APPLIED ABSTRACT ANALYSIS

JEAN-PIERRE AUBIN

Exercises by **BERNARD CORNET** and **HERVÉ MOULIN**
Translated by **CAROLE LABROUSSE**

A WILEY-INTERSCIENCE PUBLICATION

JOHN WILEY & SONS, New York · London · Sydney · Toronto

Library of Congress Cataloging in Publication Data

Aubin, Jean-Pierre.
  Applied abstract analysis.

  (Pure and applied mathematics)
  "A Wiley-Interscience publication."
  Includes index.
  1. Topology. 2. Metric spaces. 3. Game theory.
4. Mathematical optimization. I. Title.

QA611.A8513   514'.3   77-2382
ISBN 0-471-02146-6

Printed in the United States of America

10 9 8 7 6 5 4 3 2

APPLIED ABSTRACT
ANALYSIS

To my parents

PIERRE and **GEORGETTE AUBIN**

PREFACE

The object of this book is to introduce and study the principal notions of topology in the elementary framework of metric spaces rather than in the general case of topological spaces. This approach allows us to consider as soon as possible various applications (in optimization theory and game theory in particular). It is our hope to illustrate in this way the advantages of an abstract approach to problems of a more concrete origin and to encourage the student to appreciate applied mathematics.

This book will be followed by another one, entitled *Applied Functional Analysis* [AFA] to be published by Wiley, which deals in detail with those problems that are specifically linear. It is in this latter volume, moreover, that the most interesting applications will be presented.

This text grew out of two courses given to second and third year students of Mathématiques de la Décision of Université de Paris-Dauphine.

The major part of the first four chapters of this book should be considered as fundamental. The results of the fifth chapter may be omitted at a first reading and considered only as the specific need arises (in the study of applied functional analysis and game theory, for instance). Those sections marked in the contents with an asterisk (*) will be used in subsequent study or in other cases that may be considered as worked out exercises.

For the convenience of the reader, the principal propositions and theorems found in this book are grouped in a resumé at the conclusion of the text under eleven headings. This resumé serves to give a concise review of the essential results concerning each heading. Naturally, a result may appear under several headings. A terminological index appears at the end of the book.

Apart from the Brouwer theorem, which is stated without proof, all the proofs given are based on axioms or other theorems and propositions previously established. Nevertheless, it is recommended that the student be familiar with elementary differential and integral calculus of one or several variables.

A set of exercises allows the reader to apply the results proved in the book, to handle them, to single out the ones that are the most often used, and to classify them; in short, to know these results quite intimately.

I should like to express my deep gratitude to Carole Labrousse for her excellent translation of the French text, for the care she gave to her work, and for correcting many mistakes with great patience.

I also thank Bernard Cornet and Hervé Moulin who have actively contributed to teaching the properties of metric spaces and who prepared the collection of exercises. They also suggested many improvements in the first drafts of this work and pointed out parts that "did not get through." In this latter task they were efficiently helped by the students who, either by their sudden glassy stares or by the protestations of the more vocal of them, indicated the need for improving some parts of the text. Let them be thanked, for I had some pleasure in initiating them into the abstract methods and applications of analysis.

JEAN-PIERRE AUBIN

Paris, France
February 1977

CONTENTS

APPLIED ABSTRACT ANALYSIS

Introduction:
A Guide to the Reader

We summarize here the contents of the following chapters by describing the results that are essential. Since most of the terms are not always precisely defined in this introduction, these lines can be used as guidelines to replace a given result in its context.

1. In studying a mathematical object (a function, for example), one is led to isolate a restricted number of its properties and to deduce from them possible consequences. These consequences are then shared by the elements of the set E of all objects satisfying these properties.

Moreover, a classic method of *analysis* consists in "approximating" a certain object by another with *additional* properties (a polynomial, for example) that make its study easier, and then "taking the limit" (explicitly or implicitly).

The simplest and most fruitful way to undertake these approximation methods is to furnish the set E with a *distance* that makes of it a *metric space.* Indeed, we can then define the concepts of convergent sequences and continuous functions.

In the first chapter, we indicate only that every convergent sequence is a Cauchy sequence, and we define *complete spaces* as those for which the converse is true.

After accepting the axioms that characterize the field of real numbers $\mathbb{R}$, we show that $\mathbb{R}$ is a *complete* metric space. In this first chapter, we then study other examples of complete metric spaces: the extended real numbers $\bar{\mathbb{R}}$, fields with absolute value, Banach spaces and Hilbert spaces, the space of bounded functions and the space of continuous bounded functions, the space of continuous linear mappings, the spaces of sequences l^p, Fréchet spaces, and the space of functions that have derivatives of all orders.

The number and character of the examples should convince the student of the interest in discovering the common characteristics of these spaces and in developing from them consequences that are valid in every case.

1

2.   These characteristics are presented in the second chapter. They are essentially the concepts of *complete sets* (those in which every Cauchy sequence converges), *closed sets* (those that contain the limits of their convergent sequences), and especially *compact sets* (those in which every sequence has a convergent subsequence). The theorems that characterize compact sets and state their properties play an extremely important role in what follows, as the preceding definition suggests. We conclude the second chapter by establishing the topological properties of *convex sets* (those that contain the centers of gravity of their points) whose importance in analysis is considerable, in particular for vector spaces as well as for mechanics, optimization theory, mathematical economics, and so on.

3.   Continuous mappings are those that map convergent sequences onto convergent sequences. The study of their properties is the object of the third chapter. These mappings are characterized by the fact that the inverse image of every closed set is closed, and they have the interesting property of mapping compact sets onto compact sets.

The important role of linear mappings justifies our characterizing continuous linear mappings. We show, in particular, that every linear mapping from a *finite dimensional* vector space to a *finite dimensional* space is always continuous (and further that every convex function defined on a *finite dimensional* space is also continuous).

We make use on numerous occasions of a theorem on *extension "by density"* for continuous mappings.

In the case of real-valued functions we can "decompose" the notion of continuity into two parts and thus define the concepts of upper and lower semicontinuous functions. These are indispensable for studying the existence of optimal solutions in maximum and minimum problems, as well as for elementary problems in the theory of two-person games.

We conclude this chapter with the proof of the *fixed point theorem for contractions*, giving a simple sufficient condition for the existence of a solution x (called a fixed point) to the equation $x = f(x)$ when f is a contraction from a complete metric space to itself. This theorem yields numerous results affirming the existence of solutions to equations $f(x) = y$, where f is a continuous mapping from E to F and y is given in F.

We illustrate this by establishing the existence of local solutions to differential and integral equations as well as by proving the inverse function theorem. We also state, but without proof, the *Brouwer fixed point theorem*, which is the origin of most of the theorems of nonlinear analysis and whose consequences are considerable. Moreover, we demonstrate this to the student by establishing a simple sufficient condition (that the mapping be coercive) that implies that a continuous mapping f from $\mathbb{R}^n$ to itself is surjective.

4. The principal properties of the topological concepts and of continuous functions having been established, we indicate in Chapter 4 how to construct new metric spaces and how to compare them. Thus we define metric subspaces and the finite product of metric spaces. We construct a distance for which one or several mappings are continuous. We associate to every metric space E the space $\mathscr{C}_\infty(E)$ of continuous bounded functions on E and characterize the compact subsets of this space. We illustrate this by proving the existence of global solutions to differential equations.

We also construct the completion of a metric space E. Finally, the set of nonempty closed subsets of a metric space E with a suitable distance (the Hausdorff distance) enables us to study the continuity of mappings $x \mapsto A(x) \subset E$. We exploit these results to analyze the *stability of optimization problems and invariance of limit sets of dynamical systems*. We end this chapter with fixed point theorems for mappings $x \mapsto A(x) \subset E$.

5. The fifth chapter is devoted to an exposé of special properties of metric spaces. We study *locally compact spaces* and *separable spaces*, which leads to establishing the Stone–Weierstrass theorem and the Bernstein theorem on approximating a continuous function by polynomials. The *very important Baire theorem*, which is the basis of two of the three fundamental theorems of linear analysis, is proved. We make use of it to show that every lower semicontinuous seminorm on a Banach space is continuous.

We present a short study of connected sets and use them for constructing "utility functions" representing a "preference preordering".

We construct continuous partitions of unity using a theorem on the separation of two closed disjoint sets by a continuous function. These partitions of unity are then used to establish theorems in the theory of non-cooperative two-person and n-person games and in the theory of general equilibrium with which we conclude this book.

CHAPTER 1

Metric Spaces: Definitions and Examples

Introduction

In the first section of this chapter we remark that, in the final analysis, the concepts of real convergent sequences and continuous functions from $\mathbb{R}$ to $\mathbb{R}$ (that is, those functions under which the image of a convergent sequence is a convergent sequence) and most of their properties depend only on the three following properties of the distance $d(x, y) = |x - y|$ between two real numbers.

1. $d(x, y) \geq 0$ and $d(x, y) = 0$ if and only if $x = y$.
2. $d(x, y) = d(y, x)$ (symmetry).
3. $d(x, y) \leq d(x, z) + d(z, x)$ (triangle inequality).

This observation motivates the introduction in Section 2 of the general notion of a *metric space*, that is, a set E with a "*distance*" $d(x, y)$ satisfying these three properties and the verification that this distance allows us to define the notion of a convergent sequence: *A sequence $\{x_n\}$ of elements x_n of E converges to an element x of E if and only if the sequence of real numbers $d(x_n, x)$ converges to* 0.

Remark that in the case where $E = \mathbb{R}$ and $d(x, y) = |x - y|$, the *Cauchy condition plays an essential role* for it tells us that every Cauchy sequence (that is, one for which the sequence with two indices $d(x_m, x_n)$ converges to 0) converges to a limit x, which we need not know in advance. Hence in practice, we shall only use *complete spaces*, that is, those metric spaces that possess this property.

The introduction of metric spaces and their study would be pointless if there were not numerous examples useful in analysis where the metric space structure allows us to prove many results in the *simplest* and most elegant fashion as well as the most *intuitive* (that is, using geometric language). Thus

5

the remaining part of this chapter is devoted to the simplest and most important examples of metric spaces used in analysis.

After considering the metric space $\mathbb{R} = \,]-\infty, +\infty[$ in Section 1, we consider in Section 3 the extended real numbers $\bar{\mathbb{R}} = [-\infty, +\infty]$, that is, the union of $\mathbb{R}$ and the infinite points $-\infty$ and $+\infty$.

Sections 4 and 5 are devoted respectively to *fields with an absolute value* $|\cdot|$ and to *real normed vector spaces* (vector spaces with a norm $\|\cdot\|$). In these cases, the absolute value (or the norm) allows us to define a distance $d(x, y) = |x - y|$ (or $d(x, y) = \|x - y\|$) that is compatible with the algebraic structure.

We recall in Section 6 that the vector space $\mathbb{R}^n$ is a complete normed space for the norms

$$\|x\|_\infty = \sup_{1 \le i \le n} |x_i| \quad \text{and} \quad \|x\|_1 = \sum_{i=1}^n |x_i|.$$

The norms $\|x\|_p = (\sum_{i=1}^n |x_i|^p)^{1/p}$ for $p \ge 1$ and their properties are studied in Section 11 for $p = 2$ and in Section 12 in the general case.

We consider in Section 7 the first examples of infinite dimensional normed vector spaces: the space l^1 of sequences $x = (x_1, x_2, \ldots, x_n, \ldots)$ such that

$$\|x\|_1 = \sum_{k=1}^\infty |x_k| < +\infty$$

and the space l^∞ of bounded sequences x such that

$$\|x\|_\infty = \sup_{1 \le k < +\infty} |x_k| < +\infty.$$

The study of l^p spaces, the spaces of sequences x such that

$$\|x\|_p = \left(\sum_{k=1}^\infty |x_k|^p \right)^{1/p} < +\infty,$$

is made in Section 11 for $p = 2$ and in Section 12 in the general case.

We introduce in Sections 8, 9, and 10 fundamental examples of complete normed vector spaces of functions. It is for these kinds of spaces that the general study of metric spaces proves to be of the greatest importance. Note that functional analysis is essentially the study of vector spaces of functions.

In Section 8 we show that the vector space $\mathscr{U}(X)$ of *bounded*, real-valued functions on a set X with the norm

$$\|f\|_\infty = \sup_{x \in X} |f(x)|$$

is a complete metric space. We show in Section 9 that when X is a metric space, the subspace $\mathscr{C}_\infty(X)$ of *bounded, continuous*, real-valued functions on X

is also a complete metric space. We see that the convergence of a sequence of functions is precisely the notion of *uniform convergence*.

In Section 10 we prove that when E and F are normed vector spaces the space $\mathscr{L}(E, F)$ of *continuous linear* mappings from E to F is also a normed vector space that is complete if F is complete.

We show in Section 12 that when $X = [0, 1]$, we can give the space $\mathscr{C}_\infty(X)$ the norms

$$\|f\|_p = \left[\int_0^1 |f(x)|^p \, dx \right]^{1/p} \qquad \text{for } p \geq 1.$$

Unlike the norm $\|f\|_\infty$ for which $\mathscr{C}_\infty(X)$ is a complete space, the space $\mathscr{C}_\infty(X)$ with the norms $\|f\|_p$ is *not complete*. Since the use of the norms $\|f\|_p$ is very helpful, for it allows the use of results from integration theory, we are led to introduce the completions $L^p(X)$ of $\mathscr{C}_\infty(X)$ for the norms $\|f\|_p$.

Finally, we introduce in Section 13 vector spaces having a distance constructed from a *countable* family of seminorms. Thus we can, for example, describe by means of a distance the uniform convergence of a sequence of functions on all the sets K_n of a *countable* covering of a set $X = \bigcup_{n=1}^{\infty} K_n$, or further, the convergence of a sequence of functions possessing derivatives of all orders and of all its derivatives.

This section, which is more difficult than the others, may be omitted in a first reading, although it provides examples where the methods of differential and integral calculus (hard analysis) are used in the framework of abstract analysis (soft analysis). Note that we shall also study in Section 4.8 the metric space of the closed nonempty sets of a metric space.

1. Preliminaries: The Field of Real Numbers

Let us recall the definition of the field of real numbers $\mathbb{R}$.

Definition 1.   We call the "field of real numbers $\mathbb{R}$" a field with the operations $\{x, y\} \mapsto x + y$ and $\{x, y\} \mapsto xy$ and with an order relation $x \leq y$ satisfying the following properties:

  a. $\mathbb{R}$ is a "totally ordered" field, that is:

(1) $\begin{cases} \textbf{i.} & \forall x, y \in \mathbb{R},\ x \leq y \text{ or } y \leq x \text{ (the ordering is total).} \\[4pt] \textbf{ii.} & x \leq y \text{ implies that } x + z \leq y + z \text{ (the ordering is compatible} \\ & \text{with addition).} \\[4pt] \textbf{iii.} & 0 \leq x \text{ and } 0 \leq y \text{ imply that } 0 \leq xy \text{ (the ordering is compatible} \\ & \text{with multiplication).} \end{cases}$

b. $\mathbb{R}$ is an "archimedean field," that is:

(2) $\begin{cases} \forall x, y \in \mathbb{R} \text{ such that } 0 \leq x, 0 \neq x, 0 \leq y, \text{ there exists an integer } n \\ \text{such that } y \leq nx. \end{cases}$

c. $\mathbb{R}$ satisfies the nested interval property:

(3) $\qquad \bigcap_n [a_n, b_n] \neq \varnothing \qquad \text{if} \qquad \varnothing \neq [a_n, b_n] \subset [a_{n-1}, b_{n-1}] \forall n.$

Remark 1. One can either accept these axioms or derive them from the axioms of set theory. We accept the existence of the field of real numbers. ■

We shall denote by $\mathbb{R}_+$ the half line $[0, \infty[$. The total ordering on $\mathbb{R}$ allows us to define the notion of the absolute value $|x|$ of a real number x by

(4) $\qquad\qquad\qquad |x| = \sup(-x, +x)$

It is easily verified that this absolute value has the following properties:

(5) $\begin{cases} \textbf{i.} & |x| \geq 0, \quad \forall x \in \mathbb{R}, |x| = 0 \quad \text{if and only if} \quad x = 0. \\ \textbf{ii.} & |x + y| \leq |x| + |y|. \\ \textbf{iii.} & |\lambda x| = |\lambda||x|. \end{cases}$

To this absolute value, we associate the *distance*

(6) $\qquad\qquad\qquad d(x, y) = |x - y|$

between two points x and y of $\mathbb{R}$.

Let us remark that this distance serves to define the fundamental notion of analysis, that of a *convergent sequence*:

(7) $\begin{cases} \text{A sequence } \{x_n\} \text{ of real numbers } x_n \in \mathbb{R} \text{ converges to } x \text{ if } \forall \varepsilon > 0, \text{ there} \\ \text{exists } n(\varepsilon) \in \mathbb{N} \text{ such that } \forall n \geq n(\varepsilon), d(x_n, x) = |x_n - x| \leq \varepsilon; \end{cases}$

and the notion of a *continuous function* from $\mathbb{R}$ to $\mathbb{R}$:

(8) $\begin{cases} \text{A function } f: \mathbb{R} \mapsto \mathbb{R} \text{ is continuous if the image under } f \text{ of every} \\ \text{convergent sequence } \{x_n\} \text{ is a convergent sequence } \{f(x_n)\}. \end{cases}$

We shall see that most of the theorems dealing with the properties of continuous functions from $\mathbb{R}$ to $\mathbb{R}$ or with real convergent sequences depend in fact only upon the following three properties of the distance $d(x, y)$:

(9) $\begin{cases} \textbf{i.} & d(x, y) \geq 0, \quad \forall x, y \in \mathbb{R}, \quad d(x, y) = 0 \quad \text{if and only} \\ & \text{if} \quad x = y. \\ \textbf{ii.} & d(x, y) = d(y, x) \text{ (symmetry).} \\ \textbf{iii.} & d(x, y) \leq d(x, z) + d(z, y) \text{ (triangle inequality).} \end{cases}$

This thus suggests the introduction of the idea of a metric space, that is, a set E with a "distance" $d(x, y)$ and the extension of the notions of convergent sequences and continuous functions and their properties to these spaces.

First, let us recall the following proposition and the notions of least upper bound and greatest lower bound, which we shall use very often.

PROPOSITION 1.   A sequence of real numbers $\{x_n\}$ converges if and only if

(10) $\begin{cases} \forall \varepsilon > 0, \qquad \exists n(\varepsilon) \qquad \text{such that} \qquad d(x_m, x_n) = |x_m - x_n| \leq \varepsilon \\ \text{for} \qquad m, n \geq n(\varepsilon). \end{cases}$

Proof.

a.   *The condition is necessary.* If x_n converges to x, then for every $\varepsilon > 0$ there is a number $n(\varepsilon)$ such that $d(x_n, x) \leq \varepsilon/2$ and $d(x_m, x) \leq \varepsilon/2$ when m and $n \geq n(\varepsilon)$. Thus $d(x_m, x_n) \leq d(x_n, x) + d(x_m, x) \leq 2(\varepsilon/2) = \varepsilon$ when $m, n \geq n(\varepsilon)$.

b.   *The condition is sufficient.* Let $\{x_n\}$ be a sequence satisfying (10). We construct by recursion the sequence n_k $(k = 0, \ldots, m, \ldots)$ as follows: for $k = 0, n_0 = 1$; if $n_0, \ldots, n_k$ are known, we define n_{k+1} to be the smallest integer greater than n_k such that if $p \geq n_{k+1}, q \geq n_{k+1}$, then $|x_p - x_q| < 1/2^{k+2}$. This is possible according to (10) taking $\varepsilon = 1/2^{k+2}$.

Now consider the closed intervals:

(11) $$I_k = [x_{n_k} - 2^{-k}, x_{n_k} + 2^{-k}]$$

We verify that

(12) $$I_k \subset I_{k-1}, \qquad \forall k \in \mathbb{N},$$

since

(13) $$|x_{n_k} - x_{n_{k-1}}| < \frac{1}{2^k}$$

Thus if $m \geq n_k$, $x_m \in I_k \subset I_{k-1}$. According to the nested interval property, $\bigcap_{k \geq 0} I_k \neq \emptyset$ and consequently contains a point x. Since $x \in I_k$ for all k, we obtain that

(14) $$d(x, x_m) = |x - x_m| \leq \frac{1}{2^{k-1}} \qquad \text{for} \qquad m \geq n_k,$$

and this implies that $\{x_n\}$ converges to x.   ∎

GREATEST LOWER BOUNDS AND LEAST UPPER BOUNDS. Let us recall the following definitions:

Definition 2. Let $A \subset \mathbb{R}$ be a subset of the real numbers. We say that $a \in \mathbb{R}$ is the least upper bound of A, and we write $a = \sup A$ if:

$$(15) \quad \begin{cases} \text{i.} & \forall x \in A, \quad x \le a. \\ \text{ii.} & \forall n > 0, \text{ there exists } x \in A \text{ such that } a - 1/n \le x \le a. \end{cases}$$

If the least upper bound a of A belongs to A, we say that a is the largest element of A. If f is a mapping from a set E to $\mathbb{R}$, we say that $a \in \mathbb{R}$ is the supremum of f on E, and we write $a = \sup_{x \in E} f(x)$ (if it exists) if

$$(16) \qquad a = \sup f(E) = \sup_{x \in E} f(x).$$

In a similar fashion, we define the greatest lower bound of A and the infimum of f on E by

$$(17) \qquad b = \inf A = -\sup(-A), \qquad b = \inf_{x \in E} f(x) = -\sup_{x \in E}(-f(x)).$$

We say that "f attains its supremum at x_0" if

$$(18) \qquad a = \sup f(x) = f(x_0) \qquad \text{where} \qquad x_0 \in E.$$

Furthermore, let us recall the following properties.

PROPOSITION 2.

a. If f is a mapping from $E_1 \times E_2$ to $\mathbb{R}$, then

$$(19) \qquad \sup_{(x_1, x_2) \in E_1 \times E_2} f(x_1, x_2) = \sup_{x_1 \in E_1} \left(\sup_{x_2 \in E_2} f(x_1, x_2) \right)$$

b. If f_1 and f_2 are mappings from E_1 and E_2 to $\mathbb{R}$, then

$$(20) \qquad \sup_{(x_1, x_2) \in E_1 \times E_2} (f_1(x_1) + f_2(x_2)) = \sup_{x_1 \in E_1} f_1(x_1) + \sup_{x_2 \in E_2} f_2(x_2)$$

c. If f and g are mappings from E to $\mathbb{R}$ and if $f(x) \le g(x)$ for all $x \in E$, then

$$(21) \qquad \sup_{x \in E} f(x) \le \sup_{x \in E} g(x), \qquad \inf_{x \in E} f(x) \le \inf_{x \in E} g(x).$$

d. If f and g are mappings from E to $\mathbb{R}$, then

$$(22) \quad \sup_{x \in E} f(x) + \inf_{x \in E} g(x) \le \sup(f(x) + g(x)) \le \sup_{x \in E} f(x) + \sup_{x \in E} g(x).$$

Proof. The first three statements of the proposition are immediate and are left as an exercise. Let us consider the fourth statement.

Let $a = \inf_{x \in E} g(x)$. Then for all $x \in E$, $f(x) + a \leq (f(x) + g(x))$. Hence from (21), we know that

$$\sup_{x \in E}(f(x) + a) = \sup_{x \in E} f(x) + a = \sup_{x \in E} f(x) + \inf_{x \in E} g(x) \leq \sup_{x \in E}(f(x) + g(x)).$$

Moreover, if $b = \sup_{x \in E} f(x)$ and $c = \sup_{x \in E} g(x)$, then for all $x \in E$, $f(x) \leq b$ and $g(x) \leq c$; hence $f(x) + g(x) \leq b + c$. Thus

$$\sup_{x \in E}(f(x) + g(x)) \leq \sup_{x \in E} f(x) + \sup_{x \in E} g(x). \qquad \blacksquare$$

2. Metric Spaces

Let E be a set.

Definition 1.   We call a "distance" on E (or a "metric" on E) a function d from $E \times E$ to $\mathbb{R}_+$ satisfying the following three properties:

(1)
$$\begin{cases} \textbf{i.} & d(x, y) \geq 0, \quad \forall x, y \in E, \quad \text{and} \quad d(x, y) = 0 \quad \text{if and only} \\ & \text{if} \quad x = y. \\ \textbf{ii.} & d(x, y) = d(y, x) \quad \text{(symmetry).} \\ \textbf{iii.} & d(x, y) \leq d(x, z) + d(z, y) \quad \text{(triangle inequality).} \end{cases}$$

A "metric space (E, d)" is a set E with a distance d defined on E.

Let us show that this definition allows us to define the notions of convergent sequences and continuous functions.

Definition 2.   Let (E, d) be a metric space. A sequence $\{x_n\}$ of elements x_n of E "converges to x" if

(2) $\forall \varepsilon > 0, \qquad \exists n_0 \in \mathbb{N} \qquad$ such that for $\qquad n \geq n_0, \qquad d(x_n, x) \leq \varepsilon.$

Let (E, d) and (F, δ) be two metric spaces. A function $f : E \mapsto F$ is called "continuous" if for every sequence $\{x_n\}$ of elements x_n of E converging to x, the sequence $\{f(x_n)\}$ converges to $f(x)$ in F.

PROPOSITION 1.   The limit of a convergent sequence is "unique."

Proof.   Let us assume that x_n converges to x and y. Then for all ε, we have $d(x_n, x) \leq \varepsilon/2$ and $d(x_m, y) \leq \varepsilon/2$ for $n \geq n_0$, $m \geq n_1$. Hence we have $d(x, y) \leq d(x_n, x) + d(x_n, y) \leq \varepsilon$ if $n \geq \max(n_0, n_1)$. Thus $d(x, y) \leq \varepsilon$ for all $\varepsilon > 0$. This implies that $d(x, y) = 0$ and, hence, that $x = y$. $\qquad \blacksquare$

The following proposition characterizes continuous functions.

PROPOSITION 2.   Let (E, d) and (F, δ) be two metric spaces. A function f from E to F is continuous if and only if

(3) $\begin{cases} \forall x, \forall \varepsilon > 0, \qquad \exists \eta(x, \varepsilon) > 0 \qquad \text{such that} \qquad \delta(f(x), f(y)) \leq \varepsilon \\ \text{when} \qquad d(x, y) \leq \eta(x, \varepsilon). \end{cases}$

Proof.

a.   *The condition is sufficient.* If x_n converges to x, then $d(x_n, x) \leq \eta(x, \varepsilon)$ for $n \geq n_0$. Thus (3) implies that $\delta(f(x), f(x_n)) \leq \varepsilon$ when $n \geq n_0$ and therefore, that $f(x_n)$ converges to $f(x)$.

b.   *The condition is necessary.* Let us argue by contradiction. Suppose that f is continuous and (3) is false. In this case, there exist $\varepsilon > 0$ and $x \in E$ such that $\forall n \in N$, there exists $y = y_n$ satisfying

(4) $$\delta(f(x), f(y_n)) > \varepsilon \qquad \text{and} \qquad d(x, y_n) \leq \frac{1}{n}.$$

Thus the sequence y_n converges to x. Since f is continuous, $f(y_n)$ must converge to $f(x)$. But (4) implies that $\delta(f(y_n), f(x)) > \varepsilon$. This is a contradiction.   ∎

Now we are going to introduce the notions of *Cauchy sequences* and *complete metric spaces.*

PROPOSITION 3.   Let $\{x_n\}$ be a sequence of elements x_n of E. If x_n converges to x, then

(5) $\forall \varepsilon > 0, \qquad \exists n_0 \in \mathbb{N} \qquad \text{such that} \qquad \forall n, m \geq n_0, \qquad d(x_n, x_m) \leq \varepsilon.$

Proof.   Take $\varepsilon > 0$. Since x_n converges to x, there exists n_0 such that $d(x_n, x) \leq \varepsilon/2$ and $d(x_m, x) \leq \varepsilon/2$, if n and $m \geq n_0$. Thus

$$d(x_n, x_m) \leq d(x_n, x) + d(x, x_m) \leq \frac{2\varepsilon}{2} = \varepsilon.$$ ∎

The advantage of property (5) over property (1) is that *the limit x does not appear in it.* If condition (5) were sufficient for convergence, it would allow us to know that a sequence is convergent without needing to know its limit.

But we know that (5) is not always sufficient. Take the case where $E = \mathbb{Q}$, and consider the sequence 0.1, 0.101, 0.101001, 0.1010010001, ... This sequence satisfies condition (5) but does not converge in $\mathbb{Q}$. However, condition (5) is always sufficient for convergence if $E = \mathbb{R}$. (See Proposition 1.1.) This leads us to introduce the following fundamental definition.

Definition 3.   Let (E, d) be a metric space. We say that a sequence $\{x_n\}$ is a "Cauchy sequence" if it satisfies condition (5). We shall say that E is a "complete (metric) space" if every Cauchy sequence is convergent.

We have thus motivated the introduction of the fundamental concept of complete spaces and their almost exclusive use in the rest of this book. What happens when a metric space is not complete? We "complete" it using a theorem we prove later. (See Theorem 4.7.1.)   ∎

Let us now state some elementary properties of distances.

PROPOSITION 4.   If $n > 2$, if $x_1, \ldots, x_n \in E$, then

(6) $d(x_1, x_n) \leq d(x_1, x_2) + d(x_2, x_3) + \cdots + d(x_{n-1}, x_n).$

If $x, y \in E$, then

(7) $|d(x, z) - d(y, z)| \leq d(x, y).$

Proof.   Formula (6) follows from the triangle inequality by recursion. To establish (7), consider the inequalities

$$d(x, z) \leq d(y, x) + d(z, y),$$
$$d(y, z) \leq d(x, y) + d(x, z),$$

which imply that

$$-d(x, y) \leq d(x, z) - d(y, z) \leq d(x, y);$$

namely, statement (7).   ∎

PROPOSITION 5.   Let (E, d) be a metric space. For x_0 fixed, the function $x \mapsto d(x, x_0)$ is continuous from E to $\mathbb{R}$.

Proof.   Given $\varepsilon > 0$, take $\eta = \varepsilon$. We conclude then from (7) that

(8) $|d(x, x_0) - d(y, x_0)| \leq d(x, y) \leq \varepsilon,$

and thus $x \mapsto d(x, x_0)$ is continuous.   ∎

Remark 1.   To say that x_n converges to x is the same as to say that the sequence of positive real numbers $d(x_n, x)$ converges to 0.   ∎

Before going into the study of examples of metric spaces, let us consider the case of discrete spaces that will serve to establish counter examples.

   EXAMPLE 1. Discrete metric spaces.   If E is an arbitrary set, consider the function d defined by

(9) $$d(x, y) = \begin{cases} 1 & \text{if} \quad x \neq y. \\ 0 & \text{if} \quad x = y. \end{cases}$$

The properties for a distance (1) are trivially satisfied. We say that (E, d) is *a discrete space.*

3.   The Extended Real Numbers $\overline{\mathbb{R}} = [-\infty, +\infty]$

Consider the interval $]-1, +1[$ of $\mathbb{R}$ and the application φ from $]-1, +1[$ onto $\mathbb{R}$ defined by

(1) $$x \in]-1, +1[\mapsto \varphi(x) = \frac{x}{1 - |x|} \in \mathbb{R}$$

It is clear that φ is a bijection from $]-1, +1[$ onto $\mathbb{R}$, for which the inverse mapping $\psi = \varphi^{-1}$ is defined by

(2) $$y \in \mathbb{R} \mapsto \psi(y) = \frac{y}{1 + |y|} \in]-1, +1[.$$

Moreover,

(3)   the mappings φ and ψ are increasing.

Setting

(4) $$\varphi(-1) = -\infty \quad \text{and} \quad \varphi(+1) = +\infty$$

and designating by

(5) $$\overline{\mathbb{R}} = [-\infty, +\infty] = \{-\infty\} \cup \mathbb{R} \cup \{+\infty\},$$

we conclude that one can "extend" φ to a bijective mapping from $[-1, +1]$ onto $\overline{\mathbb{R}}$ whose inverse ψ is defined by

(6) $$\psi(-\infty) = -1 \quad \text{and} \quad \psi(+\infty) = +1$$

   Definition 1.   We call "the extended real numbers" $\overline{\mathbb{R}} = [-\infty, +\infty]$ the set that is the union of $\{-\infty\}$, the set $\mathbb{R}$ and $\{+\infty\}$. The two new elements $-\infty$ and $+\infty$ are called infinite points of $\overline{\mathbb{R}}$.

DISTANCE ON $\overline{\mathbb{R}}$.   We are going to give $\overline{\mathbb{R}}$ a distance $\overline{d}$ by a procedure we shall generalize later. Since $d(x, y) = |x - y|$ is a distance on the closed interval $[-1, +1]$, and since ψ is a bijection from $\overline{\mathbb{R}}$ onto $[-1, +1]$, it is natural to put

$$(7) \qquad \bar{d}(x, y) = |\psi(x) - \psi(y)| = \left| \frac{x}{1 + |x|} - \frac{y}{1 + |y|} \right|$$

In particular,

$$
(8) \quad
\begin{cases}
\bar{d}(-\infty, y) = \left| \dfrac{1 + |y| + y}{1 + |y|} \right| = \begin{cases} \dfrac{1}{1 - y} & \text{if} \quad y \leq 0 \\[3mm] \dfrac{1 + 2y}{1 + y} & \text{if} \quad y \geq 0 \end{cases} \\[10mm]
\bar{d}(x, +\infty) = \left| \dfrac{1 + |x| - x}{1 + |x|} \right| = \begin{cases} \dfrac{1 - 2x}{1 - x} & \text{if} \quad x \leq 0 \\[3mm] \dfrac{1}{1 + x} & \text{if} \quad x \geq 0 \end{cases} \\[10mm]
\bar{d}(-\infty, +\infty) = 2
\end{cases}
$$

PROPOSITION 1.   The function $\bar{d}$ from $\overline{\mathbb{R}} \times \overline{\mathbb{R}}$ to $\mathbb{R}_+$ defined by (7) is a distance on $\overline{\mathbb{R}}$.

Proof.   (*Left as an exercise.*)

Remark 1.   The distance $\bar{d}(x, y)$ applied to two elements x, y of $\mathbb{R}$ is different from the distance $d(x, y) = |x - y|$.   ■

AN ORDER RELATION ON $\overline{\mathbb{R}}$.   Since the functions φ and ψ are strictly increasing, the conditions $x \leq y$ and $\psi(x) \leq \psi(y)$ are equivalent when x and y are real numbers. Hence, we can define an order relation on $\overline{\mathbb{R}}$ by setting

$$(9) \qquad\qquad x \leq y \qquad \text{if and only if} \qquad \psi(x) \leq \psi(y).$$

In particular, if $x \in \mathbb{R}$, we obtain

$$(10) \qquad\qquad\qquad -\infty < x < +\infty,$$

which leads us to say that the real numbers are the "finite numbers" of $\overline{\mathbb{R}}$.

The notions of least upper bound and greatest lower bound and their properties (see Section 1, Definition 2) carry over to $\overline{\mathbb{R}}$, for they depend upon the order relation only.

ALGEBRAIC OPERATIONS ON $\overline{\mathbb{R}}$.   Addition and multiplication can be "extended" to $\overline{\mathbb{R}}$ by adopting the following conventions:

(11)
$$
\begin{cases}
x + \infty = \infty + x = +\infty & \text{if} \quad -\infty < x \le +\infty. \\[4pt]
x - \infty = -\infty + x = -\infty & \text{if} \quad -\infty \le x < +\infty. \\[4pt]
x\infty = \infty x = \infty, \qquad x(-\infty) = (-\infty)x = -\infty \\
\text{if} \quad 0 < x \le \infty. \\[4pt]
x\infty = \infty x = -\infty, \qquad x(-\infty) = (-\infty)x = +\infty \\
\text{if} \quad -\infty \le x < 0. \\[4pt]
0\infty = \infty 0 = 0(-\infty) = (-\infty)0 = 0.
\end{cases}
$$

Only $-\infty + \infty, +\infty - \infty, \infty/0, 0/\infty, \infty/\infty$ cannot be defined. The extended real numbers $\overline{\mathbb{R}}$ *do not form a field*.

Remark 2.   We shall see that the metric space $\overline{\mathbb{R}}$ with distance $\overline{d}$ is complete.   ∎

In fact, by definition of $\overline{\mathbb{R}}$ a sequence $\{x_n\}$ of elements x_n of $\overline{\mathbb{R}}$ is a Cauchy sequence if and only if the sequence $\{\psi(x_n)\}$ is a Cauchy sequence in $\mathbb{R}$. Hence $\{\psi(x_n)\}$ is a sequence that converges to an element a of $\mathbb{R}$. *We show later that this limit a belongs to the closed interval* $[-1, +1]$. Thus we deduce that if $\overline{x} = \varphi(a) \in \overline{\mathbb{R}}$, then $\overline{d}(x_n, \overline{x}) = |(\psi(x_n) - \psi(\overline{x})| = |\psi(x_n) - a|$ converges to 0; that is, x_n converges to $\overline{x}$ in $\overline{\mathbb{R}}$. (See Proposition 3.1.5 and Remark 3.1.5.)

4.   Fields with an Absolute Value

Suppose E is a field.

Definition 1.   If E is a field, we call an "absolute value" on E a mapping $x \mapsto |x|$ from E to $\mathbb{R}_+$ such that

(1)
$$
\begin{cases}
\textbf{i.} \quad |x| \ge 0, \qquad |x| = 0 \quad \text{if and only if} \quad x = 0. \\[4pt]
\textbf{ii.} \quad |x + y| \le |x| + |y|. \\[4pt]
\textbf{iii.} \quad |xy| = |x||y|.
\end{cases}
$$

The pair $(E, |\cdot|)$ is called "a field with absolute value."

PROPOSITION 1.   An absolute value satisfies the following conditions:

(2)
$$
\begin{cases}
\textbf{i.} \quad |+1| = 1. \\[4pt]
\textbf{ii.} \quad |-x| = |x|. \\[4pt]
\textbf{iii.} \quad \big||x| - |y|\big| \le |x - y|.
\end{cases}
$$

Proof. We have $|x| = |1 \cdot x| = |1||x|$, and if $x \neq 0$, we conclude that $|1| = 1$. Moreover, $1 = |(-1)^2| = |-1|^2$. Hence $|-1| = 1$. Consequently, $|-x| = |(-1)x| = |-1||x| = |x|$. Finally, $|x| = |x + y - y| \leq |x - y| + |y|$ and $|y| = |y - x + x| \leq |x - y| + |x|$, which implies that $-|x - y| \leq |x| - |y| \leq |x - y|$, namely (2)iii.   ∎

Having an absolute value on a field allows us to define a distance that is compatible with the algebraic operations.

PROPOSITION 2. Let $(E, |\cdot|)$ be a field with absolute value. The function d defined by

$$(3) \qquad\qquad d(x, y) = |x - y|$$

is a distance on E, which satisfies the following compatibility conditions:

$$(4) \quad \begin{cases} \textbf{i.} & d(x + z, y + z) = d(x, y). \\ \textbf{ii.} & d(xz, yz) = d(x, y)|z|. \end{cases}$$

Proof.

$d(x, y) = |x - y| \geq 0$ and $d(x, y) = |x - y| = 0$ if and only if $x - y = 0$.

$d(x, y) = |x - y| = |y - x| = d(y, x)$ by   (2)ii.

$d(x + z, y + z) = |x + z - (y + z)| = |x - y| = d(x, y).$

$d(xz, yz) = |z(x - y)| = |z|d(x, y).$ ∎

EXAMPLE 1. $\mathbb{R}$ is a complete field with absolute value.

The absolute value is $|x| = \sup(-x, x)$, which satisfies the conditions (1) and yields the classic distance $d(x, y) = |x - y|$. $\mathbb{R}$ is complete according to Proposition 1.1.

EXAMPLE 2. $\mathbb{C}$ is a field with absolute value.

We can verify that if $z = x + iy = \rho e^{i\theta} \in \mathbb{C}$, then $|z| = \rho = (x^2 + y^2)^{1/2}$ is an absolute value that makes $\mathbb{C}$ a field with absolute value.

We show that $\mathbb{C}$ is *complete*. (See Corollary 1, Section 1–11.)

5. Banach Spaces

Suppose that E is a real vector space, that is, a vector space defined on $\mathbb{R}$, the field of real numbers.

Definition 1. If E is a real vector space, a "norm" on E is a mapping $x \mapsto \|x\|$ of E in $\mathbb{R}_+$ such that

(1)
$$\begin{cases} \textbf{i.} & \|x\| \geq 0, \quad \|x\| = 0 \quad \text{if and only if} \quad x = 0. \\ \textbf{ii.} & \|x + y\| \leq \|x\| + \|y\| \quad \forall x, \quad y \in E. \\ \textbf{iii.} & \|\lambda x\| = |\lambda| \|x\| \quad \forall x \in E \quad \forall \lambda \in \mathbb{R}. \end{cases}$$

The pair $(E, \|\cdot\|)$ is called a normed space.

PROPOSITION 1. A norm has the following properties:

(2)
$$\begin{cases} \textbf{i.} & \|-x\| = \|x\|. \\ \textbf{ii.} & |\|x\| - \|y\|| \leq \|x - y\|. \end{cases}$$

Proof. (*Left as an exercise.*)

PROPOSITION 2. Let $(E, \|\cdot\|)$ be a normed space. The function d defined by

(3)
$$d(x, y) = \|x - y\|,$$

is a distance on E satisfying the following compatibility conditions:

(4)
$$\begin{cases} \textbf{i.} & d(x + z, y + z) = d(x, y) \quad \text{(invariance by translation)}. \\ \textbf{ii.} & d(\lambda x, \lambda y) = |\lambda| d(x, y). \end{cases}$$

Proof. (*Left as an exercise.*)

According to Proposition 2, *every normed space is a metric space.* The study of normed spaces constitutes a part of what is called *functional analysis.* The structure of a normed space, combining the structures of a vector space and of a metric space, has a large role both in mathematical theory and in applications. Complete normed spaces, in particular, play a very important role.

Definition 2. A "Banach space" is a complete normed vector space.

We see later how to characterize the norms on a vector space. For the moment, we only study some examples.

Remark 1. If E is a vector space on a field K and if K has an absolute value, we can define a norm, a normed space, and, if the field with absolute value is complete, a Banach space. ∎

6. The Normed Space $\mathbb{R}^n$

Consider the vector space $\mathbb{R}^n$ of sequences $x = (x_1, \ldots, x_n)$ of n scalars. We give two examples of norms.

PROPOSITION 1. The following functions are norms on $\mathbb{R}^n$

(1)
$$
\begin{cases}
\textbf{i.} & \|x\|_1 = \sum_{i=1}^n |x_i| \\[2ex]
\textbf{ii.} & \|x\|_\infty = \sup_{1 \le i \le n} |x_i|
\end{cases}
$$

satisfying the following conditions:

(2)
$$
\begin{cases}
\textbf{i.} & \left| \sum_{i=1}^n x_i y_i \right| \le \|x\|_1 \|y\|_\infty \qquad \forall x, \qquad y \in \mathbb{R}^n. \\[2ex]
\textbf{ii.} & \|x\|_\infty \le \|x\|_1, \qquad \forall x \in \mathbb{R}^n. \\[2ex]
\textbf{iii.} & \|x\|_1 \le n\|x\|_\infty, \qquad \forall x \in \mathbb{R}^n.
\end{cases}
$$

Proof.

a. It is clear that the functions $\|x\|_p$, $p = 1, \infty$, are positive. Moreover, if $\|x\|_p = 0$, this implies that the x_i vanish and, hence, that $x_i = 0$ for $i = 1, \ldots, n$. Since $\lambda x = (\lambda x_1, \ldots, \lambda x_n)$ and since $|\lambda x_i| = |\lambda||x_i|$, we conclude that $\|\lambda x\|_p = |\lambda| \|x\|_p$ for $p = 1, \infty$. Since

$$ z = x + y = (x_1 + y_1, \ldots, x_n + y_n) $$

if $x = (x_1, \ldots, x_n)$ and $y = (y_1, \ldots, y_n)$ and since $z_i = x_i + y_i$, we obtain

i. $\quad \|z\|_1 = \sum_{i=1}^n |x_i + y_i| \le \sum_{i=1}^n (|x_i| + |y_i|) \le \sum_{i=1}^n |x_i| + \sum_{i=1}^n |y_i|.$

ii. $\quad \|z\|_\infty = \sup_{i=1,\ldots,n} |x_i + y_i| \le \sup_{i=1,\ldots,n} (|x_i| + |y_i|)$

$$ \le \sup_{i=1,\ldots,n} |x_i| + \sup_{i=1,\ldots,n} |y_i|. $$

We have thus shown that $\|x\|_1$ and $\|x\|_\infty$ are norms on $\mathbb{R}^n$.

b. Moreover,

$$ \left| \sum_{i=1}^n x_i y_i \right| \le \sum_{i=1}^n |x_i y_i| = \sum_{i=1}^n |x_i||y_i| \le \sup_{i=1,\ldots,n} |y_i| \sum_{i=1}^n |x_i| = \|y\|_\infty \|x\|_1. $$

In particular, taking $x = (1, \ldots, 1)$, we conclude that

$$\|y\|_1 = \sum_{i=1}^{n} |y_i| = \sum_{i=1}^{n} |y_i| 1 \leq \|y\|_\infty \left(\sum_{i=1}^{n} 1 \right) = n\|y\|_\infty.$$

Finally, if $x = (x_1, \ldots, x_n)$, we know that for all i $|x_i| \leq \sum_{i=1}^{n} |x_i| = \|x\|_1$. Thus $\|x\|_\infty = \sup_{i=1,\ldots,n} |x_i| \leq \|x\|_1$. ∎

PROPOSITION 2.   The normed spaces $(\mathbb{R}^n, \|\cdot\|_1)$ and $(\mathbb{R}^n, \|\cdot\|_\infty)$ are Banach spaces.

Proof.   We shall give the proof for the norm $\|\cdot\|_\infty$. Let $x^m = (x_1^m, \ldots, x_n^m)$ be a Cauchy sequence in $\mathbb{R}^n$. Then for all $\varepsilon > 0$, there exists m_0 such that if $p, q \geq m_0$,

(3) $$\|x^p - x^q\|_\infty = \sup_{1 \leq i \leq n} |x_i^p - x_i^q| \leq \varepsilon.$$

This implies that the scalar sequences x_i^p are Cauchy sequences in $\mathbb{R}$ for $i = 1, \ldots, n$. Since $\mathbb{R}$ is complete, each sequence x_i^p converges to an element x_i. This means that $\forall i = 1, \ldots, n, \forall \varepsilon > 0$, there exists m_i such that if $p \geq m_i$,

(4) $$|x_i^p - x_i| \leq \frac{\varepsilon}{n}.$$

Taking $m = \sup_{i=1,\ldots,n} m_i$, we see that $\forall \varepsilon, \exists m$ such that if $p \geq m$,

(5) $$\|x^p - x\|_\infty = \sup_{1 \leq i \leq n} |x_i^p - x_i| \leq n \cdot \frac{\varepsilon}{n} = \varepsilon.$$

This implies that the sequence x^p converges to x in $\mathbb{R}^n$. ∎

Remark 1.   This kind of proof is very general, and we shall use it again often. ∎

Remark 2.   The inequalities (2)ii and (2)iii show that *the notion of a convergent sequence is the same for the norms* $\|x\|_1$ *and* $\|x\|_\infty$. Indeed, if a sequence x^m converges to x in $(\mathbb{R}^n, \|\cdot\|_\infty)$, this implies that the sequence $\|x^m - x\|_\infty$ converges to 0. Thus according to (2)iii, the sequence $\|x^m - x\|_1$ converges to 0; hence x^m converges to x in $(\mathbb{R}^n, \|\cdot\|_1)$. Using (2)ii we show that if x^m converges to x in $(\mathbb{R}^n, \|\cdot\|_1)$, x^m converges to x in $(\mathbb{R}^n, \|\cdot\|_\infty)$.

For this reason we say that the norms $\|\cdot\|_1$ and $\|\cdot\|_\infty$ are *equivalent*. We show, moreover, that all the norms on $\mathbb{R}^n$ are equivalent. (This is not true for infinite dimensional spaces.) We come back later to this question. (See Chapter 4.) ∎

ECONOMIC INTERPRETATION.   In economic theory $\mathbb{R}^n$ *is interpreted as a commodity space*: For this, we index each economic good by an appropriate index $i = 1, \ldots, n$. We associate to each good a *unit of quantity* that allows us to speak of the *quantity of good i*, which is a real number x_i. Then a commodity bundle, in short, a "commodity" $x = (x_1, \ldots, x_n)$ indicates the sequence of x_1 quantities of good $1, \ldots, x_n$ quantities of good n. The unit of quantity e^i of good i is the commodity $e^i = (0, \ldots, 1, \ldots, 0)$. Then the distance $\|x - y\|_\infty$ between two commodities is equal to

$$(6) \qquad \|x - y\|_\infty = \sup_{i = 1, \ldots, n} |x_i - y_i|.$$

A sequence x^m of commodities converges to a commodity x if each quantity x_i^m of good i converges to the quantity x_i of good i.

We interpret the dual space $\mathbb{R}^{n*}$ of $\mathbb{R}^n$ as the space of *price systems* $p = (p^1, \ldots, p^n)$, where the ith component p^i is the *unit price* of good i.

If $x \in \mathbb{R}^n$ and $p \in \mathbb{R}^{n*}$, the value of the commodity x for the price system p is equal to

$$(7) \qquad \langle p, x \rangle = \sum_{i=1}^{n} p^i x_i.$$

We agree to give to the space of price systems $\mathbb{R}^{n*}$ the norm

$$(8) \qquad \|p\|_1 = \sum_{i=1}^{n} |p^i|,$$

and we conclude from (2)i that

$$(9) \qquad |\langle p, x \rangle| \le \|p\|_1 \|x\|_\infty.$$

7.   The Spaces l^1 and l^∞

Consider the set $\mathbb{N}$ of positive integers (or the set $\mathbb{Z}$ of positive and negative integers).

Definition 1.   We denote by $l^1 = l^1(\mathbb{N})$ the set of real sequences $\{x_n\}_{n \ge 0}$ for which the associated series converges absolutely, that is, such that

$$(1) \qquad \|x\|_1 = \sum_{n=0}^{\infty} |x_n| < +\infty.$$

We denote by $l^\infty = l^\infty(\mathbb{N})$ the set of real sequences $\{x_n\}_{n \ge 0}$ that are bounded, that is, such that

$$(2) \qquad \|x\|_\infty = \sup_{0 \le n < \infty} |x_n| < +\infty.$$

PROPOSITION 1.   The spaces $(l^1, \|\cdot\|_1)$ and $(l^\infty, \|\cdot\|_\infty)$ are normed vector spaces. Moreover, if $x \in l^\infty$ and if $p \in l^1$, then the series

(3)
$$\langle p, x \rangle = \sum_{n=1}^{\infty} p_n x_n$$

is absolutely convergent and satisfies

(4)
$$|\langle p, x \rangle| \le \|p\|_1 \|x\|_\infty$$

Finally, the space l^1 is contained in l^∞ and

(5)
$$\|x\|_\infty \le \|x\|_1 \qquad \text{for all} \qquad x \in l^1.$$

Proof.

a.   Let us show that l^1 is a vector space and that $\|x\|_1$ is a norm. If $x = (x_0, \ldots, x_n, \ldots) \in l^1$ and if $\lambda \in \mathbb{R}$, then $\lambda x = (\lambda x_0, \ldots, \lambda x_n, \ldots)$ and $\|\lambda x\|_1 = \sum_{n=0}^{\infty} |\lambda x_n| = \sum_{n=0}^{\infty} |\lambda| |x_n| = |\lambda| \|x\|_1 < +\infty$. Moreover, if x and $y \in l^1$, then $x + y = (x_0 + y_0, \ldots, x_n + y_n, \ldots)$ and $\|x + y\|_1 = \sum_{n=0}^{\infty} |x_n + y_n| \le \sum_{n=0}^{\infty} (|x_n| + |y_n|) = \sum_{n=0}^{\infty} |x_n| + \sum_{n=0}^{\infty} |y_n| = \|x\|_1 + \|y\|_1$. This shows that l^1 is a vector space. To show that $\|x\|_1$ is a norm, it remains to show that if $\|x\|_1 = \sum_{n=0}^{\infty} |x_n| = 0$, then $|x_n| = 0$ for all n; hence $x_n = 0$, and so $x = 0$.

b.   Let us show that l^∞ is a vector space and that $\|x\|_\infty$ is a norm. Indeed, if $x \in l^\infty$ and $\lambda \in \mathbb{R}$, then

$$\|\lambda x\|_\infty = \sup_{n \in \mathbb{N}} |\lambda x_n| = \sup_{n \in \mathbb{N}} |\lambda| |x_n| = |\lambda| \sup_{n \in \mathbb{N}} \|x_n\| = \lambda \|x\|_\infty$$

and

$$\|x + y\|_\infty \le \sup_{n \in \mathbb{N}} |x_n + y_n| \le \sup_{n \in \mathbb{N}} (|x_n| + |y_n|).$$

Moreover, since

$$|x_n| \le \|x\|_\infty \qquad \text{and} \qquad |y_n| \le \|y\|_\infty$$

for all n,

$$|x_n| + |y_n| \le \|x\|_\infty + \|y\|_\infty,$$

and, thus,

$$\|x + y\|_\infty \le \sup_{n \in \mathbb{N}} (|x_n| + |y_n|) \le \|x\|_\infty + \|y\|_\infty.$$

Finally if $\|x\|_\infty = \sup_{n \in \mathbb{N}} |x_n| = 0$, this implies that $0 \le |x_n| \le 0$ for all n; hence $x = 0$.

c. Now let us show (3). Since $|x_n| \le \|x\|_\infty$,

$$|\langle p, x \rangle| = |\sum_{n \in \mathbb{N}} p_n x_n| \le \sum_{n \in \mathbb{N}} |p_n x_n| = \sum_{n \in \mathbb{N}} |p_n| |x_n| \le \|x\|_\infty \sum_{n \in \mathbb{N}} |p_n|$$

$$= \|x\|_\infty \|p\|_1.$$

Finally, $l^1 \subset l^\infty$, since if the series whose general term is $|x_n|$ converges, then the sequence $|x_n|$ converges to 0 and is, therefore, bounded. Moreover, for every n fixed, we have $|x_n| \le \sum_{k=0}^\infty |x_k| = \|x\|_1$. Thus $\|x\|_\infty = \sup_{n \in \mathbb{N}} |x_n| \le \|x\|_1$. ∎

Remark 1. We shall show in Sections 8 and 12 that the spaces l^∞ and l^1 arc complete. (See Corollary 8.1 and Theorem 12.3.) ∎

ECONOMIC INTERPRETATION. We can interpret the index n as a time index indicating an nth period and x_n as the quantity of an economic object at date n. Then $x \in l^\infty$ (or $x \in l^1$) describes the evolution of x_n as a function of time n from 0 to infinity.

8. The Space $\mathscr{U}(X)$ of Bounded Functions

Let X be a set and $\mathscr{U}(X)$ the set of bounded real-valued functions on X:

(1)
$$\begin{cases} f \in \mathscr{U}(X) & \text{if} \quad f : x \in X \mapsto f(x) \in \mathbb{R} \quad \text{satisfies} \\ \|f\|_\infty = \sup_{x \in X} |f(x)| < +\infty \end{cases}$$

PROPOSITION 1. The space $(\mathscr{U}(X), \|\cdot\|_\infty)$ is a Banach space.

Proof. If $f \in \mathscr{U}(X)$ and $\lambda \in \mathbb{R}$, then $\lambda f : x \mapsto \lambda f(x)$ and

$$\|\lambda f\|_\infty = \sup_{x \in X} |\lambda f(x)| = \sup_{x \in X} |\lambda| |f(x)| = |\lambda| \sup_{x \in X} |f(x)| = |\lambda| \|f\|_\infty$$

If f and $g \in \mathscr{U}(X)$, then $f + g : x \mapsto f(x) + g(x)$ and

$$\|f + g\|_\infty = \sup_{x \in X} |f(x) + g(x)| \le \sup_{x \in X} (|f(x)| + |g(x)|)$$

Moreover, $|f(x)| \le \|f\|_\infty$ and $|g(x)| \le \|g\|_\infty$ for all $x \in X$, which implies that

$$|f(x)| + |g(x)| \le \|f\|_\infty + \|g\|_\infty,$$

and thus

$$\|f + g\|_\infty = \sup_{x \in X} (|f(x)| + |g(x)|) \le \|f\|_\infty + \|g\|_\infty.$$

Consequently, $\mathscr{U}(X)$ is a vector space. Furthermore, $\|f\|_\infty$ is a norm for if $\|f\|_\infty = \sup_{x \in X} |f(x)| = 0$, we conclude that $|f(x)| = 0$ for all $x \in X$, thus $f(x) = 0 \; \forall x \in X$; that is, $f = 0$. Hence $\mathcal{U}(X)$ is a normed vector space. It remains to show that $\mathcal{U}(X)$ is complete. Let f_n be a Cauchy sequence in $\mathcal{U}(X)$: $\forall \varepsilon > 0, \exists n_0$ such that

$$(2) \qquad \| f_n - f_m \|_\infty = \sup_{x \in X} | f_n(x) - f_m(x) | \leq \varepsilon \qquad \text{for} \qquad m, n \geq n_0.$$

Consequently, the sequences $f_n(x)$ are Cauchy sequences in $\mathbb{R}$, and, since $\mathbb{R}$ is complete, $f_n(x)$ converges to a real number we call $f(x)$.

We have thus defined a function $x \mapsto f(x)$. It remains to show that f is bounded and that $\lim_{n \to \infty} \| f_n - f \|_\infty = 0$. Let us fix $\varepsilon > 0$, m and $n \geq n_0$ and $x \in X$. We conclude from (2) that

$$(3) \quad \begin{cases} | f_n(x) - f(x) | \leq | f_n(x) - f_m(x) | + | f_m(x) - f(x) | \\ \qquad \leq \| f_n - f_m \|_\infty + | f_m(x) - f(x) | \leq \varepsilon + | f_m(x) - f(x) | \end{cases}$$

Letting m go to infinity while keeping n fixed greater than n_0, we obtain

$$(4) \qquad\qquad\qquad | f_n(x) - f(x) | \leq \varepsilon + 0 = \varepsilon.$$

This inequality implies, first of all, that

$$(5) \qquad\qquad | f(x) | \leq | f_n(x) | + \varepsilon \qquad \text{for all} \qquad x \in X,$$

which implies, taking the sup over X, that

$$(6) \qquad\qquad \sup_{x \in X} | f(x) | \leq \sup_{x \in X} | f_n(x) | + \varepsilon \leq \| f_n \|_\infty + \varepsilon < +\infty.$$

Thus f is bounded.

On the other hand, taking the sup over X of the left-hand side of inequality (4), we obtain

$$(7) \qquad\qquad \| f_n - f \|_\infty = \sup_{x \in X} | f_n(x) - f(x) | \leq \varepsilon.$$

That is, we have established that to every $\varepsilon > 0$ we can associate n_0 such that for all $n \geq n_0$, $\| f_n - f \|_\infty \leq \varepsilon$. We have, therefore, shown that the Cauchy sequence f_n converges to f in the metric space $\mathcal{U}(X)$. ∎

COROLLARY 1.   The space l^∞ is a Banach space.

Proof.   In fact, $l^\infty = l^\infty(\mathbb{N}) = \mathcal{U}(\mathbb{N})$ is the space of bounded functions from $\mathbb{N}$ to $\mathbb{R}$. Thus l^∞ is complete. ∎

Definition 1.   We shall say that the norm $\| \cdot \|_\infty$ on $\mathcal{U}(X)$ is the "maximum norm" or the "norm of uniform convergence." If a sequence f_n converges to f in $\mathcal{U}(X)$ in the sense that $\| f_n - f \|_\infty$ converges to 0, we say that "f_n converges to f uniformly."

More generally, if E is a normed vector space, we define the space $\mathscr{U}(X; E)$ of *bounded functions from* X *to* E: $f \in \mathscr{U}(X; E)$ if and only if $\|f\|_\infty = \sup_{x \in X} \|f(x)\|_E < +\infty$.

PROPOSITION 2. The set $\{\mathscr{U}(X; E), \|\cdot\|_\infty\}$ is a normed space. If E is a Banach space, $\mathscr{U}(X; E)$ is also a Banach space.

9. The Space $\mathscr{C}_\infty(X)$ of Continuous Bounded Functions

We suppose now that (X, d) is a metric space for a distance d.

We denote by $\mathscr{C}_\infty(X)$ the subspace of $\mathscr{U}(X)$ of real-valued continuous functions bounded on X: $f \in \mathscr{C}_\infty(X)$ if and only if

$$(1) \qquad\qquad \|f\|_\infty = \sup_{x \in X} |f(x)| < +\infty.$$

and

$$(2) \qquad \begin{cases} \forall \varepsilon, \quad \forall x, \quad \exists \eta \quad \text{such that if} \quad d(x, y) \leq \eta \quad \text{then} \\ |f(x) - f(y)| \leq \varepsilon. \end{cases}$$

Since $\mathscr{C}_\infty(X)$ is a vector subspace of $\mathscr{U}(X)$ (each linear combination of continuous functions being a continuous function), we shall give it the norm $\|f\|_\infty$ of uniform convergence.

PROPOSITION 1. The space $\mathscr{C}_\infty(X)$ is a Banach space.

Proof. What we must show is that every Cauchy sequence $f_n \in \mathscr{C}_\infty(X)$ is convergent. However, if f_n is a Cauchy sequence in $\mathscr{C}_\infty(X)$, it is a Cauchy sequence in $\mathscr{U}(X)$; hence since $\mathscr{U}(X)$ is complete, f_n converges uniformly to f in $\mathscr{U}(X)$.

It remains to establish the following lemma.

LEMMA 1. If a sequence of continuous bounded functions f_n converges uniformly to a bounded function f, then f is continuous.

Proof. Since f_n converges to f in $\mathscr{U}(X)$, $\forall \varepsilon > 0$, $\exists n_0$ such that

$$(3) \qquad \|f_n - f\|_\infty = \sup |f_n(x) - f(x)| \leq \frac{\varepsilon}{3} \quad \text{for} \quad n \geq n_0.$$

For a fixed $x \in X$, let us show that f is continuous, that is, that

$$(4) \qquad\qquad |f(x) - f(y)| \leq \varepsilon \quad \text{when} \quad d(x, y) \leq \eta(x)$$

Since the f_n are continuous, we know that for $n = n_0$, and for $\varepsilon > 0$ there exists $\eta(n_0, x)$ such that

(5) $$|f_{n_0}(x) - f_{n_0}(y)| \le \frac{\varepsilon}{3} \qquad \text{when} \qquad d(x, y) \le \eta(n_0, x).$$

Thus using (3) and (5) we establish (4) with $\eta(x) = \eta(n_0, x)$ since

$$\begin{aligned}
|f(x) - f(y)| &= |f(x) - f_{n_0}(x) + f_{n_0}(x) - f_{n_0}(y) + f_{n_0}(y) - f(y)| \\
&\le |f(x) - f_{n_0}(x)| + |f_{n_0}(x) - f_{n_0}(y)| + |f_{n_0}(y) - f(y)| \\
&\le 3\frac{\varepsilon}{3} = \varepsilon. \qquad\blacksquare
\end{aligned}$$

More generally, if E is a normed space, we define $\mathscr{C}_\infty(X; E)$ to be the vector subspace of functions $f \in \mathscr{U}(X; E)$, which are continuous from X to E: $f \in \mathscr{C}_\infty(X; E)$ if

(6) $$\|f\|_\infty = \sup_{x \in X} \|f(x)\|_E$$

and

(7) $$\begin{cases} \forall \varepsilon, \forall x, \exists \eta(x) \qquad \text{such that} \qquad \|f(x) - f(y)\|_E \le \varepsilon \qquad \text{when} \\ d(x, y) \le \eta. \end{cases}$$

PROPOSITION 2. The space $(\mathscr{C}_\infty(X; E), \|\cdot\|_\infty)$ is a normed space. If E is a Banach space, $\mathscr{C}_\infty(X; E)$ is also a Banach space.

Remark 1. We shall see later that Lemma 1 expresses the fact that "$\mathscr{C}_\infty(X)$ is a *closed* subset of $\mathscr{U}(X)$." $\qquad\blacksquare$

Remark 2. The study of the space $\mathscr{C}_\infty(X)$ is continued in Section 4.6. $\quad\blacksquare$

10. The Space $\mathscr{L}(E, F)$ of Continuous Linear Mappings

Let $(E, \|\cdot\|_E)$ and $(F, \|\cdot\|_F)$ be two normed vector spaces and A a linear mapping from E to F; that is, A satisfies

(1) $$A\left(\sum_{i=1}^n \lambda_i x_i \right) = \sum_{i=1}^n \lambda_i A x_i$$

for every linear combination $\sum_{i=1}^n \lambda_i x_i$ of elements x_i of E.

PROPOSITION 1.   Let $(E, \|\cdot\|_E)$ and $(F, \|\cdot\|_F)$ be two normed vector spaces, and let A be a linear mapping from E to F. Then A is continuous if and only if

(2)
$$\|A\| = \sup_{\substack{x \in X \\ x \neq 0}} \frac{\|Ax\|_F}{\|x\|_E} < +\infty.$$

Proof.

a.   If A is continuous, it is continuous at the point $x = 0$. Thus $\forall \varepsilon > 0$, $\exists \eta > 0$ such that if $\|y\|_E \leq \eta$, then $\|Ay\|_F \leq \varepsilon$. Let $x \neq 0$ be an arbitrary element of E. Then $y = \eta x / \|x\|_E$ satisfies

$$\|y\|_E = \frac{\|\eta x\|_E}{\|x\|_E} = \eta \frac{\|x\|_E}{\|x\|_E} = \eta.$$

Hence

$$\|Ay\|_F = \left\| A\left(\frac{\eta x}{\|x\|_E} \right) \right\|_F = \left\| \frac{\eta}{\|x\|_E} Ax \right\|_F = \eta \frac{\|Ax\|_F}{\|x\|_E} \leq \varepsilon.$$

This implies that, for all $x \in E$,

(3)
$$\frac{\|Ax\|_F}{\|x\|_E} \leq \frac{\varepsilon}{\eta}.$$

Thus we have that

$$\|A\| = \sup_{\substack{x \in E \\ x \neq 0}} \frac{\|Ax\|_F}{\|x\|_E} \leq \frac{\varepsilon}{\eta} < +\infty.$$

b.   Conversely, if $\|A\| < +\infty$, then A is continuous. Indeed, let $\varepsilon > 0$ and choose $\eta = \varepsilon / \|A\|$ (independent of x). Then if $\|x - y\|_E \leq \eta$, we conclude that

$$\|Ax - Ay\|_F = \|A(x - y)\|_F \leq \|A\| \|x - y\|_E \leq \eta \|A\| = \varepsilon.$$

which shows that A is continuous from E to F.   ∎

Definition 1.   Let $(E, \|\cdot\|_E)$ and $(F, \|\cdot\|_F)$ be two normed spaces. We denote by $\mathscr{L}(E, F)$ the set of continuous linear mappings from E to F.

PROPOSITION 2.   The space $(\mathscr{L}(E, F), \|A\|)$ is a normed vector space. It is a Banach space if F is a Banach space.

Proof.

a. The space $\mathscr{L}(E, F)$ is obviously a vector space for the following operations

$$(4) \qquad \begin{cases} \textbf{i.} & \lambda A : x \mapsto \lambda Ax. \\ \textbf{ii.} & A + B : x \mapsto Ax + Bx. \end{cases}$$

b. We must show that $\|A\| = \sup_{x \neq 0}(\|Ax\|_F / \|x\|_E)$ is a norm. $\|A\| \geq 0$. However, if $\|A\| = 0$ and $x \neq 0$, we conclude that $\|Ax\|_F = 0$, and, consequently, that $Ax = 0$. Thus $A = 0$. Furthermore,

$$\|\lambda A\| = \sup_{x \neq 0} \frac{\|\lambda Ax\|_F}{\|x\|_E} = |\lambda| \sup_{x \neq 0} \frac{\|Ax\|_F}{\|x\|_E} = \lambda \|A\|.$$

Finally, since

$$\|(A + B)x\|_F = \|Ax + Bx\|_F \leq \|Ax\|_F + \|Bx\|_F \leq (\|A\| + \|B\|)\|x\|_E$$

we conclude that $\|A + B\| \leq \|A\| + \|B\|$.

c. It remains to show that $\mathscr{L}(E, F)$ is complete if F is complete. Let $A_n \in \mathscr{L}(E, F)$ be a Cauchy sequence: $\forall \varepsilon > 0, \exists n_0$ such that

$$(5) \qquad \sup_x \frac{\|A_n x - A_m x\|_F}{\|x\|_E} = \|A_n - A_m\| \leq \varepsilon \qquad \text{when} \qquad n, m \geq n_0.$$

Thus for each fixed x

$$(6) \qquad \|A_n x - A_m x\|_F \leq \varepsilon \|x\|_E \qquad \text{when} \qquad n, m \geq n_0,$$

which implies that $A_n x \in F$ is a Cauchy sequence in F. Since F is complete, $A_n x$ converges to an element Ax of F. Let us show first of all that $A : x \mapsto Ax$ is linear. Since

$$(7) \qquad A_n(\lambda x + \mu y) = \lambda A_n x + \mu A_n y,$$

we conclude, by taking the limit, that

$$(8) \qquad A(\lambda x + \mu y) = \lambda Ax + \mu Ay.$$

It remains to show that $\|A\| < +\infty$ and that $\|A_n - A\|$ converges to 0 when n goes to infinity. For this let us fix $\varepsilon > 0$, n and $m \geq n_0$ and $x \in E$, $x \neq 0$. We deduce from (6) the following inequality:

$$\begin{aligned} |\|A_n x\|_F - \|Ax\|_F| &\leq \|A_n x - Ax\|_F = \|A_n x - A_m x + A_m x - Ax\|_F \\ &\leq \|A_n x - A_m x\|_F + \|A_m x - Ax\|_F \\ &\leq \varepsilon \|x\|_E + \|A_m x - Ax\|_F \end{aligned}$$

Fixing $n \geq n_0$, we let m go to infinity. Since $\|A_m x - A x\|_F$ converges to 0, we obtain the inequality

$$(9) \qquad \big| \|A_n x\|_F - \|A x\|_F \big| \leq \|A_n x - A x\|_F \leq \varepsilon \|x\|_E$$

In particular, we conclude that

$$\|A x\|_F \leq \|A_n x\|_F + \varepsilon \|x\|_E \leq (\|A_n\| + \varepsilon) \|x\|_E$$

Dividing by $\|x\|_E$ and taking the sup with respect to x, we obtain

$$\|A\| = \sup_x \frac{\|A x\|_F}{\|x\|_E} \leq (\|A_n\| + \varepsilon) < +\infty.$$

Thus $\|A\| < +\infty$ and $A \in \mathscr{L}(E, F)$. Moreover, we conclude from (9), dividing by $\|x\|_E$, that

$$\frac{\|A_n x - A x\|_F}{\|x\|_E} \leq \varepsilon.$$

Taking the sup with respect to x, we have that

$$\|A_n - A\| = \sup_x \frac{\|A_n x - A x\|_F}{\|x\|_E} \leq \varepsilon,$$

which shows that A_n converges to A in $\mathscr{L}(E, F)$. ∎

Definition 2. Let $(E, \|\cdot\|_E)$ be a normed space. We call the topological dual space of E the space $E^* = \mathscr{L}(E, \mathbb{R})$ of continuous linear forms on E with the norm

$$(10) \qquad \|f\|_* = \sup_{\substack{x \in E \\ x \neq 0}} \frac{|f(x)|}{\|x\|_E}.$$

COROLLARY 1. If $(E, \|\cdot\|)$ is a normed space, the topological dual space $(E^*, \|\cdot\|_*)$ is a Banach space.

Proof. The space $(F, \|\cdot\|_F) = (\mathbb{R}, |\cdot|)$ is a complete normed space.

EXAMPLE. The case where $E = \mathbb{R}^n$ and $F = \mathbb{R}^p$. We shall show that whatever the norms defined on $E = \mathbb{R}^n$ and $F = \mathbb{R}^p$, every linear operator A from $\mathbb{R}^n$ to $\mathbb{R}^p$ is continuous (Theorem 3.3.1).

Meanwhile, let us consider, for example, the case where $\mathbb{R}^n$ and $\mathbb{R}^p$ have the norms $\|x\|_\infty$ and $\|y\|_\infty$.

PROPOSITION 3.   Every linear operator A from $\mathbb{R}^n$ to $\mathbb{R}^p$ is continuous: There exists a constant $M < +\infty$ such that

(11)
$$\|Ax\|_\infty \le M\|x\|_\infty \qquad \text{for all} \qquad x \in \mathbb{R}^n.$$

Proof.   With respect to the canonical bases for $\mathbb{R}^n$ and $\mathbb{R}^p$, the operator A is defined by

$$(Ax)_j = \sum_{k=1}^{n} a_j^k x_k \qquad \text{for} \qquad j = 1, \ldots, p,$$

where $(a_j^k)_{\substack{1 \le k \le n \\ 1 \le j \le p}}$ is the matrix of the operator A. Hence

$$|(Ax)_j| \le \sum_{k=1}^{n} |a_j^k||x_k| \le \|x\|_\infty \sum_{k=1}^{n} |a_j^k| = \|x\|_\infty \|A_j\|_1,$$

where $A_j = (a_j^1, \ldots, a_j^p)$ is the jth column vector and $\|A_j\|_1$ is its norm. Thus

$$\|Ax\|_\infty = \sup_{j=1,\ldots,p} |(Ax)_j| \le \left(\sup_{j=1,\ldots,p} \|A_j\|_1 \right)\|x\|_\infty. \qquad \blacksquare$$

Remark 1.   We can show as an exercise that A is continuous when $\mathbb{R}^n$ and $\mathbb{R}^p$ have the norms $\|x\|_1$ and $\|y\|_1$. $\qquad \blacksquare$

Remark 2.   Let us recall that if $E = \mathbb{R}^n$, $F = \mathbb{R}^p$, the vector space $\mathscr{L}(\mathbb{R}^n, \mathbb{R}^p)$ is isomorphic to the vector space $\mathbb{R}^{np}$. This isomorphism associates to $A \in \mathscr{L}(\mathbb{R}^n, \mathbb{R}^p)$ the coefficients $(a_j^k)_{\substack{1 \le k \le n \\ 1 \le j \le p}} \in \mathbb{R}^{np}$ of the matrix of A. $\qquad \blacksquare$

11.   Hilbert Spaces

Definition 1.   If E is a vector space, we call a scalar product $((x, y))$ on E a function from $E \times E$ to $\mathbb{R}_+$ such that

(1)
$$
\begin{cases}
\textbf{i.} & \left(\left(\sum_{i=1}^{n} \lambda_i x_i, y \right)\right) = \sum_{i=1}^{n} \lambda_i ((x_i, y)) \quad \text{(linearity with respect to x).} \\[2ex]
\textbf{ii.} & \left(\left(x, \sum_{i=1}^{n} \mu_i y_i \right)\right) = \sum_{i=1}^{n} \mu_i ((x, y_i)) \quad \text{(linearity wih respect to y).} \\[2ex]
\textbf{iii.} & ((x, y)) = ((y, x)) \quad \text{(symmetry).} \\[2ex]
\textbf{iv.} & ((x, x)) \ge 0 \quad \text{and} \quad ((x, x)) = 0 \quad \text{if and only if} \quad x = 0.
\end{cases}
$$

Note that (1)i and iii imply ii.

EXAMPLE 1. *Canonical scalar product on* $\mathbb{R}^n$.   Consider the case where $E = \mathbb{R}^n$ with the scalar product:

$$(2) \qquad ((x, y)) = \langle x, y \rangle = \sum_{i=1}^{n} x_i y_i.$$

The conditions (1) are trivially satisfied. In the case where $n = 2$ or $n = 3$, this gives us the usual geometric notion of the Euclidean scalar product. ■

EXAMPLE 2. *Other scalar products on* $\mathbb{R}^n$.   Let A be a matrix from $\mathbb{R}^n$ to $\mathbb{R}^n$ with coefficients (a^{ij}). Consider the function from $\mathbb{R}^n \times \mathbb{R}^n$ to $\mathbb{R}$ defined by

$$(3) \qquad ((x, y)) = \langle Ax, y \rangle = \sum_{i,j=1}^{n} a^{ij} x_j y_i.$$

The properties (1)i and ii are trivially satisfied. *The bilinear form* $((x, y))$ *is symmetric if and only if* $a^{ij} = a^{ji}$ *for every pair of indices* i, j, that is, if and only if *the matrix A is symmetric*. The property (1)iv becomes in this case

$$((x, x)) = \sum_{i,j=1}^{n} a^{ij} x_i x_j > 0 \qquad \text{for all} \qquad x \neq 0.$$

Definition 2.   We say that a matrix $A = (a^{ij})$ is "positive definite" if

$$(4) \qquad \sum_{i,j=1}^{n} a^{ij} x_i x_j > 0 \qquad \text{for all} \qquad x = (x_1, \ldots, x_n) \neq 0.$$

Thus if A is a symmetric positive definite matrix, then $((x, y)) = \langle Ax, y \rangle$ is a scalar product. Conversely consider a *bilinear* form (that is, satisfying (1)i and ii) on $\mathbb{R}^n \times \mathbb{R}^n$. Let $e^i = (0, \ldots, 1, \ldots, 0)$ be the canonical base for $\mathbb{R}^n$. Then $x = \sum_{i=1}^{n} x_i e^i$ and $y = \sum_{j=1}^{n} y_j e^j$. We conclude then from (1)i and ii that

$$((x, y)) = \left(\left(\sum_{i=1}^{n} x_i e^i, \sum_{j=1}^{n} y_j e^j \right) \right)$$

$$(5) \qquad \qquad = \sum_{i,j=1}^{n} ((e^i, e^j)) x_i y_j$$

$$= \langle Ax, y \rangle$$

where A is the matrix with coefficients $a^{ij} = ((e^i, e^j))$.

We have thus shown the following proposition to hold.

PROPOSITION 1. Every scalar product on $\mathbb{R}^n$ can be written in the form

$$(6) \qquad ((x, y)) = \langle Ax, y \rangle = \sum_{i, j = 1}^{n} a^{ij} x_i y_j$$

where $A = (a^{ij})$ is a symmetric positive definite matrix.

EXAMPLE 3. *A scalar product on* $\mathscr{C}_\infty(0, 1)$. Consider the interval $X = [0, 1]$ of $\mathbb{R}$ and the space $\mathscr{C}_\infty(0, 1)$ of real-valued continuous (thus bounded) functions on X. Then the product fg of two functions f and g of $\mathscr{C}_\infty(0, 1)$ is also a continuous and bounded function and thus integrable.

PROPOSITION 2. The bilinear form

$$(7) \qquad \langle f, g \rangle = \int_0^1 f(x)g(x)dx$$

is a scalar product on $\mathscr{C}_\infty(0, 1)$.

EXAMPLE 4. *A scalar product on* $\mathscr{D}^{(1)}(0, 1)$. Consider the interval $X = [0, 1]$ of $\mathbb{R}$ and the set $\mathscr{D}^{(1)}(0, 1)$ of all real-valued functions having continuous (thus bounded) derivatives on X. The derivative of a linear combination of functions being the linear combination of the derivatives, $\mathscr{D}^{(1)}(0, 1)$ is a vector space. If f and g belong to $\mathscr{D}^{(1)}(0, 1)$, the functions fg and $((d/dx)f)\ ((d/dx)g)$ are continuous and bounded and thus integrable.

PROPOSITION 3. The bilinear form

$$(8) \qquad ((f, g))_1 = \int_0^1 f(x)g(x)dx + \int_0^1 \frac{d}{dx} f(x) \frac{d}{dx} g(x)dx$$

is a scalar product on $\mathscr{D}^{(1)}(0, 1)$.

PROPOSITION 4. (*Cauchy–Schwarz inequality*). Every scalar product $((x, y))$ on E satisfies

$$(9) \qquad |((x, y))| \leq ((x, x))^{1/2}((y, y))^{1/2}.$$

Proof. The inequality is true when $y = 0$. Suppose $y \neq 0$. Take $\lambda \in \mathbb{R}$. We have from (1) that

$$(10) \qquad 0 \leq ((x + \lambda y, x + \lambda y)) = ((x, x)) + 2\lambda((x, y)) + \lambda^2((y, y))$$

Replacing λ by $((x, x))^{1/2}/((y, y))^{1/2}$ (which is possible because $((y, y)) > 0$ if $y \neq 0$), we obtain

$$0 \leq ((x, x)) + 2 \frac{((x, x))^{1/2}}{((y, y))^{1/2}} ((x, y)) + ((x, x)) = 2\left[((x, x)) + \frac{((x, x))^{1/2}}{((y, y))^{1/2}} ((x, y)) \right]$$

From this we conclude that

$$(11) \qquad -((x, y)) \leq \frac{((y, y))^{1/2}}{((x, x))^{1/2}} ((x, x)) = ((x, x))^{1/2}((y, y))^{1/2}.$$

Similarly, taking $\lambda = -((x, x))^{1/2}/((y, y))^{1/2}$, we obtain the inequality

$$(12) \qquad ((x, y)) \leq ((x, x))^{1/2}((y, y))^{1/2}.$$

We, therefore, obtain (9) by combining (11) and (12). ∎

PROPOSITION 5. The function $x \mapsto \|x\| = ((x, x))^{1/2}$ is a norm on E, which satisfies the following condition:

$$(13) \qquad 2(\|x\|^2 + \|y\|^2) = \|x + y\|^2 + \|x - y\|^2.$$

Proof. First of all, $\|x\| = ((x, x))^{1/2}$ is positive and if $\|x\| = 0$, then $((x, x)) = 0$ and thus $x = 0$ according to (1)iv. Moreover,

$$\|\lambda x\| = ((\lambda x, \lambda x))^{1/2} = [\lambda^2((x, x))]^{1/2} = |\lambda|((x, x))^{1/2} = \lambda \|x\|.$$

From the Cauchy–Schwarz inequality we conclude that

$$\|x + y\|^2 = ((x + y, x + y)) = \|x\|^2 + \|y\|^2 + 2((x, y))$$
$$\leq \|x\|^2 + \|y\|^2 + 2(\|x\| \, \|y\|) = (\|x\| + \|y\|)^2;$$

that is, $\|x + y\| \leq \|x\| + \|y\|$. Finally, adding the inequalities (10) for $\lambda = 1$ and $\lambda = -1$, we obtain (13). ∎

Thus having a scalar product on E allows us to consider E as a normed space (giving it the norm $\|x\| = ((x, x))^{1/2}$) and, consequently, as a metric space.

Definition 3. We shall say that the pair $(E, ((\cdot, \cdot)))$ is a "prehilbert" space. The pair $(E, ((\cdot, \cdot)))$ is called a Hilbert space if the associated normed space $(E, ((\cdot, \cdot))^{1/2})$ is complete.

COROLLARY 1. If $x \in \mathbb{R}^n$, then

$$(14) \qquad \|x\|_2 = \left(\sum_{j=1}^{n} |x_j|^2 \right)^{1/2}$$

is a norm on $\mathbb{R}^n$ for which $\mathbb{R}^n$ is a Hilbert space; it satisfies

(15) $$\begin{cases} \|x\|_2 \le n^{1/2}\|x\|_\infty; & \|x\|_\infty \le \|x\|_2 & \text{for all} & x \in \mathbb{R}^n. \\ |\langle x, y\rangle| \le \|x\|_2\|y\|_2 & \text{for all} & x, y \in \mathbb{R}^n. \end{cases}$$

For $n = 2$, we see that the metric space $\mathbb{R}^2$ coincides with the metric space $\mathbb{C}$. Thus $\mathbb{C}$ is complete.

COROLLARY 2.   If $f \in \mathbb{C}_\infty(-1, 1)$ then

(16) $$\|f\|_2 = \left(\int_{-1}^1 |f(x)|^2 \, dx \right)^{1/2}$$

is a norm such that

(17) $$\int_{-1}^1 |f(x)g(x)| \, dx \le \|f\|_2 \|g\|_2$$

and such that

(18) $$\|f\|_2 \le 2^{1/2}\|f\|_\infty \qquad \forall f \in \mathscr{C}_\infty(-1, 1).$$

Proof.   Let us show (18). We obtain from the mean value theorem that

$$\|f\|_2^2 = \int_{-1}^1 |f(x)|^2 \, dx \le \int_{-1}^1 \|f\|_\infty^2 \, 1 \, dx$$

$$= \|f\|_\infty^2 \int_{-1}^1 dx = 2\|f\|_\infty^2. \qquad \blacksquare$$

Remark 1.   We can show that the normed space $(\mathscr{C}_\infty(-1, 1), \|\cdot\|_2)$ *is not complete* (although, let us recall that $(\mathscr{C}_\infty(-1, 1), \|\cdot\|_\infty)$ is complete): To this end consider the sequence of continuous functions f_n defined on $[-1, +1]$ as follows:

(19) $$f_n(x) = \begin{cases} 0 & \text{if} & -1 \le x \le -\dfrac{1}{n}. \\[2mm] nx + 1 & \text{if} & -\dfrac{1}{n} \le x \le 0. \\[2mm] 1 & \text{if} & 0 \le x \le 1. \end{cases}$$

Since

$$f_n(x) - f_m(x) = \begin{cases} 0 & \text{if} \quad -1 \leq x \leq -\dfrac{1}{n}, \\[2mm] nx + 1 & \text{if} \quad -\dfrac{1}{n} \leq x \leq -\dfrac{1}{m}, \\[2mm] (n - m)x & \text{if} \quad -\dfrac{1}{m} \leq x \leq 0, \\[2mm] 0 & \text{if} \quad x \geq 0. \end{cases}$$

we obtain that

$$\int_{-1}^{+1} | f_n(x) - f_m(x)|^2 \, dx \leq \frac{(m - n)^2}{3m^2 n}.$$

Thus the sequence f_n is a Cauchy sequence for the norm $\| \cdot \|_2$. This sequence does not converge to a continuous function. Indeed, if f is a continuous function on $[-1, +1]$ such that

$$\int_{-1}^{+1} | f_n(x) - f(x)|^2 \, dx = \int_{-1}^{-1/n} | f(x)|^2 + \int_{-1/n}^{0} |nx + 1 - f(x)|^2 \, dx$$

$$+ \int_{0}^{1} |1 - f(x)|^2 \, dx$$

converges to 0, we would have as a consequence that

$$\int_{-1}^{0} | f(x)|^2 = 0 \qquad \text{and} \qquad \int_{0}^{1} |1 - f(x)|^2 \, dx = 0,$$

that is, $f(x) = 0$ if $-1 \leq x < 0$, $f(x) = 1$ if $0 < x \leq 1$. Hence no continuous function can be the limit of f_n. ∎

We shall denote by $L^2(-1, 1)$ the completion of $(\mathscr{C}_\infty(-1, 1), \| \cdot \|_2)$, which will, therefore, be a Hilbert space. We can characterize the elements of $L^2(-1, 1)$ by means of integration theory. (These are the "classes of measurable functions" that are square integrable.) See Section 4.7.

Definition 4. We denote by $l^2 = l^2(\mathbb{N})$ the space of square summable sequences, that is, the space of sequences $x = \{x_n\}$ such that

(20) $$\|x\|_2 = \left(\sum_{n=0}^{\infty} |x_n|^2 \right)^{1/2} < +\infty.$$

PROPOSITION 6.   If $x, y \in l^2$, then

(21)
$$\langle x, y \rangle = \sum_{n=0}^{\infty} x_n y_n$$

is a scalar product satisfying

(22)
$$|\langle x, y \rangle| \leq \|x\|_2 \|y\|_2 .$$

Proof.   If x and $y \in l^2$, set

$$x^{(n)} = (x_0, \ldots, x_n, 0, \ldots,), \qquad y^{(n)} = (y_0, \ldots, x_n, 0, \ldots,)$$

and

$$z^{(n)} = (|x_0 y_0|, \ldots, |x_n y_n|, 0, \ldots,).$$

Then $\|z^{(n)}\|_1 = \sum_{j=0}^{n} |x_j y_j|$ is an increasing sequence of real numbers. In order that this sequence converge, it is sufficient that it be bounded. We conclude from (15) that

$$\|z^{(n)}\|_1 = \sum_{j=0}^{n} |x_j y_j| \leq \|x^{(n)}\|_2 \|y^{(n)}\|_2 \leq \|x\|_2 \|y\|_2 < +\infty.$$

Thus $\|z^{(n)}\|_1$ converges to $\|z\|_1 = \sum_{j=0}^{\infty} |x_j y_j|$. Therefore, the series $\sum_{n=0}^{\infty} x_n y_n$ converges; it is clear that it is a scalar product.   ■

We shall show that l^2 is *complete*, that is, a Hilbert space, in the following section (Theorem 12.3).

Remark 2.   Hilbert spaces have a richer structure than Banach spaces or, a fortiori, than metric spaces. They possess many more properties than Banach spaces or complete metric spaces. In analysis, we try whenever possible to *work within the framework of Hilbert spaces* in order to obtain more results and with less difficulty. This is what will be done in [AFA].   ■

Remark 3.   In fact, the fundamental property of Hilbert spaces is the linearity of the Fréchet derivative of the associated quadratic function φ, defined by

(23)
$$\varphi(x) = (1/2)\|x\|^2 = (1/2)((x, x)).$$
■

PROPOSITION 7.   The norm of a Hilbert space satisfies the following condition:

(24)
$$\lim_{\|y\| \to 0} \frac{\varphi(x + y) - \varphi(x) - ((x, y))}{\|y\|} = 0.$$

Proof.   By developing the scalar product, we obtain

$$\frac{\varphi(x + y) - \varphi(x) - ((x, y))}{\|y\|} = (1/2)\,\|y\|.$$

which implies (24).   ∎

12.   The Hölder and the Minkowski Inequalities

We are going to show that the function

(1) $$\|x\|_p = \left(\sum_{k=1}^{n} |x_n|^p \right)^{1/p} \qquad \text{for} \qquad p > 1, \qquad x = (x_1, \ldots, x_n) \in \mathbb{R}^n$$

is a norm on $\mathbb{R}^n$. We have already seen that this is true for $p = 1$ (see Section 5) and for $p = 2$ (see Section 11). To this end, we first need to establish the Hölder and the Minkowski inequalities.

Definition 1.   If $p > 1$, we shall say that p^* is the conjugate index of p, where p^* is the number defined by

(2) $$\frac{1}{p^*} = 1 - \frac{1}{p} = \frac{p - 1}{p}.$$

Remark 1.   It is clear that 2 is its own conjugate and that the numbers $p = 1/\theta$ and $p^* = 1/(1 - \theta)$ are conjugates if $0 < \theta < 1$. If $p = 1$, we set $p^* = \infty$, since we can write $1 = 1/1 + 1/\infty$.   ∎

LEMMA 1.   Let a and b be two positive numbers, $p > 1$ and p^* its conjugate index. Then

(3) $$ab \le \frac{1}{p} a^p + \frac{1}{p^*} b^{p^*}$$

and the two terms are equal if $a^p = b^{p^*}$.

Proof.   Consider the function

(4) $$f(x) = \frac{x^p}{p} + \frac{1}{p^*} - x \qquad \text{for} \qquad x \ge 0.$$

Its derivative $f'(x)$ is given by $f'(x) = x^{p-1} - 1$ and vanishes for $x = 1$. Since $p > 1$, $f'(x)$ is negative for $x \le 1$ and positive for $x \ge 1$. Thus $f(x)$ attains its minimum for $x = 1$ and $f(1) = 1/p + 1/p^* - 1 = 0$. This implies that $f(x) \ge 0$ for all $x \ge 0$ and attains its minimum at $x = 1$ where it vanishes.

Taking $x = ab^{-p^*/p}$, we obtain that

$$f(ab^{-p/p^*}) = \frac{a^p b^{-p^*}}{p} + \frac{1}{p^*} - ab^{-p^*/p} \geq 0.$$

Multiplying this inequality by b^{p^*} and remarking that

$$b = b^{-p^*/p + p^*} \quad \text{since} \quad p^* - \frac{p^*}{p} = p^*\left(1 - \frac{1}{p}\right) = \frac{p^*}{p^*} = 1,$$

we have the inequality

$$ab \leq \frac{a^p}{p} + \frac{b^{p^*}}{p^*}. \qquad \blacksquare$$

PROPOSITION 1. (*Hölder inequality*). Let $x = (x_1, \ldots, x_n)$ and $y = (y_1, \ldots, y_n) \in \mathbb{R}^n$. Set

(5) $$\|x\|_p = \left(\sum_{k=1}^n |x_k|^p\right)^{1/p}, \qquad \|y\|_{p^*} = \left(\sum_{k=1}^n |y_k|^{p^*}\right)^{1/p^*}.$$

If p and p^* are conjugates, we then have the Hölder inequality:

(6) $$\max_{\substack{x \in R^n \\ y \in R^n}} \frac{\left|\sum_{k=1}^n x_k y_k\right|}{\|x\|_p \|y\|_{p^*}} = 1.$$

Proof.   Let

$$a_k = \frac{|x_k|}{\|x\|_p}, \qquad b_k = \frac{|y_k|}{\|y\|_{p^*}}.$$

From inequality (3) we have that

$$a_k b_k \leq \frac{1}{p} \frac{|x_k|^p}{\|x\|_p^p} + \frac{1}{p^*} \frac{|y_k|^{p^*}}{\|y\|_{p^*}^{p^*}}.$$

Adding these inequalities for $k = 1$ to $k = n$, we obtain

$$\sum_{k=1}^n a_k b_k \leq \frac{1}{p} \frac{\|x\|_p^p}{\|x\|_p^p} + \frac{1}{p^*} \frac{\|y\|_{p^*}^{p^*}}{\|y\|_{p^*}^{p^*}} = \frac{1}{p} + \frac{1}{p^*} = 1.$$

From this we deduce that

$$\frac{|\sum x_k y_k|}{\|x\|_p \|y\|_{p^*}} \leq \sum_{k=1}^n \frac{|x_k||y_k|}{\|x\|_p \|y\|_{p^*}} = \sum_{k=1}^n a_k b_k \leq 1$$

and, consequently, the inequality we were seeking, known as the *Hölder inequality*. If $x = y = (1, 0, \ldots, 0)$, we obtain

$$\left| \sum_{k=1}^{n} x_k y_k \right| = 1 = \|x\|_p \|y\|_{p^*} = 1. \qquad \blacksquare$$

Remark 2. For $p = 2$, we have again the Cauchy–Schwarz inequality in $\mathbb{R}^n$. $\qquad \blacksquare$

PROPOSITION 2. (*Minkowski inequality*).   Let x and $y \in \mathbb{R}^n$ and $p > 1$. Then

(7) $$\|x + y\|_p \leq \|x\|_p + \|y\|_p.$$

Proof. For $k = 1, \ldots, n$

(8) $$|x_k + y_k|^p = |x_k + y_k|^{p-1} |x_k + y_k| \leq |x_k + y_k|^{p-1}(|x_k| + |y_k|).$$

On the other hand, we deduce from the Hölder inequality that

(9) $$\sum_{k=1}^{n} |x_k + y_k|^{p-1} |x_k| \leq \left[\sum_{k=1}^{n} |x_k + y_k|^{(p-1)p^*} \right]^{1/p^*} \left(\sum_{k=1}^{n} |x_k|^p \right)^{1/p}.$$

Noting that $(p - 1)p^* = p$, we conclude from these inequalities that

(10) $$\sum_{k=1}^{n} |x_k + y_k|^p = \|x + y\|_p^p \leq \|x + y\|_p^{p/p^*}(\|x\|_p + \|y\|_p).$$

Multiplying the two sides of this inequality by $\|x + y\|_p^{-p/p^*}$ and noting that $p(1 - 1/p^*) = p/p = 1$, we obtain the desired inequality, known as the *Minkowski inequality*. $\qquad \blacksquare$

THEOREM 1.   If $p > 1$, the functions

(11) $$\|x\|_p = \left(\sum_{k=1}^{n} |x_k|^p \right)^{1/p} \qquad \text{where} \qquad x = (x_1, \ldots, x_n) \in \mathbb{R}^n.$$

are norms on $\mathbb{R}^n$ such that

(12) $$\left| \sum_{k=1}^{n} x_k y_k \right| \leq \|x\|_p \|y\|_{p^*} \qquad \text{if} \qquad \frac{1}{p} + \frac{1}{p^*} = 1.$$

Proof. If $\|x\|_p = 0$, then $|x_k|^p = 0$ for $k = 1, \ldots, n$ and thus $x = 0$. Moreover,

$$\|\lambda x\|_p = \left(\sum_{k=1}^{n} |\lambda|^p |x_k|^p \right)^{1/p} = |\lambda|^{p/p} \left(\sum_{k=1}^{n} |x_k|^p \right)^{1/p} = |\lambda| \|x\|_p.$$

Finally, the inequality $\|x + y\|_p \leq \|x\|_p + \|y\|_p$ is the Minkowski inequality and inequality (12) is the Hölder inequality. ∎

PROPOSITION 3. The norms $\|x\|_p$ on $\mathbb{R}^n$ satisfy the following inequalities: If $p < q$, then

(13)
$$
\begin{cases}
\textbf{i.} \quad \max_{x \in R^n} \dfrac{\|x\|_q}{\|x\|_p} = 1 \\[4mm]
\textbf{ii.} \quad \max_{x \in R^n} \dfrac{\|x\|_p}{\|x\|_q} = n^{(q-p)/pq}.
\end{cases}
$$

Proof. Since $q - p \geq 0$, we deduce that if $a \leq 1$, $a^{q-p} \leq 1$, that is $a^q \leq a^p$. Then if $a_k \leq 1$, we obtain that

(14)
$$
\sum_{k=1}^n a_k^q \leq \sum_{k=1}^n a_k^p.
$$

Moreover if $p \geq 1$,

(15)
$$
|x_k| = (|x_k|^p)^{1/p} \leq \left(\sum_{j=1}^n |x_j|^p \right)^{1/p} = \|x\|_p.
$$

Thus taking $a_k = |x_k| / \|x\|_p \leq 1$, we deduce from (14) that

$$
\frac{\|x\|_q^q}{\|x\|_p^q} = \sum_{k=1}^n \frac{|x_k|^q}{\|x\|_p^q} \leq \sum_{k=1}^n \frac{|x_k|^p}{\|x\|_p^p} = \frac{\|x\|_p^p}{\|x\|_p^p} = 1
$$

and, hence, that $\|x\|_q \leq \|x\|_p$ if $p \leq q$. If $x = (1, 0, \ldots, 0)$, then $\|x\|_p = \|x\|_q = 1$. We have, therefore, shown (13)i. On the other hand, applying the Hölder inequality for the conjugate indices $r = q/p \geq 1$ and $r^* = q/(q - p)$ gives

(16)
$$
\sum_{k=1}^n a_k b_k \leq \left(\sum_{k=1}^n a_k^{q/p} \right)^{p/q} \left(\sum_{k=1}^n b_k^{q/(q-p)} \right)^{(q-p)/q}
$$

With $a_k = |x_k|^p$ and $b_k = 1$, we obtain

$$
\|x\|_p^p = \sum_{k=1}^n |x_k|^p \, 1 \leq \left(\sum_{k=1}^n |x_k|^q \right)^{p/q} \left(\sum_{k=1}^n 1 \right)^{(q-p)/q}
$$

$$
= \|x\|_q^p n^{(q-p)/q}
$$

Taking $x = (1, 0, \ldots, 0)$, we deduce that $\|x\|_p = n^{1/p}$ and $\|x\|_q = n^{1/q}$ and, consequently, that $\|x\|_p / \|x\|_q = n^{(q-p)/pq}$. ∎

EXAMPLE 1.   The sequence spaces l^p.

Definition 2.   We denote by $l^p = l^p(\mathbb{N})$ the space of sequences $x = \{x_n\}$ such that

$$
(17) \qquad \|x\|_p = \left(\sum_{n=1}^{\infty} |x_n|^p \right)^{1/p} < +\infty.
$$

THEOREM 2.   If $p \geq 1$, the set l^p is a normed vector space. If $p \leq q$, then $l^p \subset l^q \subset l^\infty$ and

$$
(18) \qquad \|x\|_\infty \leq \|x\|_q \leq \|x\|_p \qquad \text{if} \qquad x \in l^1.
$$

If p and p^* are conjugate indices, if $x \in l^p$ and $y \in l^{p^*}$, then

$$
(19) \qquad \left| \sum_{k=0}^{\infty} x_k y_k \right| \leq \|x\|_p \|y\|_{p^*}.
$$

Proof.   It is clear that if $x \in l^p$, $\lambda x \in l^p$ since

$$
\|\lambda x\|_p = \left(\sum_{k=1}^{\infty} |\lambda|^p |x_k|^p \right)^{1/p} = |\lambda| \|x\|_p < +\infty.
$$

Let us show that the sum $x + y$ of two elements x and y belonging to l^p also belongs to l^p. For this it is sufficient to show that the increasing sequence

$$
\|x^{(n)} + y^{(n)}\|_p = \left(\sum_{k=1}^{n} |x_k + y_k|^p \right)^{1/p}
$$

(where $x^{(n)} = (x_1, x_2, \ldots, x_n, 0, 0, \ldots)$) is bounded above. But according to the Minkowski inequality for $\mathbb{R}^n$,

$$
\|x^{(n)} + y^{(n)}\|_p \leq \left(\sum_{k=1}^{n} |x_k|^p \right)^{1/p} + \left(\sum_{k=1}^{n} |y_k|^p \right)^{1/p}
$$

$$
(20) \qquad \qquad \leq \|x\|_p + \|y\|_p.
$$

This implies that $x + y$ belongs to l^p and that

$$
\|x + y\|_p \leq \|x\|_p + \|y\|_p.
$$

Moreover, for all k,

$$
|x_k| = (|x_k|^p)^{1/p} \leq \left(\sum_{k=1}^{\infty} |x_k|^p \right)^{1/p} = \|x\|_p.
$$

Thus

$$
\|x\|_\infty = \sup_k |x_k| \leq \|x\|_p.
$$

Furthermore, $\|x\|_p$ is the limit of the increasing sequence

$$\|x^{(n)}\|_p = \left(\sum_{k=1}^{n} |x_k|^p \right)^{1/p}.$$

Since

(21) $$\|x^{(n)}\|_q \le \|x^{(n)}\|_p \qquad \text{if} \qquad p \le q,$$

we obtain that $\|x\|_q \le \|x\|_p$ if $x \in l^p$ and hence that $l^p \subset l^q$ if $p \le q$. Since $\sum_{k=1}^{\infty} |x_k y_k|$ is the limit of the increasing sequence $\sum_{k=1}^{n} |x_k y_k|$ and since

$$\sum_{k=1}^{n} |x_k y_k| \le \left(\sum_{k=1}^{n} |x_k|^p \right)^{1/p} \left(\sum_{k=1}^{n} |y_k|^{p^*} \right)^{1/p^*} \le \|x\|_p \|y\|_{p^*}$$

if $x \in l^p$ and $y \in l^{p^*}$, we conclude that

$$\sum_{k=1}^{\infty} |x_k y_k| \le \|x\|_p \|y\|_{p^*}. \qquad \blacksquare$$

THEOREM 3. The normed spaces l^p are Banach spaces for $1 \le p \le +\infty$.

Proof. The theorem has already been shown for the case $p = \infty$. (See Corollary 8.1.) In the case where $1 \le p < +\infty$, consider a *Cauchy sequence* $\{x^n\}$ of elements $x^n = (x_1^n, \ldots, x_k^n, \ldots)$ of l^p: For every $\varepsilon > 0$, there exists $n_0(\varepsilon)$ such that

(22) $$\sum_{k=1}^{\infty} |x_k^m - x_k^n|^p = \|x^m - x^n\|_p^p \le \varepsilon^p \qquad \text{for} \qquad m, n \ge n_0(\varepsilon).$$

Fix $n \ge n_0(\varepsilon)$, and set $a_k^m = |x_k^m - x_k^n|^p$. Since, according to inequality (15),

(23) $$\forall k \in \mathbb{N}, \qquad |x_k^m - x_k^n| \le \|x^m - x^n\|_p \le \varepsilon \qquad \text{for} \qquad m, n \ge n_0(\varepsilon),$$

we deduce that the sequence of real numbers $\{x_k^m\}$ is a Cauchy sequence and, thus, that it converges to a number $x_k \in \mathbb{R}$. Let $x = (x_1, \ldots, x_k, \ldots)$, $a_k^m = |x_k^m - x_k^n|^p$ and $a_k = |x_k - x_k^n|^p$. Then a_k^m converges to a_k. Since, according to (22), $\sum_{k=1}^{\infty} a_k^m \le \varepsilon^p$ and since the a_k^m converge to a_k, we obtain that

(24) $$\sum_{k=1}^{\infty} a_k = \sum_{k=1}^{\infty} |x_k - x_k^n|^p = \|x - x^n\|_p^p \le \varepsilon^p \qquad \text{for} \qquad n \ge n_0(\varepsilon).$$

[In fact, for every integer j, the finite sums $s_j^m = \sum_{k=1}^{j} a_k^m \le \varepsilon^p$ converge to the finite sums $s_j = \sum_{k=1}^{j} a_k \le \varepsilon^p$. Consequently, the increasing sequence $\{s_j\}_{j \in \mathbb{N}}$ is bounded by ε^p; thus it converges to an element $s \le \varepsilon^p$, and this element s is precisely $\sum_{k=1}^{\infty} a_k$.] We deduce then from (24) with $n = n_0(\varepsilon)$ that

(25) $$\|x\|_p \le \varepsilon + \|x_{n_0(\varepsilon)}\|_p < +\infty,$$

that is, that x belongs to l^p. Moreover, (24) expresses the fact that the sequence x^n converges to x in l^p, which completes the proof of the theorem.   ∎

Remark 3. If $y \in l^{p^*}$, the Hölder inequality (19) shows that the linear form $x \in l^p \mapsto \langle y, x \rangle = \sum_{k=1}^{\infty} x_k y_k$ is continuous and of norm 1, that is $y \in l^{p^*}$ defines a linear form $j(y)$ belonging to the topological dual space $(l^p)^*$ of l^p. It can be shown that this mapping j is a bijection from l^{p^*} onto $(l^p)^*$, by which l^{p^*} can be identified with the topological dual space of l^p.   ∎

*THEOREM 4.   The space l^{p^*} is isomorphic to the topological dual space of l^p if $1 \le p < +\infty$ and $1/p + 1/p^* = 1$. Moreover,

$$(26) \qquad \|j(y)\|_{(l^p)^*} = \|y\|_{p^*}.$$

Proof.   Let us show that j is *surjective*. We denote by e^n the sequence $(0, 0, \ldots, 0, 1, 0, \ldots)$. Let $f \in (l^p)^*$ be a continuous linear form and set $y_n = f(e^n)$. Since $x^{(n)} = \sum_{k=1}^{n} x_k e^k$, we obtain that $f(x^{(n)}) = \sum_{k=1}^{n} y_k x_k$. Since $x^{(n)}$ converges to $x = (x_1, \ldots, x_k, \ldots) \in l^p$ in l^p, and since f is continuous we deduce that $f(x) = \lim_{n \to \infty} \sum_{k=1}^{\infty} y_k x_k$. We must show that the sequence $y = (y_1, \ldots, y_k, \ldots)$ thus constructed is in l^{p^*}. (This will imply that $f = j(y)$.) For this we associate to the sequence $y = (y_1, \ldots, y_n, \ldots)$ the sequence $z = (z_1, \ldots, z_n, \ldots)$ defined by

$$z_n = y_n |y_n|^{p^* - 2} \qquad \text{if} \qquad p > 1.$$

Since

$$z_n y_n = |y_n|^{p^*} \qquad \text{and} \qquad |z_n|^p = |y_n|^{p(p^* - 1)} = |y_n|^{p^*},$$

we obtain the relations

$$\|z^{(n)}\|_p^p = \|y^{(n)}\|_{p^*}^{p^*} \qquad \text{and} \qquad f(z^{(n)}) = \|y^{(n)}\|_{p^*}^{p^*}.$$

Since the linear form f is continuous, there exists a constant $a > 0$ such that $|f(x)| \le a \|x\|_p$ for every $x \in l^p$. Taking $x = z$, we deduce that

$$\|y^{(n)}\|_{p^*}^{p^*} = |f(z^{(n)})| \le a \|z^{(n)}\|_p = a \|y^{(n)}\|_{p^*}^{p^*/p}.$$

Since $p^* - p^*/p = 1$, this implies that

$$\|y^{(n)}\|_{p^*} = \|y^{(n)}\|_{p^*}^{p^* - p^*/p} \le a.$$

Letting n go to infinity, we obtain that $\|y\|_{p^*} \le a$, that is $y \in l^{p^*}$. If $p = 1$, since,

$$|y^n| = |f(e^n)| \le a \|e^n\|_1 = a,$$

we have that the sequence y belongs to l^∞. It is clear that (26) implies that j is *injective*; in order to show (26), we verify that if $y \in l^{p^*}$, the supremum

$$\| j(y) \|_{(l^p)^*} = \sup_{x \in l^p} \frac{|\langle y, x \rangle|}{\|x\|_p} \le \|y\|_{p^*}$$

is attained at the point $\bar{x}$ defined by

$$\bar{x}_n = y_n |y_n|^{p^* - 2} \qquad \text{if} \qquad p > 1.$$

Indeed, in this case

$$\langle y, \bar{x} \rangle = \|y\|_{p^*}^{p^*}, \qquad \|\bar{x}\|_p = \|y\|_{p^*}^{p^*/p},$$

and, consequently,

$$\| j(y) \|_{p^*} = \|y\|_{p^*}. \qquad \blacksquare$$

Inequality (26) holds also when $p = 1$.

EXAMPLE 2. *The norms* $\|f\|_p$. Consider the vector space $\mathscr{C}_\infty(-1, 1)$ of continuous functions on the interval $[-1, 1]$.

PROPOSITION 4. If $p \ge 1$, then

(27)
$$\|f\|_p = \left[\int_{-1}^{1} |f(x)|^p \, dx \right]^{1/p}$$

is a norm on $\mathscr{C}_\infty(-1, 1)$. If p and p^* are conjugate indices, then

(28)
$$\left| \int_{-1}^{1} f(x)g(x)dx \right| \le \|f\|_p \|g\|_{p^*}.$$

Finally if $p < q$,

(29)
$$\|f\|_p \le 2^{(q-p)/pq} \|f\|_q.$$

Proof. First, we shall prove inequality (28) (*called the Hölder inequality*). For every fixed x, we deduce from (3) with $a = f(x)/\|f\|_p$ and $b = g(x)/\|g\|_{p^*}$ the inequality

(30)
$$\frac{|f(x)g(x)|}{\|f\|_p \|g\|_{p^*}} \le \frac{1}{p} \frac{|f(x)|^p}{\|f\|_p^p} + \frac{1}{p^*} \frac{|g(x)|^{p^*}}{\|g\|_{p^*}^{p^*}}.$$

Integrating this inequality on $[-1, 1]$, we obtain

$$\frac{\int_{-1}^{1} |f(x)g(x)|dx}{\|f\|_p \|g\|_{p^*}} \le \frac{1}{p} \frac{\|f\|_p^p}{\|f\|_p^p} + \frac{1}{p^*} \frac{\|g\|_{p^*}^{p^*}}{\|g\|_{p^*}^{p^*}} = \frac{1}{p} + \frac{1}{p^*} = 1.$$

It remains to show that $\|f\|_p$ is a norm. If $\|f\| = 0$, $\int_{-1}^{1} |f(x)|^p \, dx = 0$. This implies that $|f(x)|^p = 0$ for all $x \in [-1, 1]$, thus that $f = 0$. Moreover,

(31)
$$\begin{cases} \|\lambda f\|_p = \left[\int_{-1}^{1} |\lambda|^p |f(x)|^p \, dx \right]^{1/p} \\ \\ = |\lambda| \left[\int_{-1}^{1} |f(x)|^p \, dx \right]^{1/p} = |\lambda| \|f\|_p \end{cases}$$

Now we must establish the inequality

(32)
$$\|f + g\|_p \leq \|f\|_p + \|g\|_p$$

(called the Minkowski inequality). As in the case for $\mathbb{R}^n$ we write the inequality

$$|f(x) + g(x)|^p \leq |f(x) + g(x)|^{p-1} |f(x)| + |f(x) + g(x)|^{p-1} |g(x)|,$$

and we apply the Hölder inequality to each of the two terms of the right-hand side. Since $p^*(p - 1) = p$, we obtain

$$\|f + g\|_p^p = \int_{-1}^{1} |f(x) + g(x)|^p \, dx$$

$$\leq \left[\int_{-1}^{1} |f(x) + g(x)|^{(p-1)p^*} \right]^{1/p^*}$$

$$\times \left(\left[\int_{-1}^{1} |f(x)|^p \, dx \right]^{1/p} + \left[\int_{-1}^{1} |g(x)|^p \, dx \right]^{1/p} \right)$$

$$= \|f + g\|_p^{p/p^*} (\|f\|_p + \|g\|_p)$$

Since $1 = p - p/p^*$, we obtain (32). Finally if $p \leq q$, by applying the Hölder inequality for the conjugate indices $r = q/p$ and $r^* = q/(q - p)$ to the functions $|f(x)|^p$ and $g(x) = 1$, we obtain

$$\|f\|_p^p = \int_{-1}^{1} |f(x)|^p 1 \, dx \leq \left[\int_{-1}^{1} |f(x)|^q \right]^{p/q} \left[\int_{-1}^{1} 1 \, dx \right]^{(q-p)/q}$$

$$= 2^{(q-p)/q} \|f\|_q^p. \qquad \blacksquare$$

Remark 4. The normed space $(\mathscr{C}_\infty(-1, 1), \|f\|_p)$ is not complete. Indeed, we can verify that the sequence of functions f_n defined by relations (11.19) is a Cauchy sequence for the norm $\|\cdot\|_p \, (1 \leq p < +\infty)$ that cannot converge to a continuous function. $\qquad \blacksquare$

We denote by $L^p(-1, 1)$ the completion of the normed space $(\mathscr{C}_\alpha(-1, 1), \|f\|_p)$, which will be defined in Section 4.7.

If a sequence of continuous functions f_n is such that $\lim_{n \to \infty} \| f_n - f \|_p = 0$, we say that f_n converges to f *in the mean of order p* (or in the L^p norm). In particular, we say that f_n converges to f *in the mean* (for $p = 1$) and that f_n converges to f *in the quadratic mean* (for $p = 2$).

13. Fréchet Spaces

We have seen in Section 5 that normed spaces (and in particular prehilbert spaces) are metric spaces for the distance $d(x, y) = \|x - y\|$ associated with the norm.

Definition 1.   We say that a vector space E is "metrizable" if there exists a distance d on E for which the mappings

(1) $\quad \begin{cases} \textbf{i.} & \{x, y\} \in E \times E \mapsto x + y \in E \\ \textbf{ii.} & \{\lambda, x\} \in \mathbb{R} \times E \mapsto \lambda x \in E \end{cases}$

are continuous. We shall say that E is a Fréchet space if it is complete as well.

We show how to construct a distance on a vector space E on which a countable family of seminorms is defined.

Definition 2.   A mapping p from E to $\mathbb{R}_+$ is a "seminorm" if

(2) $\quad \begin{cases} \textbf{i.} & p(x + y) \le p(x) + p(y) \qquad \forall x, y \in E. \\ \textbf{ii.} & p(\lambda x) = |\lambda| p(x) \qquad\qquad \forall x \in E, \qquad \lambda \in \mathbb{R}. \end{cases}$

Thus a seminorm is a norm if it has the additional property that $p(x) = 0$ implies that $x = 0$.

PROPOSITION 1.   Consider an increasing sequence $\{p_k\}_{k \ge 1}$ of seminorms p_k defined on a vector space E. Suppose that they satisfy the condition:

(3) $\qquad$ If $\qquad p_k(x) = 0 \qquad$ for all $\qquad k \ge 1, \qquad$ then $\qquad x = 0$.

The function d defined by

(4) $$d(x, y) = \sum_{k=1}^{\infty} 2^{-k} \frac{p_k(x - y)}{1 + p_k(x - y)}$$

is a distance satisfying the following conditions:

(5) $$d(x + z, y + z) = d(x, y)$$

and for every $k \geq 1$

(6)
$$
\begin{cases}
\textbf{i.} \quad d(x, y) \leq \dfrac{p_k(x - y)}{1 + p_k(x - y)} + \dfrac{1}{2^k}. \\[2em]
\textbf{ii.} \quad \dfrac{p_k(x - y)}{1 + p_k(x - y)} \leq 2^{k-1}\, d(x, y).
\end{cases}
$$

Furnished with this distance E is a metric vector space.

Before proving this theorem, let us recall that the function $\varphi : u \mapsto u/(1 + u)$ is strictly increasing (hence bijective) from $[0, \infty[$ onto $[0, 1[$ and that its inverse from $[0, 1[$ onto $[0, \infty[$ is defined by $\psi(v) = v/(1 - v)$. These functions are continuous: For every $\varepsilon > 0$,

(7)
$$
\begin{cases}
\dfrac{u}{1 + u} \leq \varepsilon \quad \text{when} \quad u \leq \psi(\varepsilon) \quad \text{and} \\[2em]
\dfrac{v}{1 - v} \leq \varepsilon \quad \text{when} \quad v \leq \varphi(\varepsilon).
\end{cases}
$$

Finally, it is easy to show that

(8)
$$
\varphi(u_1 + u_2) \leq \varphi(u_1) + \varphi(u_2)
$$

Proof. First of all, since $0 \leq p_k(x - y)/(1 + p_k(x - y)) \leq 1$, the series defining $d(x, y)$ is convergent for

$$
0 \leq d(x, y) \leq \sum_{k=1}^{\infty} 2^{-k} = 1.
$$

If $d(x, y) = 0$, then $p_k(x - y) = 0$ for all k. Condition (3) then implies that $x = y$. It is obvious that $d(x, y) = d(y, x)$. Since the function $\varphi : u \mapsto u/(1 + u)$ is strictly increasing from $[0, \infty[$ onto $[0, 1[$ and since $p_k(x - y) \leq p_k(x - z) + p_k(z - y)$, we conclude from (8) that

$$
\frac{p_k(x - y)}{1 + p_k(x - y)} \leq \frac{p_k(x - z)}{1 + p_k(x - z)} + \frac{p_k(z - y)}{1 + p_k(z - y)}.
$$

Multiplying these inequalities by 2^{-k} and summing them, we obtain the triangle inequality

$$
d(x, y) \leq d(x, z) + d(z, y).
$$

Since $p_k(x - y) = p_k(x + z - (y + z))$, it is clear that $d(x, y) = d(x + z, y + z)$. Since $p_j(x - y)/(1 + p_j(x - y)) \leq p_k(x - y)/(1 + p_k(x - y))$ if $j \leq k$ and since $p_k(x - y)/(1 + p_k(x - y)) \leq 1$ if $j \geq k + 1$, we obtain the following upper estimate for $d(x, y)$:

$$d(x, y) \leq \left(\sum_{j=1}^{k} 2^{-j} \right) \frac{p_k(x - y)}{1 + p_k(x - y)} + \sum_{j-k+1}^{\infty} 2^{-j} \leq \frac{p_k(x - y)}{1 + p_k(x - y)} + \frac{1}{2^k}.$$

We obtain a lower estimate for $d(x, y)$ as follows:

$$d(x, y) \geq \sum_{j=1}^{k-1} 2^{-j} \frac{p_j(x - y)}{1 + p_j(x - y)} + \left(\sum_{j=k}^{\infty} 2^{-j} \right) \frac{p_k(x - y)}{1 + p_k(x - y)}$$

$$\geq \left(\sum_{j=k}^{\infty} 2^{-j} \right) \frac{p_k(x - y)}{1 + p_k(x - y)} = 2^{-k+1} \frac{p_k(x - y)}{1 + p_k(x - y)}.$$

Since, using the triangle inequality and (5), we have

$$d(x_0 + y_0, x + y) \leq d(x_0 + y_0, x + y_0) + d(x + y_0, x + y)$$
$$= d(x_0, x) + d(y_0, y),$$

we obtain the continuity of addition $\{x, y\} \in E \times E \mapsto x + y \in E$. To show the continuity of multiplication by scalars $\{\lambda, x\} \mapsto \lambda x$ at $\{\lambda_0, x_0\}$, take $\varepsilon > 0$ and let k be an integer such that $1/2^k \leq \varepsilon/2$. According to (7),

$$(9) \qquad \frac{p_k(\lambda_0 x_0 - \lambda x)}{1 + p_k(\lambda_0 x_0 - \lambda x)} \leq \frac{\varepsilon}{2} \qquad \text{when} \qquad p_k(\lambda_0 x - \lambda x) \leq \alpha = \psi\left(\frac{\varepsilon}{2}\right).$$

Now we verify that

$$(10) \qquad \begin{cases} p_k(\lambda_0 x_0 - \lambda x) \leq \alpha \qquad \text{when} \qquad |\lambda - \lambda_0| \leq \gamma \qquad \text{and} \\ p_k(x - x_0) \leq \gamma, \end{cases}$$

where

$$\gamma = \min\left(1, \frac{\alpha}{|\lambda_0| + p_k(x_0) + 1}\right).$$

(Indeed,

$$p_k(\lambda_0 x_0 - \lambda x) \leq |\lambda| p_k(x_0 - x) + |\lambda_0 - \lambda| p_k(x_0)$$
$$\leq (|\lambda_0| + 1)\gamma + \gamma p_k(x_0) \leq \alpha.)$$

Moreover, we obtain from (7) that

$$(11) \qquad \frac{2^{k-1} d(x, y)}{1 - 2^{k-1} d(x, y)} \leq \gamma \qquad \text{when} \qquad 2^{k-1} d(x, y) \leq \varphi(\gamma) < 1.$$

Thus by taking $|\lambda - \lambda_0| \leq \gamma$ and $d(x_0, x) \leq 2^{-k+1} \varphi(\gamma)$, we obtain from (6)ii that $p_k(x - y) \leq 2^{k-1} d(x, y)/(1 - 2^{k-1} d(x, y)) \leq \varepsilon$, from (10) that $p_k(\lambda_0 x_0 - \lambda x) \leq \alpha$ and from (9) and (6)i that

$$d(\lambda_0 x_0, \lambda x) \leq \frac{p_k(\lambda_0 x_0 - \lambda x)}{1 + p_k(\lambda_0 x_0 - \lambda x)} + \frac{1}{2^k} \leq 2\frac{\varepsilon}{2} \leq \varepsilon.$$

We have thus shown the continuity of multiplication by scalars, and, therefore, E is a metrizable vector space. ∎

In practice, we shall generally use the following characterization of convergent sequences and continuous functions.

PROPOSITION 2. Consider an increasing sequence of seminorms $\{p_k\}_{k \geq 1}$ satisfying condition (3) and the associated distance defined by (4). A sequence x_n converges to x if and only if

(12)
$$\begin{cases} \forall \varepsilon > 0, & \forall p_k, & \exists N(k, \varepsilon) & \text{such that} \\ p_k(x_n - x) \leq \varepsilon & \text{for} & n \geq N(k, \varepsilon). \end{cases}$$

Let (F, δ) be a metric space. A mapping f from E to F is continuous if and only if

(13)
$$\begin{cases} \forall \varepsilon > 0, & \exists p_k, & \exists \eta & \text{such that} \\ p_k(x - y) \leq \eta \Rightarrow \delta(f(x), f(y)) \leq \varepsilon. \end{cases}$$

A mapping g from F to E is continuous at $u \in F$ if

(14)
$$\begin{cases} \forall \varepsilon > 0, & \forall p_k, & \exists \eta & \text{such that} \\ \delta(u - v) \leq \eta \Rightarrow p_k(g(u) - g(v)) \leq \varepsilon. \end{cases}$$

Proof.

a. Let us consider the characterization of convergent sequences. Suppose that the sequence x_n converges to x in the metric space (E, d), and let us establish (12). To this end, we fix $\varepsilon > 0$ and a seminorm p_k. According to (7), there exists N such that $2^{k-1} d(x_n, x) \leq \varphi(\varepsilon)$ when $n \geq N$. Hence, by (6)ii, $p_k(x_n - x) \leq \psi(2^{k-1} d(x_n, x)) \leq \psi\varphi(\varepsilon) = \varepsilon$ when $n \geq N$.

Conversely, suppose that (12) is satisfied. We can associate to $\varepsilon > 0$ an integer k such that $1/2^k \leq \varepsilon/2$, and according to (12), an integer N such that $p_k(x_n - x) \leq \psi(\varepsilon/2)$ when $n \geq N$. By (6)i we deduce that $d(x_n, x) \leq \varphi(p_k(x_n - x)) + 2^{-k} \leq \varphi\psi(\varepsilon/2) + \varepsilon/2 = \varepsilon$ when $n \geq N$.

b. We shall characterize continuous functions in the same fashion. Suppose that f is a continuous function from E to F, and let us establish (13).

Take $\varepsilon > 0$. There exists γ such that $\delta(f(x), f(y)) \leq \varepsilon$ when $d(x, y) \leq \gamma$. Choose a seminorm p_k such that $1/2^k \leq \gamma/2$ and y such that $p_k(y - x) \leq \eta = \psi(\gamma/2)$. Then, by (6)i, $d(x, y) \leq \varphi(p_k(x - y)) + \gamma/2 \leq \gamma$ and, consequently, $\delta(f(x), f(y)) \leq \varepsilon$.

Conversely, suppose that statement (13) holds. Associating to $\varepsilon > 0$, the seminorm p_k and the number γ such that $\delta(f(x), f(y)) \leq \varepsilon$ when $p_k(x - y) \leq \gamma$, we conclude from (6)ii and from (7) that $p_k(x - y) \leq \gamma$ when $d(x, y) \leq 2^{-k+1}\varphi(\gamma)$. Thus f is continuous.

   c. Suppose that g is continuous from F to E. Choose $\varepsilon > 0$ and a seminorm p_k. Let $\beta = \varphi(\varepsilon)$ be the number that, according to (7), implies that $2^{k-1} d(g(u), g(v))/(1 - 2^{k-1} d(g(u), g(v))) \leq \varepsilon$ when $d(g(u), g(v)) \leq 2^{-k+1}\beta$. Then we deduce from (6)ii that $p_k(g(u) - g(v)) \leq \varepsilon$ when $d(g(u), g(v)) \leq 2^{-k+1}\beta$. Since g is continuous, we know that this last inequality holds if $\delta(u, v) \leq \alpha$ for a suitable α. Hence (14) is satisfied.

   Conversely, if (14) holds, if k is such that $1/2^k \leq \varepsilon/2$ and if $\alpha = \psi(\varepsilon/2)$ is defined by (7), we conclude from (14) that $\delta(u, v) \leq \eta$ implies that $p_k(g(u) - g(v)) \leq \alpha$, thus that $p_k(g(u) - g(v))/(1 + p_k(g(u) - g(v))) \leq \varepsilon/2$ and, consequently, that $d(g(u), g(v)) \leq 2\varepsilon/2 = \varepsilon$ using (6)i.   ∎

Remark 1. We can express in another fashion the results stated in Proposition 2. To say that *a sequence of elements x_n converges to x in (E, d) is the same as saying that for all $k \geq 1$, the sequence of real numbers $p_k(x_n - x)$ converges to* 0.   ∎

Remark 2. It is clear by Proposition 2 that *the seminorms p_k are continuous on the metric space (E, d).*   ∎

EXAMPLE 1. We write a set X as a *countable union of an increasing sequence of subsets K_n of X*:

$$(15) \qquad\qquad X = \bigcup_{n=1}^{\infty} K_n.$$

Now we consider the set

$(16) \qquad \mathscr{A}(X)$ of real-valued functions bounded on each set K_n.

The set $\mathscr{A}(X)$ is clearly a vector space on which we can define the functions p_n by

$$(17) \qquad\qquad p_n(f) = \sup_{x \in K_n} |f(x)|.$$

Obviously, these functions p_n are seminorms. They satisfy condition (3): Indeed, suppose that $p_n(f) = 0$ for all n. Take $x \in E$. Using (15), x belongs to one of the K_n's and thus $|f(x)| \leq \sup_{y \in K_n} |f(y)| = p_n(f) = 0$. Hence $f(x) = 0$. Consequently, $\mathscr{A}(X)$ is a *metrizable vector space*. Proposition 2 implies that a sequence of functions $f_j \in \mathscr{A}(X)$ converges to $f \in \mathscr{A}(X)$ if and only if $\sup_{y \in K_n} |f_j(x) - f(x)|$ converges to 0 for all n, that is, if and only if f_j converges uniformly to f on each K_n. We say that the associated *distance d on $\mathscr{A}(X)$ is the distance of uniform convergence on each set K_n.*

THEOREM 1.   The space $\mathscr{A}(X)$ is a Fréchet space.

Proof.   We must show that $\mathscr{A}(X)$ is complete. Consider a Cauchy sequence of functions f_j of $\mathscr{A}(X)$. We derive from properties (6) and (7) that

(18) $\forall n, \quad \sup_{y \in K_n} |f_j(y) - f_l(y)| = p_n(f_j - f_l) \leq \dfrac{2^{n-1} \, d(f_j, f_l)}{1 - 2^{n-1} \, d(f_j, f_l)} \leq \varepsilon$

when $d(f_j, f_l) \leq 2^{-n+1} \varphi(\varepsilon)$. Since we are dealing with a Cauchy sequence, we know that this inequality is satisfied when $i, j \geq N(2^{-n+1} \varphi(\varepsilon))$. Thus for every fixed x in X, there exists n such that $x \in K_n$ (according to (15)). Thus (18) implies that the sequence $f_j(x)$ is a Cauchy sequence of real numbers that converge to a number $f(x)$. On the other hand, for all $x \in K_n$, $i, j \geq N(2^{-n+1} \varphi(\varepsilon))$, we obtain the inequality

$$|f_j(x) - f(x)| \leq \sup_{y \in K_n} |f_j(y) - f_l(y)| + |f_l(x) - f(x)|$$
$$\leq \varepsilon + |f_l(x) - f(x)|$$

using (18). Letting l go to infinity, we find that $|f_j(x) - f(x)| \leq \varepsilon$. Consequently, f is bounded on K_n, and $p_n(f_j - f) \leq \varepsilon$ when $j \geq N(\varepsilon)$. Since this is true for all n, the function f belongs to $\mathscr{A}(X)$ and f_j converges to f uniformly on all the K_n. ∎

Remark 3.   We can, therefore, describe uniform convergence on the members of a countable covering of X by a distance. In the case of uniform convergence on the members of an uncountable covering, it is necessary to leave the framework of metric spaces and take up the study of more general topological spaces. However, we can obtain enough results and study enough examples to justify deferring the study of general topological spaces. ∎

EXAMPLE 2.   An important example is that where

(19) $$X = \mathbb{R}; \qquad K_n = [-n, +n].$$

The space $\mathscr{C}(\mathbb{R})$ of continuous functions is contained in $\mathscr{A}(\mathbb{R})$, since every continuous function is bounded on every closed and bounded interval. *It is a Fréchet space.* More generally, we can take

(20) $$X = \mathbb{R}^p; \qquad K_n = [-n, +n]^p.$$

EXAMPLE 3.   Consider the following important example:

(21) $\quad\begin{cases} E = \mathscr{E}(\mathbb{R}) \text{ is the vector space of all real-valued functions possessing} \\ \text{derivatives of all orders on } \mathbb{R}. \end{cases}$

This is clearly a vector space. Since $\mathbb{R} = \bigcup_{n=1}^{\infty} [-n, +n]$, and since every continuous function is bounded on the closed and bounded intervals $[-n, +n]$, we can define the functions $p_{n,m}$ on $\mathscr{E}(\mathbb{R})$ by

(22) $$p_{n,m}(f) = \sup_{|x| \le n} \sup_{p \le m} |f^{(p)}(x)|.$$

These are clearly seminorms that satisfy condition (3). Therefore, they define a distance on the vector space $\mathscr{E}(\mathbb{R})$. Proposition 2 shows that for this distance a sequence of functions $f_j \in \mathscr{E}(\mathbb{R})$ converges to a function $f \in \mathscr{E}(\mathbb{R})$ if and only if *all the derivatives $f_j^{(m)}$ converge uniformly to the derivatives $f^{(m)}$ on every closed and bounded interval*. Hence this distance takes into account the convergence of the functions f_j and of *all their derivatives*.

THEOREM 2. $\mathscr{E}(\mathbb{R})$ is a Fréchet space.

Proof.   Consider a Cauchy sequence of functions $f_j \in \mathscr{E}(\mathbb{R})$. We deduce that for every m, the sequence of derivatives $f_j^{(m)}$ is a Cauchy sequence in the space $\mathscr{A}(\mathbb{R})$ of functions bounded on every interval $[-n, +n]$. Theorem 1 implies then that $f_j^{(m)}$ converges uniformly to a function $f_m \in \mathscr{A}(\mathbb{R})$ on every interval $[-n, +n]$. Since the functions $f_j^{(m)}$ are continuous and bounded on $[-n, +n]$, Lemma 9.1 implies that the limit f_m is a continuous function. Thus it is sufficient to verify that f_1 is the derivative of f_0, that f_2 is the derivative of f_1, and so on, that is, that $f \in \mathscr{E}(\mathbb{R})$. This is a consequence of the following lemma.

LEMMA 1.   If the sequences f_j and f'_j of continuous functions converge uniformly to continuous functions f and f_1 on an interval $[-n, +n]$, then at every $x \in]-n, +n[$, $f_1(x) = f'(x)$ is the derivative of f at x.

Proof.   Let $x_0 \in]-n, +n[$. We can then write that

(23) $$f_j(x) - f_j(x_0) = \int_{x_0}^{x} f'_j(y)dy.$$

Since the functions f'_j converge uniformly to f_1 on $[-n, +n]$, we conclude that $\int_{x_0}^{x} f'_j(y)dy$ converges to $\int_{x_0}^{x} f_1(y)dy$. Thus, by taking the limit, we obtain

(24) $$f(x) - f(x_0) = \int_{x_0}^{x} f_1(y)dy.$$

This equality expresses the fact that f_1 is the derivative of f.   ■

CHAPTER 2

Topological Properties
of Metric Spaces

Introduction

We present in this chapter the principal elementary topological properties of metric spaces. Since the structure of a metric space allows us to define the notion of a convergent sequence, we are led to distinguish among the subsets A of a metric space E those which are:

1. *Complete*, that is, such that every Cauchy sequence $\{x_n\}$ of elements $x_n \in A$ converges to an element x of A.

2. *Closed*, that is, such that A contains the limits $x = \lim_{n \to \infty} x_n$ of the sequences $\{x_n\}$ of elements x_n of A.

3. *Compact*, that is, such that one can extract from every sequence $\{x_n\}$ of elements x_n of A a subsequence $\{x_{n_k}\}$ that converges to an element x of A.

We shall characterize these sets and give their properties. In Chapter 3, we use these topological notions for the study of continuous mappings.

It is also convenient to introduce the subsets A of E which are:

4. *Open*, that is, such that A is the complement of a closed set.

5. *Neighborhoods* of an element x of E, such that A contains an open set B containing x,

and, among the neighborhoods of x, the

6. *Balls* $B(x, \varepsilon)$ of center x and radius ε, that is, the sets $A = B(x, \varepsilon) = \{y \in E \text{ such that } d(x, y) \leq \varepsilon\}$, which allow us to use the familiar geometric language.

Thus we shall begin by defining these balls in the first section and then study successively closed sets (Section 2), open sets (Section 3), and neighborhoods (Section 4). Finally, after having defined the notions of subsequence and cluster point of a sequence (Section 5), we consider the study of compact subsets (Section 6).

53

Except for Theorem 6.1 characterizing compact sets, most of the results follow directly from the definitions. This chapter is, therefore, essentially devoted to terminology. The introduction of these topological notions, which we could do without "in theory," allows us to clarify the properties most useful in applications and, once this new vocabulary is assimilated, to avoid complicated and roundabout expressions.

We conclude this chapter with the study of the topological properties of convex sets that play an important role in linear and convex analysis.

1.   Balls and Diameters

Definition 1.   Consider a metric space (E, d). The "open ball with center x and radius ε" is the set

$$(1) \qquad \mathring{B}(x, \varepsilon) = \{y \in E \text{ such that } d(x, y) < \varepsilon\},$$

and the "closed ball with center x and radius ε" is the set

$$(2) \qquad B(x, \varepsilon) = \{y \in E \text{ such that } d(x, y) \leq \varepsilon\}.$$

If A and B are two sets, the "distance from A to B" is the number

$$(3) \qquad d(A, B) = \inf_{x \in A} \inf_{y \in B} d(x, y),$$

and we set

$$(4) \qquad d(x, B) = d(\{x\}, B) = \inf_{y \in B} d(x, y).$$

The "diameter of A" is the finite or infinite number

$$(5) \qquad \delta(A) = \sup_{x \in A} \sup_{y \in A} d(x, y),$$

and we say that a "set A is bounded" if A is nonempty and $\delta(A) < \infty$.

Remark 1.   The terminology "distance from A to B," athough well established by usage, is unfortunate since the function d defined by (4) is not a distance in the sense that we defined it in Chapter 1. We shall see in Chapter 4 a (true) distance between sets, called the Hausdorff distance (Section 4.8).   ■

EXAMPLES 1.

a.   In the case of the metric space $(\mathbb{R}, |x|)$, the balls $\mathring{B}(x, \varepsilon)$ and $B(x, \varepsilon)$ are the intervals

$$]x - \varepsilon, x + \varepsilon[\qquad \text{and} \qquad [x - \varepsilon, x + \varepsilon].$$

b.   In the case of $\mathbb{R}^n$, consider the balls

$$B_p(x, \varepsilon) = \{y \in \mathbb{R}^n \text{ such that } \|x - y\|_p \leq \varepsilon\}.$$

In the case where $n = 2$, $x = 0$, $\varepsilon = 1$, the balls $B_p(0, 1)$ with center 0 and radius 1 are defined by

$$B_1(0, 1) = \{y = (y_1, y_2) \text{ such that } |y_1| + |y_2| \leq 1\}.$$
$$B_2(0, 1) = \{y = (y_1, y_2) \text{ such that } |y_1|^2 + |y_2|^2 \leq 1\}.$$
$$B_\infty(0, 1) = \{y = (y_1, y_2) \text{ such that } \sup(|y_1|, |y_2|) \leq 1\}.$$

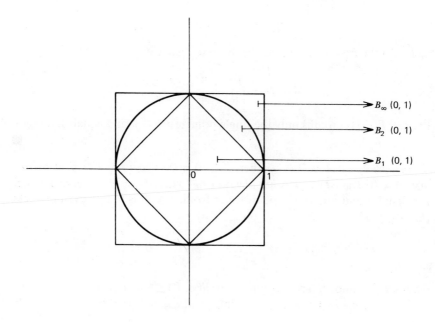

c.   In the case of the Banach space $\mathscr{C}_\infty(0, 1)$ the ball $B(f, \varepsilon)$ with center f and radius ε is the set of functions g such that

$$\sup_{x \in [0, 1]} |f(x) - g(x)| \leq \varepsilon,$$

that is, the set of continuous functions g whose graphs are contained in the shaded band.

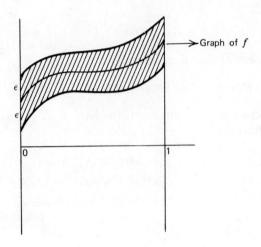

Graph of f

d.   In the case where E has the discrete distance

$$d(x, y) = \begin{cases} 0 & \text{if} \quad x = y, \\ 1 & \text{if} \quad x \neq y, \end{cases}$$

the ball $B(x, \varepsilon)$ contains only the single point x if $\varepsilon < 1$, and equals E if $\varepsilon \geq 1$.   ∎

Remark 2.   If $A \cap B \neq \varnothing$, then $d(A, B) = 0$ since if $x \in A \cap B, d(A, B) \leq d(x, x) = 0$. *The converse is not always true*: There exist sets A and B, which are disjoint and for which the distance from A to B is 0. For example, take $(E, d) = (\mathbb{R}, |\cdot|)$,

$$A = \mathbb{N}_* = \{1, 2, \ldots, n, \ldots\} \quad \text{and} \quad B = \left\{ n - \frac{1}{n} \right\}_{n \in \mathbb{N}_*}.$$

The sets A and B are disjoint, but $d(A, B) \leq d(n, n - 1/n) = 1/n$ since $n \in A, n - 1/n \in B$. Thus $d(A, B) \leq \inf_{n \in \mathbb{N}_*} d(n, n - 1/n) = 0$.   ∎

PROPOSITION 1.   If $y \notin B(x, r)$, then

(6) $d(y, B(x, r)) \geq d(x, y) - r$.

Moreover,

(7) $\delta(B(x, r)) \leq 2r$.

Finally,

(8) $$|d(x, A) - d(y, A)| \le d(x, y).$$

Proof.

a. If $y \notin B(x, r)$, then $d(x, y) - r > 0$. Moreover, if $z \in B(x, r)$, then the triangle inequality implies that

(9) $$d(y, z) \ge d(y, x) - d(x, z) \ge d(x, y) - r \qquad \forall z \in B(x, r)$$

and, consequently, that

$$d(y, B(x, r)) = \inf_{z \in B(x, r)} d(y, z) \ge d(x, y) - r.$$

b. If y and $z \in B(x, r)$, then

(10) $$d(y, z) \le d(x, y) + d(x, z) \le 2r,$$

and, thus,

$$\delta(B(x, r)) = \sup_{y \in A} \sup_{z \in A} d(y, z) \le 2r$$

c. If $z \in A$, the triangle inequality implies that

(11) $$\begin{cases} \textbf{i.} & d(x, z) \le d(x, y) + d(y, z). \\ \textbf{ii.} & d(y, z) \le d(x, y) + d(x, z). \end{cases}$$

Thus

$$d(x, A) = \inf_{z \in A} d(x, z) \le d(x, y) + \inf_{z \in A} d(y, z) = d(x, y) + d(y, A)$$

and similarly

$$d(y, A) \le d(x, y) + d(x, A). \qquad \blacksquare$$

PROPOSITION 2.   The union of two bounded sets is bounded.

Proof.   Let A and B be two bounded sets, $a \in A$, $b \in B$ and x, y two points of $A \cup B$. If x and y belong to A, then $d(x, y) \le \delta(A)$. If x and y belong to B, then $d(x, y) \le \delta(B)$. Now suppose that $x \in A$ and $y \in B$. Then if $a \in A$ and $b \in B$ are arbitrary elements, we obtain

(12) $$d(x, y) \le d(x, a) + d(a, b) + d(b, y) \le d(a, b) + \delta(A) + \delta(B).$$

This inequality being true for any $a \in A$, $b \in B$, we obtain

(13) $$d(x, y) \le \delta(A) + \delta(B) + \inf_{a \in A} \inf_{b \in B} d(a, b) = \delta(A) + \delta(B) + d(A, B).$$

Thus

$$(14) \qquad \delta(A \cup B) = \sup_{x \in A \cup B} \sup_{y \in A \cup B} d(x, y) \le \delta(A) + \delta(B) + d(A, B),$$

which implies that $A \cup B$ is bounded when A and B are bounded.   ∎

2.   Closure and Closed Sets

PROPOSITION 1.   Let A be a subset of a metric space (E, d) and let $x \in E$. The following conditions on x are equivalent:

$$(1) \quad \begin{cases} \textbf{i.} \quad x = \lim_{n \to \infty} x_n \quad \text{where} \quad x_n \in A. \\[2mm] \textbf{ii.} \quad d(x, A) = 0. \\[2mm] \textbf{iii.} \quad \forall n, B\left(x, \dfrac{1}{n}\right) \cap A \ne \varnothing. \end{cases}$$

Proof.   Let us show that i $\Rightarrow$ ii $\Rightarrow$ iii $\Rightarrow$ i.

a.   If $x = \lim_{n \to \infty} x_n$ where $x_n \in A$, then

$$(2) \qquad d(x, A) = \inf_{y \in A} d(x, y) \le d(x, x_n) \qquad \text{for all } n.$$

Since x_n converges to x, $d(x, x_n)$ converges to 0, and hence $d(x, A) = 0$.

b.   By definition of $d(x, A)$, for each n there exists $x_n \in A$ such that

$$(3) \qquad d(x, x_n) \le \inf_{y \in A} d(x, y) + \frac{1}{n} = d(x, A) + \frac{1}{n}$$

Thus if $d(x, A) = 0$, this implies that for every n there exists x_n belonging to A and to the ball $B(x, 1/n)$.

c.   By hypothesis, for every n there exists $x_n \in B(x, 1/n) \cap A$, that is, $x_n \in A$ such that $d(x, x_n) \le 1/n$. This implies, therefore, that $x = \lim x_n$.   ∎

Definition 1.   If $x \in E$ satisfies one of the three equivalent conditions (1), we say that "x belongs to the closure of A." We say that the set $\bar{A}$ of such points is the "closure" of A. A set A is "closed" if $\bar{A} = A$. We say that a subset "A is dense in B" if $B \subset \bar{A}$, that is, if every element of B can be approximated by elements of A.

Remark 1.   To say that a set A is *closed* is the same as to say that A "encloses" all its limit points or that "A is stable under taking limits."   ∎

To show that A is *dense* in B is the same as *establishing an approximation result* allowing an element of B to be replaced by an element of A within an error that can be made arbitrarily small.

EXAMPLES 1.   The following propositions provide examples of closed sets.

PROPOSITION 2.   Every closed ball $B(x, \varepsilon)$ is closed.

Proof.   Let $y \in \overline{B(x, \varepsilon)}$. Since for every $z \in B(x, \varepsilon)$

(4) $$d(x, y) \le d(x, z) + d(y, z) \le \varepsilon + d(y, z),$$

and since $d(y, B(x, \varepsilon)) = \inf_{z \in B(x, \varepsilon)} d(y, z) = 0$ by hypothesis, we deduce from (4) that $d(x, y) \le \varepsilon$, that is that $y \in B(x, \varepsilon)$.   ∎

PROPOSITION 3.   The set E is closed.

Proof.   (*Obvious.*)
By convention we shall say that the empty set $\varnothing$ is closed.   ∎

PROPOSITION 4.   If E has the discrete distance, every subset $A \subset E$ is closed.

Proof.   If $d(x, A) = 0$, then $\inf_{z \in A} d(x, z) = 0$. There exists, therefore, $z \in A$ such that $d(x, z) \le 1/2$, which implies that $x = z \in A$.   ∎

PROPOSITION 5.   Every set that consists of a single point is closed.

Proof.   A point x belongs to the closure of $\{x_0\}$ if and only if $d(x, x_0) = 0$, that is, $x = x_0$.   ∎

PROPOSITION 6.   The closure $\overline{A}$ of a set A satisfies the following conditions:

(5) $\begin{cases} \textbf{i.} & A \subset \overline{A}. \\ \textbf{ii.} & A \subset B \quad \text{implies that} \quad \overline{A} \subset \overline{B}. \\ \textbf{iii.} & \overline{A \cup B} = \overline{A} \cup \overline{B}. \end{cases}$

Proof.

a.   The properties (5)i and ii are obvious.

b. $\overline{A} \cup \overline{B} \subset \overline{A \cup B}$ since $A \subset A \cup B$ and $B \subset A \cup B$ imply that $\overline{A} \subset \overline{A \cup B}$ and $\overline{B} \subset \overline{A \cup B}$, thus $\overline{A} \cup \overline{B} \subset \overline{A \cup B}$.

c.   Suppose that x is a point of closure of $A \cup B$, and let $\{x_n\}$ be a sequence of points $x_n \in A \cup B$ converging to x. Set $x_n = a_n$ if $x_n \in A$ and $x_n = b_n$ if $x_n \in B$.

The sequences $\{a_n\}$ and $\{b_n\}$ cannot both be finite sequences. One of them, $\{a_n\}$ for example, is a subsequence of $\{x_n\}$, which converges to x. Since A is closed, this implies that $x \in \bar{A} \subset \bar{A} \cup \bar{B}$. ∎

THEOREM 1.

a. Every intersection of closed sets is closed.
b. A "finite" union of closed sets is closed.

Proof.

a. Let $A = \bigcap_{i \in I} A_i$ be an intersection of closed sets A_i. If $A = \varnothing$, A is closed. Otherwise, suppose that x is the limit of a sequence $\{x_n\}$ of elements $x_n \in A$. Since $\{x_n\} \subset A_i$ for every $i \in I$, then $x \in A_i$. Thus $x \in \bigcap_{i \in I} A_i = A$.
b. According to (5)iii the union $A_1 \cup A_2$ of two closed sets is closed since $\overline{A_1 \cup A_2} = \bar{A}_1 \cup \bar{A}_2 = A_1 \cup A_2$. Thus by recursion we obtain that the union of a finite sequence of closed sets is closed. ∎

Definition 2. A subset A of a metric space E is "complete" if every Cauchy sequence $\{x_n\}$ of elements $x_n \in A$ has a limit $x \in A$.

PROPOSITION 7.

a. If E is a complete metric space, every closed subset $A \subset E$ is complete.
b. If E is a metric space and if $A \subset E$ is complete, then A is closed.

Proof.

a. Let $\{x_n\}$ be a Cauchy sequence of elements x_n of A. Then x_n converges to x in E, and since A is closed, x belongs to A. Thus A is complete.
b. Let $\{x_n\}$ be a sequence of elements x_n of A converging to an element x of E. Thus $\{x_n\}$ is a Cauchy sequence in E and in A, and since A is complete, $\{x_n\}$ converges to an element x belonging to A. Thus A is closed. ∎

THEOREM 2. Let E be a complete metric space. Let $\{A_n\}$ be a decreasing sequence of nonempty closed subsets A_n of E such that

$$(6) \qquad \lim_{n \to \infty} \delta(A_n) = 0.$$

Then

$$(7) \qquad \bigcap_n A_n = \{a\} \qquad \text{where} \qquad a \in E.$$

Proof. For each n choose an element a_n belonging to A_n. Since $A_n \subset A_p$ if $p \le n$, then $a_n \in A_p \forall p \le n$. The sequence $\{a_n\}$ is a Cauchy sequence in E: For since $\delta(A_n)$ tends to 0, we can associate to every $\varepsilon > 0$ an integer $n_0(\varepsilon)$

such that $\delta(A_p) \leq \delta(A_{n_0(\varepsilon)}) \leq \varepsilon$ when $p \geq n_0(\varepsilon)$. Thus if $p, q \geq n_0(\varepsilon)$, $d(a_p, a_q)$ $\leq \delta(A_{n_0(\varepsilon)}) \leq \varepsilon$. Since E is complete, the sequence a_n converges to an element a of E. Moreover, for each fixed n, the sequence a_{n+p} converges to a as p goes to infinity. Since $a_{n+p} \in A_n$ for every p and since A_n is closed, we conclude that $a \in \bar{A}_n = A_n$. Thus a belongs to the intersection $A = \bigcap_n A_n$ of the A_n's. Since $\delta(A) \leq \delta(A_n)$, for $A = \bigcap_m A_m \subset A_n$ for every n and since $\delta(A_n)$ tends to 0, then $\delta(A) = 0$, which implies that A contains at most a single point. ∎

3. Interior and Open Sets

Let A be a subset of E and B its complement.

PROPOSITION 1.   The following conditions on x are equivalent

(1) $\quad \begin{cases} \textbf{i.} & \text{There exists } \varepsilon > 0 \text{ such that } B(x, \varepsilon) \subset A. \\ \textbf{ii.} & x \text{ does not belong to the closure of the complement of } A. \end{cases}$

Proof.   Indeed, the negation of (1)i is equivalent to

$$(2) \qquad\qquad \forall \varepsilon > 0, \qquad B(x, \varepsilon) \cap \complement A \neq \varnothing,$$

which is the same as saying that x belongs to the closure of the complement of A. ∎

Definition 1.   If $x \in A$ satisfies one of the two equivalent conditions of (1), we shall say that x is an "interior point" of A. We shall say the set $\mathring{A}$ of interior points of A is the "interior" of A and that A is an "open" set if $A = \mathring{A}$.

PROPOSITION 2.   If B is the complement of A, then $\mathring{A} = \complement \bar{B}$. A set is open if and only if its complement is closed.

Proof.   The first statement is a consequence of Proposition 1. Thus $A = \mathring{A}$ if and only if $B = \complement A = \complement \mathring{A}$ is equal to $\bar{B}$. ∎

The properties for closed sets imply symmetric properties for open sets by taking complements.

PROPOSITION 3.   The interior $\mathring{A}$ of a set A satisfies the following properties:

(3) $\quad \begin{cases} \textbf{i.} & \mathring{A} \subset A. \\ \textbf{ii.} & \text{If } A \subset B, \quad \text{then} \quad \mathring{A} \subset \mathring{B}. \\ \textbf{iii.} & \widehat{A \cap B} = \mathring{A} \cap \mathring{B}. \end{cases}$

Moreover, Theorem 2.1 implies the following.

THEOREM 1.

a.   The set E and the empty set $\varnothing$ are open.
b.   Every union of open sets is open.
c.   A "finite" intersection of open sets is open.

PROPOSITION 4.   Every open ball is open.

Proof.   Indeed, if $y \in \mathring{B}(x, \varepsilon)$, then $d(x, y) < \varepsilon$. Thus

$$B\left(y, \frac{\varepsilon - d(x, y)}{2}\right) \subset \mathring{B}(x, \varepsilon) \qquad \text{since if} \qquad z \in B\left(y, \frac{\varepsilon - d(x, y)}{2}\right),$$

$$d(x, z) \leq d(x, y) + d(y, z) \leq (1/2)(d(x, y) + \varepsilon) < \varepsilon. \qquad \blacksquare$$

EXAMPLE 1.   Let E be a vector space. If A and B are two subsets of E, we denote by $A + B$ *the subset of sums* $a + b$ *as a runs over A and b runs over B*:

$$(4) \qquad\qquad A + B = \{a + b\}_{\substack{a \in A \\ b \in B}} \subset E.$$

PROPOSITION 5.   Let A and B be two nonempty subsets of a normed vector space E. If A is open, then $A + B$ is open.

Proof.   First remark that for every fixed $b \in B$, the subset

$$(5) \qquad\qquad A + b = A + \{b\} = \{a + b\}_{a \in A}$$

is open, since if $B(a, \varepsilon)$ is a ball with center a and radius ε contained in A, then $B(a + b, \varepsilon)$ is a ball with center $a + b$, radius ε and contained in $A + b$. Then since

$$(6) \qquad\qquad A + B = \bigcup_{b \in B} (A + b)$$

is a union of open sets, we conclude from Theorem 3.1 that $A + B$ is open.
$\blacksquare$

4.   Neighborhoods

Definition 1.   Let $A \subset E$ be a nonempty subset of a metric space E. A "neighborhood" V of A is any set containing an open set containing A. We denote by $\mathscr{V}(A)$ the set of neighborhoods of A. If $A = \{x\}$, we denote by $\mathscr{V}(x)$ the set of neighborhoods of the point x.

The following propositions provide some *examples* of neighborhoods.

PROPOSITION 1.   If $A \subset E$ is a nonempty set, the balls

(1) $$\mathring{B}(A, \varepsilon) = \{x \in E \text{ such that } d(x, A) < \varepsilon\}$$

are open neighborhoods of A, and the balls

(2) $$B(A, \varepsilon) = \{x \in E \text{ such that } d(x, A) \leq \varepsilon\}$$

are closed neighborhoods of A.

Proof.   Given $\varepsilon > 0$, we have the following inclusions:

(3) $$A \subset \mathring{B}\left(A, \frac{\varepsilon}{2}\right) \subset \mathring{B}(A, \varepsilon) \subset B(A, \varepsilon).$$

It is sufficient to show that $\mathring{B}(A, \varepsilon/2)$ is an open set. Let $y \in \mathring{B}(A, \varepsilon/2)$, that is $d(y, A) < \varepsilon/2$. Then $B(y, (\varepsilon - 2d(y, A))/4) \subset \mathring{B}(A, \varepsilon/2)$ since if $z \in B(y, (\varepsilon - 2d(y, A))/4)$, we obtain for every $x \in A$

(4) $$d(x, z) \leq d(x, y) + d(y, z),$$

and, consequently, taking the inf as x runs over A,

(5) $$d(z, A) \leq d(y, A) + d(y, z) \leq \frac{\varepsilon}{4} + \frac{d(y, A)}{2} < \frac{\varepsilon}{2}.$$

We can prove that $B(A, \varepsilon)$ is closed as in Proposition 2.2.   ■

Remark 1.   The set E is a neighborhood of every nonempty subset A since E is open.   ■

THEOREM 1.   The set $\mathscr{V}(x)$ of neighborhoods of x has the following properties:

(6)
   i.   If $B \supset V$ and if $V \in \mathscr{V}(x)$, then $B \in \mathscr{V}(x)$.
   ii. $x \in V$ for every $V \in \mathscr{V}(x)$.
   iii.   If $V_1, \ldots, V_n \in \mathscr{V}(x)$, then $V = V_1 \cap \cdots \cap V_n \in \mathscr{V}(x)$.
   iv.   If $V \in \mathscr{V}(x)$, there exists $W \in \mathscr{V}(x)$ such that, for every $y \in W$, $V \in \mathscr{V}(y)$.

Proof.   The properties (6)i and (6)ii are obvious. Property (6)iii is a consequence of the fact that every finite intersection of open sets is open. Property (6)iv expresses the fact that *a neighborhood V of x is also a neighborhood of all those points y of W sufficiently close to x*: It is a compatibility property for the families $\mathscr{V}(x)$ of neighborhoods as x runs over E. In order to prove (6)iv, it is sufficient to take W to be an open set containing x and contained in V.   ■

THEOREM 2.   In order for a set A to be open, it is necessary and sufficient that A be a neighborhood of each of its points.

Proof.

a.   It is clear that if A is open, it is a neighborhood of each of its points.

b.   Conversely, suppose that A is a neighborhood of each of its points. Then for every $x \in A$ there exists an open set $U(x)$ such that $x \in U(x) \subset A$. Thus

$$A = \bigcup_{x \in A} \{x\} \subset \bigcup_{x \in A} U(x) \subset A,$$

which shows that $A = \bigcup_{x \in A} U(x)$ is an open set since it is the union of a family of open sets.   ∎

PROPOSITION 2.   Let E by a metric space. For every neighborhood $V \in \mathscr{V}(x)$, there exists $n > 0$ such that $B(x, 1/n) \subset V$.

Proof.   Since $V \supset U \ni x$ where U is an open set, then x is an interior point of U and there exists a sphere $B(x, \varepsilon)$ contained in $U \subset V$. It is sufficient to choose n greater than $1/\varepsilon$.   ∎

PROPOSITION 3.   In a metric space E every closed set is the intersection of a decreasing sequence of open sets, and every open set is the union of an increasing sequence of closed sets.

Proof.   Let A be a closed set and consider the decreasing sequence of open neighborhoods $U_n = \mathring{B}(A, 1/n)$ containing A. Hence $A \subset \bigcap_{n \geq 1} U_n$. Conversely, if $x \in \bigcap_{n \geq 1} U_n$, then $d(x, A) \leq 1/n$ for every n and thus $d(x, A) = 0$. Since A is closed, this implies that $x \in A$.

We obtain the second assertion of the proposition from the first by taking complements: If $B = \complement A$ is closed,

$$(7) \qquad B = \complement A = \complement \bigcap_{n \geq 1} U_n = \bigcup_{n \geq 1} \complement U_n = \bigcup_{n \geq 1} F_n,$$

where the F_n's form an increasing sequence of closed sets.   ∎

PROPOSITION 4.   Let A and B be two closed subsets of a metric space E. If $A \cap B = \varnothing$, there exists $U \in \mathscr{V}(A)$ and $V \in \mathscr{V}(B)$ such that $U \cap V = \varnothing$.

Proof.   Take

$$(8) \quad U = \{x \mid d(x, A) < d(x, B)\} \qquad \text{and} \qquad V = \{y \mid d(y, A) > d(y, B)\}$$

a. $A \subset U$ and $B \subset V$. Indeed if $x \in A, d(x, A) = 0$. Moreover, x belongs to the complement $\complement B$ of the closed set B, which is open. Thus $d(x, B) > 0 = d(x, A)$. Hence $A \subset U$, and we can show similarly that $B \subset V$.

b. $U \cap V = \emptyset$. This is obvious by the definition of U and V.

c. U and V are open. Indeed, let $x_0 \in U$ and $\varepsilon = d(x_0, B) - d(x_0, A) > 0$. Then $B(x_0, \varepsilon/3) \subset U$ since, if $y_0 \in B(x_0, \varepsilon/3)$, the inequalities

(9)
$$
\begin{cases}
\textbf{i.} \quad d(y_0, A) - d(x_0, A) \le d(x_0, y_0) \le \dfrac{d(x_0, B) - d(x_0, A)}{3} \\[4mm]
\textbf{ii.} \quad d(x_0, B) - d(y_0, B) \le d(x_0, y_0) \le \dfrac{d(x_0, B) - d(x_0, A)}{3}
\end{cases}
$$

imply that

(10) $d(y_0, A) - d(y_0, B) \le -1/3(d(x_0, B) - d(x_0, A)) < 0,$

that is, that $y_0 \in U$. In the same way we show that V is open.   ∎

Remark 2.   There exist sets which are *both open and closed*; for example, if E is a discrete metric space, since every subset A of E is closed, then every subset is *both open and closed* (because $\complement A$ is closed).

There exist sets that are *neither open nor closed*. For example, If $E = \mathbb{R}$ and if $A = \,]a, b]$, this set is not closed since a belongs to the closure of A and does not belong to A, and is not open since b is not an interior point of A, b belonging to the closure of the complement of A.   ∎

Remark 3.   We have defined the notion of closure by means of a distance and, thereafter, the notions of closed sets, open sets, and neighborhoods using only the notion of closure.   ∎

We can obtain these concepts without using distance. For example, consider this definition.

Definition 2.   We shall say the a space E is a "topological space" if we can associate to every $x \in E$ a family $\mathscr{V}(x)$ of neighborhoods of x satisfying the conditions (6).

Using this, we can define convergent sequences, closure, and interior of a subset.

Definition 3.   We shall say that a sequence $\{x_n\}$ of elements x_n of a topological space E converges to x if for every $V \in \mathscr{V}(x)$, there exists n_0 such that $x_n \in V$ for every $n \ge n_0$. A point $x \in E$ is called a point of closure of a subset A of E if

(11) $\forall V \in \mathscr{V}(x), \quad A \cap V \ne \emptyset,$

and $x \in A$ is called an interior point of A if

(12) $\exists V \in \mathscr{V}(x) \quad$ such that $\quad V \subset A.$

Let us remark, however, that *the notion of a Cauchy sequence has no meaning in a topological space.*

Many properties of metric spaces remain true for topological spaces. Let us observe, nevertheless, that *we are obliged to abandon the metric space framework* in order to consider notions of convergence that cannot be described by a distance, in particular, *pointwise convergence* of a sequence f_n of functions. The concepts of *convergence in the mean of order p for $p \geq 1$* and of *uniform convergence* can be described by the distances $\|f - g\|_p$ for $1 \leq p \leq +\infty$, as well as the concept of *uniform convergence on the sets of a countable covering.* However, we can describe the notion of pointwise convergence within a richer structure than that of a topological space, namely one *defined by a family of distances* (rather than a single distance, as in the case of metric spaces).

5.   Cluster Points of a Sequence

Consider the sequence of real numbers defined by

$$
(1) \qquad x_n = \begin{cases} \dfrac{1}{k} & \text{if} \quad n = 2k. \\[2mm] k & \text{if} \quad n = 2k + 1. \end{cases}
$$

Although the sequence x_n does not converge to any point (it is not a Cauchy sequence), the subsequence $y_k = x_{2k} = 1/k$ extracted for the sequence x_n does converge.

Definition 1.   We denote by "subsequence" any strictly increasing mapping $k \mapsto n_k$ from $\mathbb{N}$ to $\mathbb{N}$ and by "subsequence of a sequence $\{x_n\}$" the sequence $k \mapsto x_{n_k}$. In other words, a sequence $\{x_n\}$ being a mapping $n \mapsto x_n$ from $\mathbb{N}$ to E, a subsequence obtained from x_n is the composed mapping $k \mapsto n_k \mapsto x_{n_k}$ where $k \mapsto n_k$ is strictly increasing.

PROPOSITION 1.   Every subsequence of a sequence $\{x_n\}$ converging to x converges to x.

Proof.   Since x_n converges to x, $d(x, x_n) \leq \varepsilon$ for $n \geq n(\varepsilon)$. Taking $k(\varepsilon)$ such that $n_{k(\varepsilon)} \geq n(\varepsilon)$, we obtain that $d(x, x_{n_k}) \leq \varepsilon$ for $k \geq k(\varepsilon)$.   ∎

PROPOSITION 2.   The following three properties are equivalent:

$$
(2) \qquad \begin{cases} \textbf{i.} & x \text{ is the limit of a subsequence } \{x_{n_k}\}_k \text{ of the sequence } \{x_n\}_n, \\[1mm] \textbf{ii.} & \forall \varepsilon > 0 \text{ and } \forall n, \text{ there exists } m \geq n \text{ such that } d(x, x_m) \leq \varepsilon, \\[1mm] \textbf{iii.} & \forall n, x \text{ belongs to the closure of the set } A_n = \{x_m\}_{m \geq n}. \end{cases}
$$

Proof.

a.   It is clear that (2)i implies (2)ii: Since $d(x, x_{n_k}) \le \varepsilon$ for $k \ge k(\varepsilon)$, it suffices to take $m = n_k$ for $k \ge k(\varepsilon)$, which satisfies $n_k \ge n$.

b.   It is clear that (2)ii implies (2) iii.

c.   Let us show that (2)iii implies (2)i.

We know that for every $\varepsilon = 1/k$ and for every $n \ge 1$, there exists a number $n(k)$ such that $d(x, x_{n(k)}) \le 1/k$. We define the subsequence n_k by taking n_k to be the smallest integer strictly greater than n_{k-1} and verifying $d(x, x_{n_k}) \le 1/k$. The subsequence x_{n_k} converges to x since for every $\varepsilon > 0$, there exists $k_0 \ge 1/\varepsilon$ such that $d(x, x_{n_k}) \le 1/k \le 1/k_0 \le \varepsilon$ when $k \ge k_0$.   ∎

Definition 2.   We say that "x is a cluster point" of the sequence $\{x_n\}$ if one of the three equivalent conditions (2) is satisfied.

COROLLARY 1.   If x is the limit of a sequence $\{x_n\}$, x is the unique cluster point of $\{x_n\}$.

Proof.   (*Obvious.*)

Remark 1.   The sequence $\{x_n\}$ defined by (1) has exactly one cluster point (which is 0) but this does not imply that 0 is the limit of the sequence $\{x_n\}$.   ∎

PROPOSITION 3.   If a Cauchy sequence $\{x_n\}$ has a cluster point x, then x is the limit of $\{x_n\}$.

Proof.   We know that for $\varepsilon > 0$, there exist k_0 and n_0 such that $d(x, x_{n_k}) \le \varepsilon/2$ for $k \ge k_0$ and $d(x_p, x_q) \le \varepsilon/2$ for $p, q \ge n_0$. Thus by taking $n_1 = \max(n_0, n_{k_0})$, we obtain

$$d(x, x_p) \le d(x, x_{n_k}) + d(x_{n_k}, x_p) \le \varepsilon$$

if $p \ge n_1$ (taking $k \ge k_0$ such that $x_{n_k} \ge n_0$).   ∎

6.   Compact Sets

Given E a metric space, we are going to characterize the subsets K of E having the following property:

Definition 1.   We shall say that a subset K of E is "compact" if every infinite sequence $\{x_n\}$ of elements x_n of K has at least one cluster point belonging to K.

We see the importance of compact sets in all problems involving taking limits: If we know that a set is compact, then every sequence has a convergent subsequence.

In order to characterize compact sets, we need to know some of their elementary properties.

PROPOSITION 1.   Let K be a compact subset of E. Then

(1)
$$\begin{cases} \textbf{i.} & K \text{ is closed.} \\ \textbf{ii.} & K \text{ is complete.} \\ \textbf{iii.} & K \text{ is bounded.} \\ \textbf{iv.} & \text{There exists a sequence } D = \{a_1, \ldots, a_n, \ldots\} \text{ that is dense in } K. \\ \textbf{v.} & \text{If } x \text{ is the unique cluster point of a sequence } \{x_n\} \text{ of elements} \\ & x_n \text{ of } K, \text{ then } x \text{ is the limit of } \{x_n\}. \end{cases}$$

Proof.

a. K is closed since it contains the cluster points of its sequences and, therefore, the limits of its convergent sequences.

b. K is complete for if $\{x_n\}$ is a Cauchy sequence of elements x_n of K, then it has a cluster point $x \in K$, which implies that x is the limit of $\{x_n\}$ (see Proposition 5.3).

c. K is bounded; that is, its diameter $\delta = \sup_{x, y \in K} d(x, y)$ is finite. Indeed, if this were not the case, there would exist sequences $\{x_n\}$ and $\{y_n\}$ of elements x_n and y_n of K such that $d(x_n, y_n)$ would approach infinity. But these sequences have convergent subsequences x_{n_k} and y_{n_k} converging respectively to elements x and y of K. Thus $d(x, y) = \lim_{k \to \infty} d(x_{n_k}, y_{n_k}) = \infty$, which is impossible.

d.   We shall construct the sequence $\{a_n\}$ by recursion. Take $a_0 = x_0 \in E$. We construct a_1 such that

(2)
$$d(a_1, a_0) \geq \frac{d_0}{2} \quad \text{where} \quad d_0 = \sup_{x \in K} d(a_0, x)$$

and knowing the a_i's for $i \leq n - 1$, we construct a_n such that

(3)
$$\min_{0 \leq i \leq n-1} d(a_n, a_i) \geq \frac{d_{n-1}}{2} \quad \text{where} \quad d_{n-1} = \sup_{x \in K} \min_{0 \leq i \leq n-1} d(x, a_i).$$

It is clear that $d_0 \geq d_1 \geq \cdots \geq d_n \geq \cdots$. The sequence d_n converges to 0. Indeed, there exists a subsequence $\{a_{n_k}\}$ that is convergent and, therefore, is a Cauchy sequence: For every $\varepsilon > 0$, there is k_0 such that

$$d_{n_k - 1} \leq 2d(a_{n_{k_0}}, a_{n_k}) \leq \varepsilon \quad \text{for} \quad k \geq k_0.$$

Thus the subsequence d_{n_k} converges to 0, and since $\{d_n\}$ is decreasing, the sequence d_n converges to 0. Thus for every $x \in K$ and for every ε, there exists $n(\varepsilon)$ such that for $n \geq n(\varepsilon)$ we have

(4) $$\min_{0 \leq i \leq n} d(x, a_i) \leq \sup_{x \in K} \min_{0 \leq i \leq n} d(x, a_i) = d_n \leq \varepsilon.$$

Hence there exists $a_i(x, \varepsilon)$ such that $d(x, a_i(x, \varepsilon)) \leq \varepsilon$, which implies that $D = \{a_0, \ldots, a_n, \ldots\}$ is dense in K.

e. Let x be a cluster point of $\{x_n\}$, which is not the limit of the sequence $\{x_n\}$. There exist a number $\varepsilon > 0$ and an infinite subsequence x_{n_k} such that $d(x, x_{n_k}) \geq \varepsilon$. But the sequence $\{x_{n_k}\}$ has a cluster point y since K is compact, and we deduce, therefore, that $d(x, y) \geq \varepsilon$. This implies that y is a cluster point distinct from x, which contradicts the hypothesis that x was the unique cluster point of $\{x_n\}$. ∎

Now we shall introduce the following notions.

Definition 2.

a. If $K \subset E$ is the union of a family $\{V_i\}_{i \in I}$ of open sets V_i, we say that $\{V_i\}_{i \in I}$ is an "open covering of K." We say it is a "finite covering" if the family V_i is finite. A set K has the "Borel–Lebesgue property" if and only if we can extract a finite covering from every open covering.
b. A family $\{F_i\}_{i \in I}$ of closed sets F_i of K has the "finite intersection property" if and only if $\bigcap_{i \in J} F_i \neq \varnothing$ for every finite sequence $J \subset I$.

THEOREM 1. Given K a subset of a metric space, the following statements are equivalent:

(5)
> **i.** K is compact.
> **ii.** K is complete and for every $\varepsilon > 0$, K can be covered by a finite sequence of balls of radius ε.
> **iii.** K is closed and has the Borel–Lebesgue property.
> **iv.** K is closed and if $\{F_i\}_{i \in I}$ is a family of closed sets F_i with the finite intersection property, then $\bigcap_{i \in I} F_i \neq \varnothing$.

Proof. We are going to show that properties (5)i and (5)ii are equivalent, that properties (5)iii and (5)iv are equivalent, that property (5)iv implies (5)i and that property (5)i implies (5)iii.

a. (5)i IMPLIES (5)ii. First of all, K is complete according to Proposition 1. Suppose that (5)ii is false: There exists $\varepsilon > 0$ such that the intersection of K and the complement of every finite union of balls of radius ε is nonempty.

Then let us construct the following sequence by recursion. The point x_1 is chosen arbitrarily. Since $K \cap \complement B(x_1, \varepsilon) \neq \emptyset$, there exists $x_2 \in K$ such that $d(x_1, x_2) > \varepsilon$. Suppose that we have constructed $x_1, \ldots, x_n$ such that $d(x_i, x_j) > \varepsilon$ for all $i, j = 1, \ldots, n$, where $i \neq j$. Since

$$K \cap \complement \left(\bigcup_{i=1}^{n} B(x_i, \varepsilon) \right) = K \cap \bigcap_{i=1}^{n} \complement B(x_i, \varepsilon) \neq \emptyset,$$

there exists $x_{n+1} \in K$ such that $d(x_{n+1}, x_i) > \varepsilon$ for all $i = 1, \ldots, n$.

The sequence $\{x_n\}$ thus constructed has no cluster points: Indeed, if $x \in K$ were a cluster point, there would exist a subsequence x_{n_k} converging to x, and thus such that $d(x, x_{n_k}) \leq \varepsilon/3$ when $k \geq k_\varepsilon$. If h and $k \geq k_\varepsilon$, we would obtain the contradiction

$$\varepsilon < d(x_{n_h}, x_{n_k}) \leq d(x_{n_h}, x) + d(x, x_{n_k}) \leq \frac{2\varepsilon}{3}.$$

b. (5)ii IMPLIES (5)i. By hypothesis for every $n \geq 1$ there exists a finite sequence a_i^n such that $K \subset \bigcup_{\text{finite}} B(a_i^n, 1/n)$. Let $\{x_n\}$ be an infinite sequence of elements x_n of K. We show that this sequence contains a Cauchy subsequence, which, since K is complete, is convergent; this establishes that K is compact.

For $n = 1$, $K = \bigcup_{\text{finite}} B(a_i^1, 1)$ and there exists an infinite subsequence $\{x_n^1\}$ *entirely contained* in one of the balls $B(a_i^1, 1)$. (Otherwise, each sphere $B(a_i^1, 1)$ would contain a finite number of elements x_n of the sequence $\{x_n\}$, which would imply that $\{x_n\}$ is a finite sequence.)

Similarly for $n = 2$ there exists an infinite subsequence $\{x_n^2\}$ extracted from $\{x_n^1\}$ and entirely contained in a ball of radius $1/2$; thus we obtain infinite subsequences $\{x_n^k\}$ extracted from $\{x_n^{k-1}\}$ and contained in a ball of radius $1/k$.

Now set $y_k = x_k^k$. By construction, the sequence $\{y_k\}$ is a Cauchy sequence since $d(y_k, y_l) \leq 2(1/k)$ for $l \geq k$. Since K is complete, $\{y_k\}$ is thus a convergent subsequence of $\{x_n\}$.

c. (5)ii AND (5)iv ARE EQUIVALENT. To say that K has the Borel–Lebesgue property is equivalent to saying that if $\{V_i\}_{i \in I}$ is an open covering,

$$(6) \qquad K \subset \bigcup_{i \in I} V_i \Rightarrow \exists J \subset I, \quad J \text{ finite}, \quad \text{such that } K \subset \bigcup_{i \in J} V_i.$$

Setting $F_i = \complement V_i$, statement (6) is equivalent to

$$(7) \qquad \begin{cases} \forall J \subset I, \quad J \text{ finite}, \quad \text{such that} \\ K \cap \bigcap_{i \in J} F_i \neq \emptyset \Rightarrow K \cap \bigcap_{i \in I} F_i \neq \emptyset. \end{cases}$$

That is to say, if the sequence of closed sets $K \cap F_i$ has the finite intersection property, then $\bigcap_{i \in I} K \cap F_i \neq \emptyset$.

d. (5)iv IMPLIES (5)i.   Let $\{x_n\}$ be a sequence of elements $x_n \in K$, and consider the closed sets $\bar{A}_n = \overline{\{x_m\}_{m \geq n}}$. These closed sets form a decreasing sequence and thus have the finite intersection property. Hence there exists $x \in \bigcap_{n \geq 1} \bar{A}_n$, which implies that x belongs to the closure of all the $\bar{A}_n$'s. Then according to Proposition 5.2, x is a cluster point of the sequence $\{x_n\}$.

e. (5)i IMPLIES (5)iii.   Let $\{V_i\}_{i \in I}$ be an open covering of K. First, we shall show that $K = \bigcup_{n, k \geq 1} \mathring{B}(x_n, 1/k)$ where $\mathring{B}(x_n, 1/k) \subset V_i$ for a suitable i (that is, we shall extract a countable covering). Let $D = \{a_1, \ldots, a_n, \ldots\}$ be a sequence that is dense in K. If $x \in K$, it belongs to one of the V_i's, and since V_i is open, there exists a ball $\mathring{B}(x, \varepsilon_i)$ contained in V_i. Moreover, if $k > 2/\varepsilon_i$, there exists an element $a_n \in D$ such that $d(x, a_n) < 1/k$. Thus

$$(8) \qquad x \in \mathring{B}\left(a_n, \frac{1}{k}\right) \subset \mathring{B}(x, \varepsilon_i) \subset V_i$$

since if $y \in \mathring{B}(a_n, 1/k)$,

$$(9) \qquad d(x, y) \leq d(x, a_n) + d(a_n, y) < \frac{2}{k} < \varepsilon_i.$$

It suffices now to extract a finite covering from the covering of K by the countable family of balls $\mathring{B}(x_n, 1/k)$. Consider the increasing sequence of open sets:

$$C_j = \bigcup_{\substack{n \leq j \\ k \leq j}} \mathring{B}\left(a_n, \frac{1}{k}\right).$$

We must show that one of the C_j's is equal to K. If not, for every j there would exist $x_j \in K$ and not belonging to C_j and a subsequence x_{j_k} converging to $x \in K$. The sets $F_j = \complement C_j$ are closed and decreasing, the elements $x_j \in F_j$, and, hence, $x \in \cap F_j = \complement \cup C_j = \complement K = \emptyset$. We have thus arrived at at contradiction. ∎

PROPOSITION 2.

a.   If E is compact and if $K \subset E$ is closed, then K is compact.
b.   Every intersection of compact sets is compact.
c.   Every finite union of compact sets is compact.

Proof.   Statements a and b are obvious. For statement c, let K_1 and K_2 be two compact sets. Then $K_1 \cup K_2$ is closed. Let $\{V_i\}_{i \in I}$ be an open covering of $K_1 \cup K_2$. Since K_1 and K_2 are compact, there exist two finite subsets J_1 and J_2 of I such that $K_1 \subset \bigcup_{i \in J_1} V_i$ and $K_2 \subset \bigcup_{i \in J_2} V_i$. Therefore, $K_1 \cup K_2 \subset \bigcup_{i \in J} V_i$, where $J = J_1 \cup J_2$ is a finite subset of I.   ∎

Definition 3.   We say that a subset A of E is "relatively compact" if its closure $\bar{A}$ is compact and that A is "precompact" or "totally bounded" if its completion is compact.

PROPOSITION 3. (Borel–Lebesgue).   A subset K of the space $\mathbb{R}^p$ is compact if and only if it is closed and bounded.

Proof.

a. If K is compact, we know that it is closed and bounded by Proposition 1.

b.   *Conversely,* let K be a closed and bounded subset of $\mathbb{R}^p$. K is complete since it is closed in the complete space $\mathbb{R}^p$. Since K is bounded, it is contained in a product of intervals $[a_i, b_i]$. For every integer n, the spheres $\mathring{B}(x_k, 1/n) = \prod_{i=1}^{p}]x_{k_i} - 1/n, x_{k_i} + 1/n[$ form a covering of $\prod_{i=1}^{p} [a_i, b_i]$, where $x_{k_i} = a_i + (k_i(b_i - a_i)/n)\,(0 \le k_i \le n)$. Thus $K \subset \bigcup_{k=0}^{n} \mathring{B}(x_k, 1/n)$. Using statement ii of Theorem 1, we, therefore, conclude that K is compact.   ∎

Remark 1.   In infinite dimensional normed vector spaces, a set can be closed and bounded but not compact. We shall characterize the compact sets of the space $\mathscr{C}_\infty(E)$ in Section 4.6.   ∎

PROPOSITION 4.   Let E be a discrete metric space. Then E is compact if and only if E is a finite set.

Proof.

a.   Suppose E is compact. Since $\{x\}$ is open in E and since $E = \bigcup_{x \in E} \{x\}$, we can extract a finite covering, which implies that $E = \bigcup_{i=1,\ldots,n} \{x_i\} = \{x_1, \ldots, x_n\}$.

b.   If E is finite, E is compact since there are only a finite number of open sets in E.   ∎

We shall now give some supplementary properties of compact sets.

PROPOSITION 5.   Given A and B two subsets of a normed vector space, if A is compact and B is closed, then $A + B$ is closed.

Proof.   Let $\{x^n\}$ be a sequence of elements $x^n = a^n + b^n$ of $A + B$ (where $a^n \in A$ and $b^n \in B$), which converges to an element x of E. Let us show that $x = a + b$, where $a \in A$ and $b \in B$. Since A is compact, a subsequence a^{n_k}

converges to an element $a \in A$. Thus the subsequence $b^{n_k} = x^{n_k} - a^{n_k}$ converges to an element $b = x - a$, which belongs to B since B is closed. Hence $A + B$ is closed.   ∎

Remark 2.   A similar argument would show that the sum $A + B$ of two compact sets A and B is compact.   ∎

7.   Convex Sets

The convex subsets of a vector space play a fundamental role in analysis.

Definition 1.   Let E be a real vector space. A subset K of E is called "convex" if for every $x, y \in K$ and every $\lambda \in [0, 1]$, $\lambda x + (1 - \lambda)y \in K$. A subset K of E is a "convex cone" with vertex 0 if for every $x, y \in K$ and every $\lambda, \mu \geq 0$, $\lambda x + \mu y \in K$. We say that $x = \sum_{i=1}^{n} \lambda_i x_i$ is a "convex combination of elements x_i" if $\lambda_i \geq 0$ and $\sum_{i=1}^{n} \lambda_i = 1$.

We agree to say that the empty set is convex.

The following proposition summarizes the elementary properties of convex sets.

PROPOSITION 1.

a.   A subset $K \subset E$ is convex if and only if every convex combination of elements $x_i \in K$ belongs to K.

b.   Under a linear mapping, the image and the inverse image of a convex set are convex.

c.   The intersection of any family of convex sets is convex.

d.   Every product of convex sets is convex.

e.   If K and L are convex sets, $K + L$ is convex.

Proof.   (*Left as an exercise.*)

PROPOSITION 2.   Every vector subspace K of E is convex. If E is a normed space, the balls $\mathring{B}(x, \varepsilon)$ and $B(x, \varepsilon)$ are convex.

Proof.   (*Left as an exercise.*)

We shall now give some topological properties of convex sets.

THEOREM 1.   Let K be a convex subset of a normed space E. Suppose that $x_0 \in \mathring{K}$ and $x_1 \in \overline{K}$. Then

$$(1) \qquad \forall \lambda \in \,]0, 1[, \qquad x = \lambda x_0 + (1 - \lambda)x_1 \in \mathring{K}.$$

Proof. Let $\lambda \in]0, 1[$ and $x_\lambda = \lambda x_0 + (1 - \lambda)x_1$. We are going to show that x_λ belongs to an open set $\tilde{B}$ contained in K. Since $x_0 \in \overset{\circ}{K}$, there exists $\varepsilon > 0$ such that $\overset{\circ}{B}(x_0, \varepsilon) \subset K$. Since $x_1 \in \overline{K}$, there exists $y \in K$ such that $y \in \overset{\circ}{B}(x_1, \varepsilon\lambda/(1 - \lambda))$. Consider the open ball

$$\text{(2)} \qquad \qquad \tilde{B} = \overset{\circ}{B}(\lambda x_0 + (1 - \lambda)y, \lambda\varepsilon).$$

Then x_λ belongs to $\tilde{B}$ since

$$\|x_\lambda - \lambda x_0 - (1 - \lambda)y\| = (1 - \lambda)\|x_1 - y\| \le \frac{(1 - \lambda)\varepsilon\lambda}{(1 - \lambda)} = \varepsilon\lambda.$$

Moreover, $\tilde{B} \subset K$: Indeed, if $z \in \tilde{B}$, we conclude that

$$\left\|\left(\frac{1}{\lambda}\right)(z - (1 - \lambda)y) - x_0\right\| < \frac{\lambda\varepsilon}{\lambda} = \varepsilon,$$

and, consequently, that $(1/\lambda)(z - (1 - \lambda)y) \in \overset{\circ}{B}(x_0, \varepsilon) \subset K$. Since $y \in K$ and since K is convex, we obtain that

$$\lambda\left(\frac{1}{\lambda}\right)(z - (1 - \lambda)y) + (1 - \lambda)y = z \in K. \qquad \blacksquare$$

THEOREM 2. Let K be a convex subset of a normed space E. The closure $\overline{K}$ of K and the interior $\overset{\circ}{K}$ of K are convex. Moreover, if $\overset{\circ}{K} \ne \varnothing$, $\overline{K}$ is the closure of $\overset{\circ}{K}$ and $\overset{\circ}{K}$ is the interior of $\overline{K}$.

Proof.

a. Let us show that $\overline{K}$ is convex. Let λ be given in $]0, 1[$, and take $x = \lim x_n$ and $y = \lim y_n$ in the closure $\overline{K}$ of K (where x_n and y_n belong to K). Since $\lambda x + (1 - \lambda)y$ is the limit of $\lambda x_n + (1 - \lambda)y_n$ (which belongs to K since K is convex), we conclude that $\lambda x + (1 - \lambda)y \in \overline{K}$. Thus $\overline{K}$ is convex.

b. If x_0 and $x_1 \in \overset{\circ}{K}$, the elements $x_\lambda = \lambda x_0 + (1 - \lambda)x_1$ belong to $\overset{\circ}{K}$ when $\lambda \in]0, 1[$ according to Theorem 1. Hence the interior $\overset{\circ}{K}$ of K is convex.

c. If $\overset{\circ}{K} \ne \varnothing$, $\overline{K}$ is the closure of $\overset{\circ}{K}$ since if $x_0 \in \overset{\circ}{K}$ and if $x_1 \in \overline{K}$, $x_1 = \lim_{n \to \infty} x_n$, where $x_n = (1/n)x_0 + (1 - 1/n)x_1 \in \overset{\circ}{K}$ according to Theorem 1.

d. Let us show that if $\overset{\circ}{K} \ne \varnothing$, then $\overset{\circ}{K}$ is the interior of $\overline{K}$. Since $\overset{\circ}{K} \subset \overset{\circ}{\overline{K}}$ it is sufficient to show that if $x_0 \in \overset{\circ}{\overline{K}}$, then $x_0 \in \overset{\circ}{K}$. By hypothesis, there exists $\varepsilon > 0$ such that $B(x_0, \varepsilon) \subset \overline{K}$. Moreover, since $x_0 \in \overline{K}$ and since $\overset{\circ}{K} \ne \varnothing$, then $\overline{K} = \overset{\circ}{\overline{K}}$ according to c and there exists x_1 belonging to $\overset{\circ}{K} \cap B(x_0, \varepsilon)$. Thus, since $x_1 \in B(x_0, \varepsilon)$, $2x_0 - x_1$ belongs to $B(x_0, \varepsilon) \subset \overline{K}$ since $\|2x_0 - x_1 - x_0\| = \|x_0 - x_1\| \le \varepsilon$. Thus $x_0 = (1/2)x_1 + (1/2)(2x_0 - x_1)$, where $x_1 \in \overset{\circ}{K}$ and $2x_0 - x_1 \in \overline{K}$, and, therefore, $x_0 \in \overset{\circ}{K}$ using Theorem 1. $\blacksquare$

Definition 2.   Let K be a subset of a normed space E. We call the "convex hull" of K the intersection of all the convex sets containing K and the "closed convex hull" of K the intersection of all the closed convex sets containing K.

In other words, the convex hull co(K) of K is the smallest convex set containing K, and the closed convex hull $\overline{co}(K)$ of K is the smallest closed convex set containing K.

PROPOSITION 3.   Let K be a subset of a normed space E. Then the convex hull co(K) of K is the set of convex combinations $\sum \lambda_i x_i$ of elements x_i of K and the closed convex hull $\overline{co}(K)$ is the closure of co(K).

Proof.

a.   Let us denote by $\tilde{K}$ the set of convex combinations of elements of K. Then $\tilde{K}$ is convex for if $x = \sum_{i \in I} \lambda_i x_i$ and $y = \sum_{j \in J} \mu_j y_j$ are convex combinations of elements $x_i \in K$ and $y_j \in K$ and if $\alpha \in]0, 1[$, then

$$\alpha x + (1 - \alpha)y = \sum_{i \in I} (\alpha \lambda_i) x_i + \sum_{j \in J} (1 - \alpha)\mu_j y_j$$

is a convex combination of elements x_i and y_j of K since $\alpha \lambda_i$ and $(1 - \alpha)\mu_j$ belong to $[0, 1]$ and since

$$\sum_{i \in I} \alpha \lambda_i + \sum_{j \in J} (1 - \alpha)\mu_j = \alpha \left(\sum_{i \in I} \lambda_i \right) + (1 - \alpha) \left(\sum_{j \in J} \mu_j \right) = \alpha + (1 - \alpha) = 1.$$

Moreover, if C is a convex set containing K, it is clear that $\tilde{K} \subset C$. Thus $\tilde{K} \subset \text{co}(K)$. But since $\tilde{K}$ itself is a convex set containing K, we conclude that $\tilde{K} = \text{co}(K)$.

b.   Consider the closed convex hull $\overline{co}(K)$ of K. If C is a closed convex set containing K, then the closure $\overline{co(K)}$ of co(K) is contained in C; hence $\overline{co(K)} \subset \overline{co}(K)$. But since $\overline{co(K)}$ is a closed convex set containing K (according to Theorem 2), we conclude that $\overline{co}(K) \subset \overline{co(K)}$.   ∎

PROPOSITION 4.   Let us denote by

(3) $$S^n = \left\{ \lambda = \{\lambda_i\}_{i=1,\ldots,n} \in \mathbb{R}^n \text{ such that } \lambda_i \geq 0 \ \forall_i \text{ and } \sum_{i=1}^{n} \lambda_i = 1 \right\}.$$

Consider a sequence of n convex sets K_i. Then the convex hull of the union of the K_i's is the set of convex combinations $\sum_{i=1}^{n} \lambda_i x_i$ as λ runs over S^n and x_i runs over K_i for all i:

(4) $$\text{co}\left(\bigcup_{i=1}^{n} K_i \right) = \left\{ \sum_{i=1}^{n} \lambda_i x_i \right\}_{\substack{\lambda \in S^n \\ x_i \in K_i}}.$$

Proof.   Indeed the set C of these linear combinations is clearly contained in any convex set containing the sets K_i. Moreover, $K_i \subset C$ for every i. Thus the proof depends on showing that C is convex. Let $x = \sum_{i=1}^{n} \lambda_i x_i$ and $y = \sum_{j=1}^{n} \mu_j y_j$ be two elements of C (where λ and μ belong to S^n). Then $\alpha x + (1 - \alpha)y$ can be written in the form $\sum_{i=1}^{n} v_i z_i$, where $v = \{v_i\}_{i=1,\dots,n} = \alpha\lambda + (1 - \alpha)\mu \in S^n$ and where $z_i = (\alpha\lambda_i x_i + (1 - \alpha)\mu_i y_i)/(\alpha\lambda_i + (1 - \alpha)\mu_i)$ belongs to K_i, which is convex. Thus $\alpha x + (1 - \alpha)y$ belongs to C.   ∎

PROPOSITION 5.   Let K_i $(1 \leq i \leq n)$ be n compact convex sets of a normed space E. Then the convex hull $\text{co}(\bigcup_{i=1}^{n} K_i)$ of the union of the K_i's is compact (thus equal to the closed convex hull of the union).

Proof.   Let $x^k = \sum_{i=1}^{n} \lambda_i^k x_i^k$ be a sequence of

$$\text{co}\left(\bigcup_{i=1}^{n} K_i\right) = \left\{\sum_{i=1}^{n} \lambda_i x_i\right\}_{\substack{\lambda \in S^n \\ x_i \in K_i}}.$$

Since S^n is compact (for it is a closed and bounded subset of $\mathbb{R}^n$) and since the K_i's are compact, we can extract subsequences λ^{k_p} and $x_i^{k_p}$ converging to $\lambda \in S^n$ and $x_i \in K_i$, respectively. Then $x^{k_p} = \sum_{i=1}^{n} \lambda_i^{k_p} x_i^{k_p}$ converges to $x = \sum_{i=1}^{n} \lambda_i x_i \in \text{co}(\bigcup_{i=1}^{n} K_i)$. Thus this set is compact.   ∎

COROLLARY 1.   The convex hull of a finite set is compact.

More generally, the convex hull of a compact convex subset is compact. To show this we need the following theorem of Carathéodory.

PROPOSITION 6.   Let K be a subset of $\mathbb{R}^n$. The convex hull $\text{co}(K)$ of K is the set of elements of the form

$$x = \sum_{i=1}^{n+1} \lambda_i x_i \qquad \text{where} \qquad \lambda \in S^{n+1} \qquad \text{and} \qquad x_i \in K \ (i = 1, \dots, n+1).$$

Proof.   Every element $x \in \text{co}(K)$ is of the form $x = \sum_{i=1}^{k} \lambda_i x_i$ where $\lambda_i \geq 0$ and $\sum_{i=1}^{k} \lambda_i = 1$. Let us show that if $k > n + 1$, we can find among the representations of this type one for which $\lambda_i = 0$ for an index i. Indeed if $k > n + 1$, the elements $x_1 - x_k$, $x_2 - x_k$, $\dots$, $x_{k-1} - x_k$ are linearly *dependant*. Thus there exist $\mu_1, \dots, \mu_{k-1}$ *not all zero* such that

$$\mu_1(x_1 - x_k) + \mu_2(x_2 - x_k) + \cdots + \mu_{k-1}(x_{k-1} - x_k) = 0.$$

Set $\mu_k = -\sum_{i=1}^{k-1} \mu_i$. Then we have that $\sum_{i=1}^{k} \mu_i x_i = 0$ and $\sum_{i=1}^{k} \mu_i = 0$. Let I be the (nonempty) set of indices i such that $\mu_i > 0$. Set

$$t = \min_{i \in I} \frac{\lambda_i}{\mu_i} = \frac{\lambda_{i_0}}{\mu_{i_0}}, \qquad v_i = \lambda_i - t\mu_i.$$

The scalars v_i are positive or zero, $v_{i_0} = \lambda_{i_0} - t\mu_{i_0} = 0$ and

$$\sum_{i=1}^{k} v_i = \sum_{i=1}^{k} \lambda_i - t \sum_{i=1}^{k} \mu_i = 1 - 0 = 1.$$

Moreover,

$$x = \sum_{i=1}^{k} \lambda_i x_i - t \sum_{i=1}^{k} \mu_i x_i = \sum_{i=1}^{k} v_i x_i.$$

Consequently, we have shown that x is a convex combination of $k - 1$ points.

By repeating this procedure, we can successively write x as a convex combination of $k - 1, k - 2, \ldots, n + 1$ points of k, which proves the proposition. ∎

PROPOSITION 7.   The convex hull of a compact subset of a finite dimensional space is compact.

Proof.   The proof is similar to that of Proposition 5. We consider an infinite sequence of elements $x^k \in co(K)$. Applying Proposition 6, we can write $x^k = \sum_{i=1}^{n+1} \lambda_i^k x_i^k$ where $\lambda^k \in S^{n+1}$ and $x_i^k \in K$. The sets S^{n+1} and K being compact, we can extract from these sequences subsequences λ^{k_p} and $x_i^{k_p}$, which converge, respectively, to $\lambda \in S^{n+1}$ and $x_i \in K$. Thus the sequence of elements $x^{k_p} = \sum_{i=1}^{n+1} \lambda_i^{k_p} x_i^{k_p}$ converges to $x = \sum_{i=1}^{n+1} \lambda_i x_i \in co(K)$. We have, therefore, shown that $co(K)$ is compact. ∎

CHAPTER 3

Continuous Functions

Introduction

The aim of this chapter is to study the continuous functions from a metric space E to a metric space F, that is, those functions under which the image of a convergent sequence in E is a convergent sequence in F.

In the first section, we characterize continuous functions with the aid of the topological notions introduced in the previous chapter (open and closed sets and neighborhoods) and introduce the notion of uniform continuity. We show that composition of functions and the elementary algebraic operations preserve continuity.

We indicate in the second section some examples (in particular continuous multilinear functions). In the third section we characterize continuous linear forms on a normed vector space E: These are those linear forms f such that $f^{-1}\{0\}$ is closed. This characterization allows us to show that a linear mapping from a finite dimensional vector space to another is *always* continuous. The fourth section is important for it studies the properties of continuous functions on compact sets: The image $f(K)$ of a compact set under a continuous mapping f is compact and f is, therefore, uniformly continuous on this compact set. Moreover, if f is bijective, it is bicontinuous.

In particular, a continuous function on a compact set has the following property, which is very useful in approximation theory: If the sequence $f(x_n)$ converges to y in F, then a subsequence x_{n_k} converges to a solution $x \in K$ of the equation $f(x) = y$. The continuous functions on a (noncompact) metric space E, which have this important property are called *proper functions*.

In Section 5 we study the important properties of the extension of continuous functions: If two continuous functions are equal on a *dense* subset, they are everywhere equal, and if a function is uniformly continuous from a dense subset A of E to a complete space, then it can be extended to a uniformly continuous function on the space E.

In Section 6, we study the limits of functions with respect to a subset and the upper and lower limits of real-valued functions. This leads us to introduce the notions of upper and lower semicontinuous real-valued functions (Section 7).

We study the continuity properties of convex functions in Section 8. For example, we show that every convex function defined on an open set of $\mathbb{R}^n$ is continuous. We apply the preceding results to prove the existence and uniqueness of a solution to a minimization problem in Section 9 and to begin a study of two-person games in Section 10 (which will be pursued at the end of Chapter 5).

In Section 11, we introduce the very important concept of contraction f from a metric space E to itself to prove the existence of a fixed point x of f, that is, a solution to the equation $f(x) = x$. We apply this theorem to prove the existence of solutions of integral equations and the Cauchy–Lipschitz theorem on the existence of solutions of differential equations. In Section 12, we deduce from the fixed point theorem the important inverse function theorem, which gives a sufficient condition for a mapping to be a homeomorphism on a neighborhood of a point. We state without proof the Brouwer fixed point theorem, which is the origin of all the deep results of nonlinear analysis. (This theorem states that every continuous mapping from a compact convex subset of $\mathbb{R}^n$ to itself has a fixed point.) As a first application we show that every coercive mapping from $\mathbb{R}^n$ to itself is surjective.

We end this chapter by proving convergence of algorithms for solving equations $f(x) = y$.

1.   Continuous and Uniformly Continuous Functions

Let (E, d_E) and (F, d_F) be two metric spaces.

PROPOSITION 1.   Let f be a function from E to F, and let $x \in E$. The following conditions are equivalent:

(1)

> **i.**   For every sequence x_n converging to x, $f(x_n)$ converges to $f(x)$.
>
> **ii.** $\forall \varepsilon > 0, \exists \eta(\varepsilon, x) = \eta$ such that if $d_E(x, y) \leq \eta$, then $d_F(f(x), f(y)) \leq \varepsilon$.
>
> **iii.**   For every neighborhood $V \in \mathscr{V}(f(x))$ of $f(x)$ there exists a neighborhood $U \in \mathscr{V}(x)$ of x such that $f(U) \subset V$.
>
> **iv.**   For every neighborhood $V \in \mathscr{V}(f(x))$, $f^{-1}(V) \in \mathscr{V}(x)$ is a neighborhood of x.

Proof.

a.   We have already shown the equivalence of properties (1)i and (1)ii (see Proposition 1.2.2).

b. (1)ii implies (1)iii: Every neighborhood $V \in \mathscr{V}(f(x))$ contains a ball $B(f(x), \varepsilon)$ of center $f(x)$ and of radius ε; hence it is sufficient to take for the neighborhood $U \in \mathscr{V}(x)$ the ball $B(x, \eta)$ of center x and of radius η in order to obtain that $f(U) \subset B(f(x), \varepsilon) \subset V$.

c. (1)iii implies (1)iv: If $f(U)$ is contained in V, then $U \subset f^{-1}(V)$, and since $U \in \mathscr{V}(x)$, $f^{-1}(V)$ is a neighborhood of x.

d. (1)iv implies (1)ii: Take $V = B(f(x), \varepsilon) \in \mathscr{V}(f(x))$. Then $U = f^{-1}(V)$ is a neighborhood of x and, therefore, contains a ball $B(x, \eta)$ of radius η, which implies (1)ii.   ∎

Definition 1.   If a mapping f from E to F satisfies the equivalent conditions (1), we shall say that "f is continuous at x." We shall say that "f is continuous on E" if f is continuous at each point x of E.

Remark 1.   Conditions (1)iii and (1)iv show that one can define the notion of a continuous function at a point x of a topological space.

One can say in an intuitive fashion that a function f is continuous at x if "$f(y)$ *is arbitrarily close to* $f(x)$ *when* y *is sufficiently close to* x."   ∎

If f is continuous on E, we obtain the following additional character-izations.

THEOREM 1.   Let f be a mapping from E to F. The following conditions are equivalent:

(2)
$$\begin{cases} \textbf{i.} & f \text{ is continuous on } E. \\ \textbf{ii.} & \text{The inverse image } f^{-1}(U) \text{ of every open set } U \text{ of } F \text{ is open in } E. \\ \textbf{iii.} & \text{The inverse image } f^{-1}(B) \text{ of every closed set } B \text{ of } F \text{ is closed in } E. \\ \textbf{iv.} & \text{For every subset } A \text{ of } E, f(\bar{A}) \subset \overline{f(A)}. \end{cases}$$

Proof.

a. (2)i implies (2)iv: If $x \in \bar{A}$, then $x = \lim x_n$ where $\{x_n\}$ is a sequence of elements $x_n \in A$. Then $f(x_n) \in f(A)$ and $f(x_n)$ converges to $f(x)$, which implies that $f(x) \in \overline{f(A)}$.

b. (2)iv implies (2)iii: Let B be a closed set of F and $A = f^{-1}(B)$. Since $f(f^{-1}(B)) \subset B$, and since $f(\bar{A}) \subset \overline{f(A)}$, we conclude that $f(\bar{A}) \subset \bar{B} = B$ and, therefore, that $\bar{A} \subset f^{-1}(B) = A \subset \bar{A}$. Thus $A = f^{-1}(B)$ is closed.

c. (2)iii implies (2)ii: If $U = \complement B$ is an open set of F, thus the complement of a closed set B of F, then $f^{-1}(U) = f^{-1}(\complement B) = \complement f^{-1}(B)$ is an open set of E.

d. (2)ii implies (2)i: Let $x \in E$ and take for $V \in \mathscr{V}(f(x))$ an open neigh-borhood of $f(x)$. Then $U = f^{-1}(V)$ is an open set containing x, thus an (open) neighborhood of x, which shows that f is continuous at x.   ∎

Remark 2.   If f is continuous, the image $f(U)$ of an open set U of E *is not necessarily open in F* and the image $f(A)$ of a closed set A of E *is not necessarily closed* in F.

For example, take $f : x \mapsto x^2$ from $\mathbb{R}$ to $\mathbb{R}$. If U is the open set $]-1, +1[$, $f(U) = [0, 1[$ is not open. Consider $f : x \mapsto 1/x$ from $A = [1, +\infty[$ to $\mathbb{R}$; $f(A) =]0, 1]$ is not closed in $\mathbb{R}$ although A is closed in $\mathbb{R}$.

EXAMPLE 1.   Let (E, d_E) be a discrete metric space. Every function from E to a metric space F is continuous since $f^{-1}(U)$ is open for every open set U of F, every subset of E being open.   ■

The notion of continuous function is "local" in the sense that it is a notion defined at each point x of E. Hence for the property (1)ii the radius $\eta = \eta(\varepsilon, x)$ of the ball of center x *depends on ε and on x.* This remark leads us to introduce the notion of uniform continuity.

Definition 2.   A function f from a metric space E to a metric space F is said to be "uniformly continuous" if

(3) $\begin{cases} \forall \varepsilon > 0, \ \exists \eta = \eta(\varepsilon) \text{ depending on } \varepsilon \text{ and } independent \text{ of } x \text{ such that} \\ d_F(f(x), f(y)) \le \varepsilon \text{ when } d_E(x, y) \le \eta. \end{cases}$

PROPOSITION 2.

a.   Every uniformly continuous function is continuous.
b.   Every uniformly continuous function maps Cauchy sequences onto Cauchy sequences.

Proof.   (*Left as an exercise.*)

PROPOSITION 3.   If E is a metric space and if $A \subset E$ is nonempty, the function $x \mapsto d(x, A)$ is uniformly continuous from E to $\mathbb{R}$.

Proof.   The proposition is a consequence of the inequality:

$$|d(x, A) - d(y, A)| \le d(x, y).$$ ■

Remark 3.   This suggests the introduction of the following definition.

Definition 3.   We shall say that a function f from a metric space E to a metric space F is "Lipschitz" if there exists a constant $\lambda > 0$ such that

(4) $$d_F(f(x), f(y)) \le \lambda d_E(x, y) \qquad \forall x, y \in E.$$

We say that a function f is a "contraction" if, in addition, $\lambda < 1$.

For example, the function $x \mapsto d(x, A)$ is Lipschitz with $\lambda = 1$. We remark from this definition that the following proposition holds.

PROPOSITION 4.   Every Lipschitz function is uniformly continuous.

EXAMPLE 2.   Take $E = [0, 1]$, $F = \mathbb{R}$, and $f(x) = x^2$. Then f is Lipschitz with $\lambda = 2$ since

(5) $|f(x) - f(y)| = |x^2 - y^2| = |(x - y)(x + y)| \leq 2|x - y|.$

However, if $E = \mathbb{R}$, $F = \mathbb{R}$, and $f(x) = x^2$, the function f is then continuous *but not uniformly continuous*. Indeed, the equality $|f(x) - f(y)| = |x - y||x + y|$ shows that $|f(x) - f(y)| \leq \varepsilon$ when $|x - y| \leq \eta(\varepsilon, x)$, where $\eta(\varepsilon, x)$ is the positive solution of $\eta^2 + \eta|x| - \varepsilon = 0$, since in this case $\eta(\varepsilon, x)|x + y| \leq \eta(\varepsilon, x)(|x| + \eta(\varepsilon, x)) \leq \varepsilon$. Thus f is continuous, but it is not uniformly continuous, for $|f(x) - f(y)| = \eta|x + y|$ takes on infinitely large values as x tends to infinity, where $\eta = |x - y|$.

EXAMPLE 3.   Let f be a differentiable function from $\mathbb{R}$ to $\mathbb{R}$. If the derivative $Df(x) = f'(x)$ is bounded on the interval $[a, b]$, the function f is Lipschitz on this interval since the inequality

(6) $|f(x) - f(y)| = \left| \int_y^x f'(t)dt \right| \leq \left(\sup_{z \in [a, b]} |f'(z)| \right) |x - y|.$

implies that

(7) $|f(x) - f(y)| \leq \|f'\|_\infty |x - y|.$

Definition 4.   Let E and F be two metric spaces and f, a mapping from E onto F. We say that f is a "homeomorphism" (or that f is "bicontinuous") if

(8) $\begin{cases} \textbf{i.} & f \text{ is bijective.} \\ \textbf{ii.} & f \text{ and } f^{-1} \text{ are continuous.} \end{cases}$

We say that f is an "isometry" from E to F if

(9) $d_F(f(x), f(y)) = d_E(x, y) \forall x, y \in E.$

We say that f has a right inverse if

(10) $\begin{cases} \textbf{i.} & f \text{ is } surjective. \\ \textbf{ii.} & f \text{ is } continuous. \\ \textbf{iii.} & \text{There exists a mapping } g \text{ from } F \text{ to } E, \text{ which is continuous and} \\ & \text{satisfies } f[g(y)] = y \text{ for every } y \in F. \end{cases}$

Remark 4.   We can motivate these definitions in the following way. If f is a mapping from E to F, we seek to solve the problem:

(11)   to find $x \in E$ satisfying $f(x) = y$ where y is given in F

To say that f is *injective* is to say that this problem has *at most* one solution for any $y \in F$, that is, *the uniqueness of the solution is assured.* To say the f is *surjective* is to say that the problem has *at least* one solution for any $y \in F$, that is, *the existence of a solution is assured.* If f is *bijective, the existence and uniqueness of the solution of* (11) *are assured.* To say that f^{-1} is *continuous* is to say that the solution is *stable* in the following sense: If $x_1 = f^{-1}(y_1)$ and $x_2 = f^{-1}(y_2)$ are the respective solutions for y_1 and y_2 in F, then for every $\varepsilon > 0$, there exists $\eta(\varepsilon, y_1)$ such that $d_E(x_1, x_2) \leq \varepsilon$ when $d_F(y_1, y_2) \leq \eta$. *In other words, every "small perturbation" of the right-hand side produces a "small perturbation" of the solution of the problem.*

In the case where f is surjective, that is when the existence of a solution is assured, to say that f *has a right inverse* is the same as to say that we can choose a solution $x = g(y)$ that *depends continuously on* y.

PROPOSITION 5.

a.   Every isometry f from E to F is injective.
b.   If E is complete, the image $f(E)$ under an isometry f is a complete set in F.
c.   Every isometry f from E onto F is bicontinuous.

Proof.

a.   Suppose that $f(x) = f(y)$. Then $d_E(x, y) = d_F(f(x), f(y)) = 0$, which implies that $x = y$.
b.   Let $\{f(x_n)\}$ be a Cauchy sequence in $f(E)$. Then $\{x_n\}$ is a Cauchy sequence in E since $d_E(x_n, x_m) = d_F(f(x_n), f(x_m)) \leq \varepsilon$ when $n, m \geq n(\varepsilon)$. Thus the sequence $\{x_n\}$ converges to an element x of E since E is complete, which implies that $f(x_n)$ converges to $f(x)$ since f is continuous (and even Lipschitz). Hence $f(E)$ is complete (and, therefore, closed).
c.   If f is surjective, f^{-1} is also an isometry from F onto E; consequently, f is bicontinuous.   ∎

Remark 5.   Proposition 5 implies that the extended real numbers $\overline{\mathbb{R}}$ is a complete metric space, since $\overline{\mathbb{R}}$ is isometric to $[-1, +1]$ (see Remark 1.3.2).   ∎

PROPOSITION 6.   Consider three metric spaces (E, d_E), (F, d_F) and (G, d_G), a mapping f from E to F and a mapping g from F to G. If the mappings

f and g are continuous (respectively, uniformly continuous) then $g \circ f$ is continuous (respectively, uniformly continuous) from E to G.

Proof. Given $\varepsilon > 0$, since g is continuous

$$d_G(g(u), g(v)) \le \varepsilon \quad \text{if} \quad d_F(u, v) \le \eta = \eta(\varepsilon, u)$$

and since f is continuous

$$d_F(f(x), f(y)) \le \eta \quad \text{if} \quad d_E(x, y) \le \delta = \delta(\eta, x).$$

Thus taking $u = f(x)$ and $v = f(y)$, we obtain that

$$d_G(g \circ f(x), g \circ f(y)) \le \varepsilon \quad \text{if} \quad d_E(x, y) \le \beta = \beta(\varepsilon, x).$$

If f and g are uniformly continuous, $\eta = \eta(\varepsilon)$ and $\delta = \delta(\eta)$ do not depend on u or x; hence $g \circ f$ is uniformly continuous. ∎

PROPOSITION 7. Let (E, d) be a metric space, $(F, \|\cdot\|)$ a normed space and f and g two continuous mappings (respectively, uniformly continuous) from E to F. Then the mappings $f + g$ and λf defined by

$$(12) \qquad f + g : x \mapsto f(x) + g(x) \qquad \text{and} \qquad \lambda f : x \mapsto \lambda f(x)$$

are continuous (respectively, uniformly continuous).

Proof. This is a consequence of Proposition 6 and of the continuity of the mappings $\{u, v\} \in F \times F \mapsto u + v \in F$ and $\{\lambda, u\} \in \mathbb{R} \times F \mapsto \lambda u \in F$. ∎

PROPOSITION 8 Given a metric space (E, d), if f and g are two real-valued functions defined on (E, d), then

$$(13) \quad \begin{cases} \textbf{i.} & fg : x \mapsto f(x)g(x) \text{ is continuous.} \\ \textbf{ii.} & \sup(f, g) : x \mapsto \sup(f(x), g(x)) \text{ is continuous.} \\ \textbf{iii.} & \text{The function } x \mapsto 1/f(x) \text{ is continuous at every point } x \text{ for} \\ & \text{which } f(x) \ne 0. \end{cases}$$

Proof. The proposition is a consequence of Proposition 6 and the continuity of the functions $\{u, v\} \in \mathbb{R} \times \mathbb{R} \mapsto uv \in \mathbb{R}$, $\{u, v\} \in \mathbb{R} \times \mathbb{R} \mapsto \sup(u, v) \in \mathbb{R}$ and $u \in \mathbb{R} - \{0\} \mapsto (1/u) \in \mathbb{R}$. ∎

2.   Examples of Continuous and Uniformly Continuous Functions

EXAMPLE 1. *Real-valued functions on E.* If (E, d) is a metric space, a *real-valued function on E* is a mapping from E to $\mathbb{R}$. A real-valued function is continuous if one of the following equivalent conditions is satisfied:

(1)
- **i.** $\forall \varepsilon, \forall x \in E, \exists \eta = \eta(\varepsilon, x)$ such that $d(x, y) \leq \eta$ implies that $|f(x) - f(y)| \leq \varepsilon$.
- **ii.**   The sets $f^{-1}(]a, b[)$ are open for every $a \leq b, a, b \in \mathbb{R}$.
- **iii.**   The sets $f^{-1}(]\infty, a])$ and $f^{-1}([b, \infty[)$ are closed for every $a, b \in \mathbb{R}$.

A real-valued function on E is uniformly continuous if

(2) $\forall \varepsilon, \exists \eta = \eta(\varepsilon)$ such that $d(x, y) \leq \eta$ implies $|f(x) - f(y)| \leq \varepsilon$.

SPECIAL CASES.   If $E = \mathbb{R}$, we say that f is a *real-valued function of a real variable* $x \in \mathbb{R}: x \mapsto f(x)$. If $E = \mathbb{C}$, we say that f is a *real-valued function of a complex variable* $z \in \mathbb{C}: z \mapsto f(z)$. If $E = \mathbb{R}^n$, we say that f is a *real-valued function of n real variables*:

(3) $(x_1, \ldots, x_n) \mapsto f(x_1, \ldots, x_n).$

If we give $\mathbb{R}^n$ the norm $\|x\|_\alpha$, f is continuous at $x = (x_1, \ldots, x_n)$ if and only if

(4) $\begin{cases} \forall \varepsilon > 0, |f(x_1, \ldots, x_n) - f(y_1, \ldots, y_n)| \leq \varepsilon & \text{when} \\ |x_i - y_i| \leq \varepsilon & \text{for all} \quad i = 1, \ldots, n. \end{cases}$

PROPOSITION 1.   If a real-valued function f of n real variables is continuous, then the functions of one variable $x_i \mapsto f(x_1, \ldots, x_i, \ldots, x_n)$ are continuous.

Proof.   It suffices to take $y = (x_1, \ldots, x_{i-1}, y_i, x_{i+1}, \ldots, x_n)$ in (4). The converse is in general not true.   ∎

EXAMPLE 2. *Vector functions.*   Let (E, d) be a metric space. *A vector function on E is a mapping from E to $\mathbb{R}^p$, associating to each $x \in E$ the element:*

(5) $f(x) = (f_1(x), \ldots, f_i(x), \ldots, f_p(x))$ of $\mathbb{R}^p$.

We give $\mathbb{R}^p$ the norm $\|x\|_\alpha$.

PROPOSITION 2.   A vector function $f = (f_1, \ldots, f_p)$ is continuous if and only if the components $f_i : x \mapsto f_i(x)$ are continuous.

Proof.   Indeed, to say that f is continuous at x is the same as to say that $\forall \varepsilon, \exists \eta = \eta(\varepsilon, x)$ such that $d(x, y) \leq \eta$ implies

(6) $\sup_{i=1,\ldots,p} |f_i(x) - f_i(y)| \leq \varepsilon.$

Thus if f is continuous, its components f_i are continuous. *Conversely,* if the components f_i are continuous at x, there exists $\eta_i = \eta_i(\varepsilon, x)$ such that

$|f_i(x) - f_i(y)| \le \varepsilon$ when $d(x, y) \le \eta_i$. Taking $\eta = \eta(\varepsilon) = \min_{i=1,\ldots,p} \eta_i(\varepsilon)$, we conclude that f is continuous from E to $\mathbb{R}^p$ at x. ∎

SPECIAL CASES.   If $E = \mathbb{R}$, we say that f is a vector function of a real variable. If $E = [a, b] \subset \mathbb{R}$, we say that $f([a, b])$ is a *curve in* $\mathbb{R}^p$. It is the subset of elements $\{f_1(x), \ldots, f_p(x)\}$ as x runs over $[a, b]$. If $E = \mathbb{R}^n$, we say that f is a vector function of n real variables:

$$f:(x_1, \ldots, x_n) \mapsto \{f_1(x_1, \ldots, x_n), \ldots, f_i(x_i, \ldots, x_n), \ldots, f_p(x_1, \ldots, x_n)\}$$

If $E = \Omega \subset \mathbb{R}^2$ is a closed subset of $\mathbb{R}^2$, we say that $f(\Omega)$ is *a surface in* $\mathbb{R}^p$.

3.   Linear and Multilinear Mappings

Let $(E, \|\cdot\|_E)$ and $(F, \|\cdot\|_F)$ be two normed vector spaces. A function f from E to F is continuous at x if $\forall \varepsilon > 0, \exists \eta = \eta(\varepsilon, x)$ such that $\|f(x) - f(y)\|_F \le \varepsilon$ when $\|x - y\|_E \le \eta$. Consider the case where $f = A$ is a linear mapping from E to F. In this case, we obtain the following proposition.

PROPOSITION 1.   Let A be a mapping from a normed space E to a normed space F. Then the following conditions are equivalent:

(1)
$\begin{cases}
\textbf{i.} & A \text{ is continuous at } 0. \\
\textbf{ii.} & A \text{ is continuous from } E \text{ to } F. \\
\textbf{iii.} & A \text{ is uniformly continuous.} \\
\textbf{iv.} & \|A\| = \sup_{\substack{x \in E \\ x \ne 0}} \dfrac{\|Ax\|_F}{\|x\|_E} < +\infty. \\
\textbf{v.} & A \text{ is Lipschitz.}
\end{cases}$

Proof. We have already shown the equivalence of conditions (1)i, (1)ii, and (1)iv (see Proposition 1.10.1). It is clear that (1)iv implies (1)v and, consequently, (1)iii and (1)i. ∎

More generally, consider n normed vector spaces E_i and a normed vector space F.

Definiton 1.   A mapping A from $E_1 \times \cdots \times E_n$ to F is called "multilinear" if for every $i = 1, \ldots, n$ and for every finite sequence $\lambda_k (k = 1, \ldots, p)$, we have

$$(2) \qquad A\left(x_1, \ldots, \sum_{k=1}^{p} \lambda_k x_i^k, \ldots, x_n\right) = \sum_{k=1}^{p} \lambda_k A(x_1, \ldots, x_i^k, \ldots, x_n).$$

In other words, A is multilinear if the mappings $x_i \mapsto A(x_1, \ldots, x_i, \ldots, x_n)$ from E_i to F are linear. In particular, we say that A is *bilinear* if $n = 2$, *trilinear* if $n = 3$. A bilinear mapping is not linear: If $n = 2$

$$A(\alpha x_1 + \beta y_1, \alpha x_2 + \beta y_2) = \alpha^2 A(x_1, x_2) + \alpha\beta(A(x_1, y_2) + A(y_1, x_2))$$
$$+ \beta^2 A(y_1, y_2)$$

is not equal to

$$\alpha A(x_1, x_2) + \beta A(y_1, y_2)!$$

PROPOSITION 2. Let A be a multilinear mapping from $E_1 \times E_2 \times \cdots \times E_n$ to F. We give $E_1 \times \cdots \times E_n$ the norm

(3)
$$\|x\| = \sup_{i=1,\ldots,n} \|x_i\|_{E_i}.$$

The following conditions are then equivalent:

(4)
$$\begin{cases} \textbf{i.} \quad A \text{ is continuous.} \\[6pt] \textbf{ii.} \quad A \text{ is continuous at the point } (0, \ldots, 0) \\[6pt] \textbf{iii.} \quad \|A\| = \sup_{\substack{x_i \neq 0 \\ x_i \in E_i}} \dfrac{\|A(x_1, \ldots, x_n)\|_F}{\|x_1\|_{E_1} \cdots \|x_n\|_{E_n}} < +\infty. \end{cases}$$

Proof.

a. It is clear that (4)i implies (4)ii. Let us show this latter condition implies (4)iii. Given $\varepsilon > 0$, there exists $\eta > 0$ such that if $\|y_i\|_{E_i} \le \eta$ for $i = 1, \ldots, n$,

$$\|A(y_1, \ldots, y_n)\|_F = \|A(y_1, \ldots, y_n) - A(0, \ldots, 0)\|_F \le \varepsilon.$$

If $x = (x_1, \ldots, x_n)$ is an arbitrary element of $E_1 \times \cdots \times E_n$, let us consider the element

$$y = \left(\eta \frac{x_1}{\|x_1\|_{E_1}}, \ldots, \eta \frac{x_n}{\|x_n\|_{E_n}}\right).$$

Then the $y_i = \eta x_i / \|x_i\|_{E_i}$ satisfy $\|y_i\|_{E_i} = \eta$ and we obtain that

(5)
$$\|A(y_1, \ldots, y_n)\|_F = \frac{\eta^n}{\|x_1\|_{E_1} \cdots \|x_n\|_{E_n}} \|A(x_1, \ldots, x_n)\|_F \le \varepsilon,$$

which implies that $\|A\| \le (\varepsilon/\eta^n) < +\infty$.

b. Let us show that (4)iii implies (4)i. We write that

$$A(x_1, \ldots, x_n) - A(y_1, \ldots, y_n) = A(x_1 - y_1, x_2, \ldots, x_n)$$
$$+ A(y_1, x_2 - y_2, x_3, \ldots, x_n)$$
$$+ \cdots + A(y_1, \ldots, y_{n-1}, x_n - y_n),$$

which implies that

$$\|A(x_1, \ldots, x_n) - A(y_1, \ldots, y_n)\|_F \leq \|x_1 - y_1\|_{E_1} \|x_2\|_{E_2} \cdots \|x_n\|_{E_n}$$
$$+ \|y_1\|_{E_1} \|x_2 - y_2\|_{E_2} \cdots \|x_n\|_{E_n}$$
$$+ \cdots + \|y_1\|_{E_1} \cdots \|y_{n-1}\|_{E_{n-1}} \|x_n - y_n\|_{E_n}.$$

Consequently, if $\|x_i - y_i\|_{E_i} \leq \eta$, then

$$\|y_i\|_{E_i} \leq \eta + \|x_i\|_{E_i} \leq \eta + \|x\|,$$

and we derive from this the upper estimate:

$$\|A(x_1, \ldots, x_n) - A(y_1, \ldots, y_n)\|_F \leq n(\eta + \|x\|^{n-1})\eta.$$

Thus if $x \in E_1 \times \cdots \times E_n$ and $\varepsilon > 0$ are given, by choosing η such that $n(\eta + \|x\|^{n-1})\eta \leq \varepsilon$, which is always possible, we conclude that $\|A(x_1, \ldots, x_n) - A(y_1, \ldots, y_n)\|_F \leq \varepsilon$ when $\|x - y\| \leq \eta$. ∎

Remark 1. Continuous multilinear mappings are not uniformly continuous. ∎

Remark 2. The space $\mathscr{L}(E_1, \ldots, E_n; F)$ of multilinear mappings from $E_1 \times \cdots \times E_n$ to F is a normed vector space with

$$\|A\| = \sup \frac{\|A(x_1, \ldots, x_n)\|_F}{\|x_1\|_{E_1} \cdots \|x_n\|_{E_n}}. \qquad \blacksquare$$

*PROPOSITION 3. For every $A \in \mathscr{L}(E_1, E_2; F)$ and for every x_1 in E_1, we denote by $A_1 x_1$ the linear mapping $x_2 \mapsto A(x_1, x_2)$. Then the mapping $A_1 : x_1 \mapsto A_1 x_1$ from E_1 to $\mathscr{L}(E_2, F)$ is a continuous linear mapping from E_1 to $\mathscr{L}(E_2, F)$ satisfying

(6) $$\|A_1\|_{\mathscr{L}(E_1, \mathscr{L}(E_2, F))} = \|A\|_{\mathscr{L}(E_1, E_2; F)}.$$

Proof. It is clear that A_1 is linear since

$$A_1 \left(\sum_{k=1}^{p} \mu_k x_1^k \right)(x_2) = A\left(\sum_{k=1}^{p} \mu_k x_1^k, x_2 \right) = \sum_{k=1}^{p} \mu_k A(x_1^k, x_2)$$

$$= \left(\sum_{k=1}^{p} \mu_k A_1^k \right)(x_2).$$

Moreover, A_1 is continuous since

$$\|A_1 x_1\|_{\mathscr{L}(E_2, F)} = \sup_{\|x_2\|_{E_2} \leq 1} \|(A_1 x_1)(x_2)\|_F$$

$$= \sup_{\|x_2\|_{E_2} \leq 1} \|A(x_1, x_2)\|_F;$$

consequently,

$$\|A_1\|_{\mathscr{L}(E_1, \mathscr{L}(E_2, F))} = \sup_{\|x_1\|_{E_1} \leq 1} \|A_1 x_1\|_{\mathscr{L}(E_2, F)}$$

$$= \sup_{\|x_1\|_{E_1} \leq 1} \sup_{\|x_2\|_{E_2} \leq 1} \|A(x_1, x_2)\|_F$$

$$= \|A\|_{\mathscr{L}(E_1, E_2; F)}. \qquad \blacksquare$$

EXAMPLE 1. Take $E = F = l^\infty(\mathbb{N})$ and consider the operators τ_k defined by

(7) $(\tau_k x)_n = (x_{n+k})$ if $x = \{x_n\}_{n \geq 0} \in l^\infty(\mathbb{N})$.

The operators $A = \sum_{k=0}^{p} a_k \tau_k$ are continuous linear operators from $l^\infty(\mathbb{N})$ to $l^\infty(\mathbb{N})$ since

(8) $$\|Ax\|_\infty = \sup_{n \geq 0} \left| \sum_{k=1}^{p} a_k x_{n+k} \right| \leq \left(\sum_{k=1}^{p} |a_k| \right) \|x\|_\infty.$$

EXAMPLE 2. Take $E = \mathscr{D}^{(1)}(0, 1)$, the vector space of continuously differentiable functions on the interval $[0, 1]$, with the norm

$$\|f\|_{\infty, 1} = \sup_{x \in [0, 1]} |f(x)| + \sup_{x \in [0, 1]} |Df(x)|$$

where $Df(x) = f'(x) = (d/dx)f(x)$. Take $F = \mathscr{C}_\infty(0, 1)$, the vector space of continuous functions with the norm $\|f\|_\infty = \sup_{x \in [0, 1]} |f(x)|$. Consider the operator A from $\mathscr{D}^{(1)}(0, 1)$ to $\mathscr{C}_\infty(0, 1)$ defined by $Af(x) = a(x)Df(x) + b(x)f(x)$ for all $f \in \mathscr{D}^{(1)}(0, 1)$ where a and $b \in \mathscr{C}_\infty(0, 1)$. The operator A is continuous since

$$\|Af\|_\infty = \sup_{x \in (0, 1)} |a(x)Df(x) + b(x)f(x)|$$

$$\leq \sup|a(x)|\sup|Df(x)| + \sup|b(x)|\sup|f(x)|$$

$$\leq \max(\|a\|_\infty, \|b\|_\infty)\|f\|_{\infty, 1}.$$

EXAMPLE 3. Let $E = \mathscr{C}_\infty(0, 1)$. The linear forms $f \mapsto \int_0^1 f(x)dx$ and $f \mapsto \sum_{k=1}^{p} \alpha_k f(x_k)$ where $\alpha_k \in \mathbb{R}$, $x_k \in [0, 1]$ are continuous on E.

Proof.

a. According to well-known results we have

$$\left| \int_0^1 f(x)dx \right| \le \int_0^1 |f(x)|dx \le \sup_{x \in [0,1]} |f(x)| \int_0^1 dx = \|f\|_\alpha .$$

b.

$$\left| \sum_{k=1}^p \alpha_k f(x_k) \right| \le \sum_{k=1}^p |\alpha_k| |f(x_k)|$$

$$\le \sup_{x \in [0,1]} |f(x)| \left(\sum_{k=1}^p |\alpha_k| \right) \le \left(\sum_{k=1}^p |\alpha_k| \right) \|f\|_\alpha . \qquad \blacksquare$$

We are going to characterize *the continuous linear forms* on a normed space and show that every linear mapping from a finite dimensional space to another is continuous.

PROPOSITION 4.   Let E be a normed space and f a linear form on E. Then f is continuous if and only if the hyperplane $H = \{x \in E$ such that $f(x) = 0\}$ is closed.

Proof.   If f is continuous, then H is closed since it is the inverse image under f of the subset $\{0\}$ that is closed. *Conversely*, let us suppose that H is closed. Given $\varepsilon > 0$, we must show that there exists a ball $B(\eta)$ of radius η such that $|f(x)| \le \varepsilon$ when $x \in B(\eta)$. Let $x_0 \in E$ satisfy $f(x_0) = \varepsilon$. Since H is closed, the subset $K = x_0 + H$ is also closed and $0 \notin K$. Hence there exists a ball $B(\eta)$ of radius η and with center 0 such that $B(\eta) \cap K = \emptyset$. Then $|f(x)| \le \varepsilon$ when $x \in B(\eta)$. If not, there would exist $x \in B(\eta)$ for which $|f(x)| > \varepsilon$. Set $y = \varepsilon x/f(x)$. Then $y \in B(\eta)$ since $\|y\| = (\varepsilon/|f(x)|)\|x\| \le 1.\eta = \eta$; moreover, $y \in K$ since $f(y - x_0) = \varepsilon(f(x)/f(x)) - f(x_0) = \varepsilon - \varepsilon = 0$. But this is impossible since $B(\eta) \cap K = \emptyset$. $\qquad \blacksquare$

Now let us consider a vector space E of dimension n. Being given a base $\{e_1, \ldots, e_n\}$ for E is the same as having an isomorphism B from $\mathbb{R}^n$ onto E defined by

(9)
$$Bx = \sum_{i=1}^n x_i e_i \quad \text{if} \quad x = (x_1, \ldots, x_n) \in \mathbb{R}^n.$$

The inverse B^{-1} of B is the linear mapping from E onto $\mathbb{R}^n$ defined by

(10)
$$\begin{cases} x = B^{-1}u \text{ if and only if } x_i = e_i^*(u) \text{ for every } i \text{ where } \{e_1^*, \ldots, e_n^*\} \text{ is} \\ \text{the dual base of the algebraic dual space } E^* \text{ of } E. \end{cases}$$

PROPOSITION 5.   Let $(E, \|\cdot\|)$ be a normed vector space of dimension n, and let B be the isomorphism of $(\mathbb{R}^n, \|\cdot\|_\infty)$ onto E associated with the base $\{e_1, \ldots, e_n\}$ of E. Then B is bicontinuous: There exist constants m and M such that

$$(11) \qquad m\|x\|_\infty \leq \|Bx\| \leq M\|x\|_\infty \qquad \forall x \in \mathbb{R}^n.$$

Proof.   We are going to establish this result by induction on the dimension n of the space. For $n = 1$, the result is evident since $B_1 x = ax$, where $a \in E$. Hence $\|B_1 x\| = |a|\,\|x\|$, which shows that B_1 is bicontinuous. Suppose that the theorem is true for every space of dimension $n - 1$, and let us show that it is true for a space E of dimension n.

First of all, B_n is continuous since

$$(12) \qquad \begin{cases} \|B_n x\| = \left\|\sum x_i e_i\right\| \leq \sum |x_i|\,\|e_i\| \leq M\|x\|_\infty \\ \text{where} \qquad M = \sum_{i=1}^{n} \|e_i\|. \end{cases}$$

To show that B_n^{-1} is continuous we must show that the forms e_i^* are continuous. Now the subspaces $H_i = e_i^{*-1}(0)$ are generated by the elements of the base of E apart from the ith element; hence, H_i is a space of dimension $n - 1$. Therefore, the induction hypothesis implies that H_i is a *complete* subspace of E: Indeed, if $\{u^k\}_k \in H_i$ is a Cauchy sequence in H_i, the sequence $B_{n-1}^{-1} u^k$ is a Cauchy sequence in $\mathbb{R}^{n-1}$ since $\|B_{n-1}^{-1}(u^k - u^l)\|_\infty \leq m^{-1}\|u^k - u^l\|$. Thus this sequence converges to an element x of $\mathbb{R}^{n-1}$, $(\mathbb{R}^{n-1}, \|\cdot\|_\infty)$ being complete. Hence u_k converges to $u = B_{n-1}x$ in E since

$$\|u^k - u\| = \|B_{n-1}(B_{n-1}^{-1}u^k - x)\| \leq M\|B_{n-1}^{-1}u^k - x\|_\infty.$$

Therefore, H_i is closed which implies that e_i^* is a continuous linear form on E according to Proposition 4.2. There exist constants M_i such that $|e_i^*(u)| \leq M_i\|u\|$ for all $u \in E$. Thus

$$(13) \qquad \begin{cases} \|B_n^{-1}(u)\|_\infty = \sup_{1 \leq i \leq n} |e_i^*(u)| \leq m^{-1}\|u\| \\ \text{where} \qquad m^{-1} = \sup_{1 \leq i \leq n} M_i. \end{cases} \qquad \blacksquare$$

PROPOSITION 6.   If E is a finite dimensional vector space, if $\|\cdot\|_1$ and $\|\cdot\|_2$ are two norms on E, then there exist constants m and M such that

$$(14) \qquad m\|u\|_1 \leq \|u\|_2 \leq M\|u\|_1 \qquad \text{for all} \qquad u \in E.$$

Proof. According to Proposition 5, there exist constants m_1, m_2, M_1, and M_2 such that

$$(15) \qquad m_i \|B^{-1}u\|_\infty \le \|u\|_i \le M_i \|B^{-1}u\|_\infty \qquad \text{for} \qquad i = 1, 2.$$

Thus we obtain (14) with $m = m_2/M_1$ and $M = M_2/m_1$. ∎

Finally, we show from this result the following fundamental theorem.

THEOREM 1. If E and F are two finite dimensional spaces, every linear mapping from E to F is continuous.

Proof. Let $\{e_1, \ldots, e_n\}$ and $\{f_1, \ldots, f_p\}$ be the bases for E and F, respectively, and let B_n and B_p be the isomorphisms from $\mathbb{R}^n$ to E and from $\mathbb{R}^p$ to F. If A is a linear mapping from E to F, its matrix $M(A)$ is the operator $M(A) = B_p^{-1}AB_n$ from $\mathbb{R}^n$ to $\mathbb{R}^p$; it is continuous from $(\mathbb{R}^n, \|\cdot\|_\infty)$ to $(\mathbb{R}^p, \|\cdot\|_\infty)$ according to Proposition 1.10.3. Thus $A = B_p M(A) B_n^{-1}$, being the composition of three continuous mappings, is itself continuous. ∎

COROLLARY 1. If E is a finite dimensional vector space, every injective (or surjective) linear mapping A is a bicontinuous isomorphism from E onto E.

Proof. Indeed, we know that if E and F are finite dimensional spaces, a linear mapping A is bijective when it is injective or surjective. Thus A and A^{-1} are continuous. ∎

In other words, in the case of linear mappings from a finite dimensional vector space to itself, the uniqueness of the solution x of the equation $Ax = f$ for every $f \in F$ implies the existence of this solution and its continuous dependence on the right-hand side of the equation.

4. Proper Functions

PROPOSITION 1. Let (E, d_E) and (F, d_F) be two metric spaces and f a continuous mapping from E to F. If $K \subset E$ is compact, then $f(K)$ is compact (hence closed) in F, and f is a uniformly continuous mapping from K to F.

Proof.

a. Let us show that $f(K)$ is compact. Let $\{f(x_n)\}$ be a sequence in $f(K)$, and let us show that it has a convergent subsequence. Since the x_n's belong to the compact set K, a subsequence $\{x_{n_k}\}$ converges to x_* in K. Thus f being continuous, $f(x_{n_k})$ converges to $f(x_*) \in f(K)$.

b.   If f were not uniformly continuous, there would exist $\varepsilon > 0$ and two sequences $\{x_n\}$ and $\{y_n\}$ such that

(1) $$d_E(x_n, y_n) \leq \frac{1}{n} \quad \text{and} \quad d_F(f(x_n), f(y_n)) > \varepsilon.$$

Since K is compact, these sequences have convergent subsequences $\{x_{n_k}\}$ and $\{y_{n_k}\}$ converging to the same point x_* of K. Since f is continuous, we have for $k \geq k_0(\varepsilon)$

$$d_F(f(x_{n_k}), f(y_{n_k})) \leq d_F(f(x_{n_k}), f(x_*)) + d_F(f(x_*), f(y_{n_k})) \leq \frac{\varepsilon}{2},$$

which is a contradiction.   ∎

PROPOSITION 2.   Let f be a continuous bijective mapping from E onto F. If E is compact, f is bicontinuous.

Proof.   Let us show that f^{-1} transforms every sequence $\{y_n\}$ of elements $y_n \in F$ converging to an element y_* of F onto a sequence $x_n = f^{-1}(y_n)$ converging to $f^{-1}(y_*)$. Indeed, E being compact, $\{x_n\}$ has a subsequence $\{x_{n_k}\}$ that converges to an element x_* of E. Since f is continuous, the sequence $\{f(x_{n_k})\}$ converges to $f(x_*)$ and also converges to y_*. Thus $y_* = f(x_*)$, and we remark that x_* is the unique cluster point of $\{x_n\}$ since f is injective. Thus applying (2.6.1)v, the sequence $\{x_n\}$ converges to x_*.   ∎

Remark 1.   When E is not compact, the inverse image $f^{-1}(K)$ of a compact set K of F under a continuous mapping f is not necessarily compact.   ∎

In this connection, we prove the following theorem.

THEOREM 1.   Let E and F be two metric spaces and f a continuous mapping from E to F. The following two conditions are equivalent

(2)
> **i.**   Every sequence $\{x_n\}$ of elements $x_n \in E$ for which $f(x_n)$ converges to an element y of F has a subsequence $\{x_{n_k}\}$, which converges to an element $x \in E$.
>
> **ii.**   The sets $f^{-1}(y)$ are compact in E, and f sends every closed set A of E to a closed set $f(A)$ of F.

Proof.

a.   Let us show that (2)i implies (2)ii: The set $K = f^{-1}(y)$ is compact since if $\{x_n\}$ is a sequence of elements x_n of K, the sequence $f(x_n) = y$ converges to y; consequently, (2)i implies that we can extract a subsequence

$\{x_{n_k}\}$, which converges to an element x_* of F. Since f is continuous, $f(x_{n_k})$ converges to $f(x_*)$ and to y; hence $f(x_*) = y$. This implies that $x_* \in K = f^{-1}(y)$. Moreover, let A be a closed set in E, and let $y \in \overline{f(A)}$ be the limit of a sequence $\{f(x_n)\}$ of elements $f(x_n) \in f(A)$. According to (2)i, we can extract a subsequence $\{x_{n_k}\}$ of elements $x_{n_k} \in A$, which converges to x_*. Since A is closed, x_* belongs to A. But, f being continuous, $f(x_{n_k})$ converges to $f(x_*)$ and to y; hence $y = f(x_*) \in f(A)$, which implies that $f(A)$ is closed.

b.   Let us show that (2)ii implies (2)i. Consider a sequence $\{x_n\}$ of elements $x_n \in E$ for which $f(x_n)$ converges to y in F, and let us show that there exists a cluster point x of $\{x_n\}$, that is as we shall show the existence of a point x belonging to the intersection $\bigcap_n \overline{A}_n$, where $A_n = \{x_m\}_{m \geq n}$. Let $K = f^{-1}(y)$. Since $f(\overline{A}_n) = \overline{f(A_n)}$ (the image of a closed set being closed), and since $f(x_n)$ converges to y, then

$$(3) \qquad \{y\} = \bigcap_n \overline{f(A_n)} = \bigcap_n f(\overline{A}_n).$$

In other words, if $x \in K$, $f(x) \in \bigcap_n f(\overline{A}_n)$, which implies that the closed subsets $F_n = K \cap \overline{A}_n$ of K are nonempty. Since K is compact and since the closed sets F_n have the finite intersection property, we conclude that $\bigcap_n F_n = K \cap \bigcap_n \overline{A}_n$ is nonempty; consequently, there exists a cluster point x belonging to K.   ∎

Definition 1.   A continuous mapping f from a metric space E to a metric space F satisfying one of the two equivalent properties (2) is called a "proper mapping."

PROPOSITION 3.   If f is a proper mapping from E to F, then the inverse image $f^{-1}(K)$ of every compact set $K \subset F$ is compact in E.

Proof.   Let $\{x_n\}$ be a sequence of elements $x_n \in f^{-1}(K)$. Then since $\{f(x_n)\}$ is a sequence of elements $f(x_n)$ in the compact set K, we can extract a subsequence $\{f(x_{n_k})\}$ that converges to $y \in K$. Applying (2)i, we can then extract from $\{x_{n_k}\}$ a subsequence $\{x_p\}$ of elements $x_p \in f^{-1}(K)$ that converges to $x_* \in E$. Thus $f(x_p)$ converges both to $f(x_*)$ (because f is continuous) and to $y \in K$. Hence $y = f(x_*)$, which implies that $x_* \in f^{-1}(K)$. Therefore, $f^{-1}(K)$ is compact.   ∎

PROPOSITION 4.   If E is compact, every continuous mapping f from E to F is proper.

Proof.   According to (2)ii, since the inverse image $f^{-1}(y)$ is closed in the compact space E, it is compact. Moreover, if A is closed in E and, therefore, compact, the image $f(A)$ is compact in F and, therefore, closed.   ∎

PROPOSITION 5.   Let f be a continuous injective mapping from E to F. The following conditions are then equivalent.

(4) $\begin{cases} \textbf{i.} & f \text{ is proper.} \\ \textbf{ii.} & \text{The image under } f \text{ of every closed set of } E \text{ is closed in } F. \\ \textbf{iii.} & f \text{ is a bicontinuous mapping from } E \text{ onto } f(E). \end{cases}$

Proof.

a.   (4)i implies (4)ii by definition (see (2)ii).

b.   The subset $f(E)$ is closed since E is closed. Hence since f is bijective from E onto $f(E)$ and since the inverse image $f[(A)] = (f^{-1})^{-1}(A)$ under f^{-1} of every closed set A of E is closed, this implies that f^{-1} is continuous.

c.   It is clear that (4)iii implies (4)i from (2)i.   ∎

PROPOSITION 6.   Let f be a continuous mapping from E to F and g a continuous mapping from F to G. Then

(5)   if f and g are proper, $g \circ f$ is proper,

(6)   if $g \circ f$ is proper, then f is proper,

(7)   if $g \circ f$ is proper and f surjective, then g is proper.

Proof.   (*Left as an exercise.*)

Remark 2.   To show that a mapping is proper is one of the principal tasks of approximation theory: Let f be a continuous mapping from E to F. Given $y \in F$, we seek to solve the equation

(8) $f(x) = y.$

Suppose we approximate the right-hand side of the equation y by a sequence of elements $y_n \in F$, and suppose there exist solutions $x_n \in E$ of the approximating equations

(9) $f(x_n) = y_n.$

Saying that f is proper is the same as saying that we can extract a subsequence of approximating solutions x_{n_k} of equation (9), which converges to a solution of equation (8).   ∎

We shall now give some examples of proper mappings. To this end we introduce the following definition.

Definition 2.   We say that a subset X of $\mathbb{R}^n$ is bounded b   :here exists a vector $a \in \mathbb{R}^n$ such that for every $i = 1, \ldots, n$ and for ev $\in X, a_i \leq x_i$.

PROPOSITION 7. Let X and Y be two closed subsets that are bounded below. Then the mapping $\{x, y\} \in X \times Y \mapsto x + y \in \mathbb{R}^n$ is proper.

Proof. Consider two sequences of elements $x^m \in X$ and $y^m \in Y$ such that $z^m = x^m + y^m$ converges to $z \in \mathbb{R}^n$. Suppose that X and Y are bounded below by the vectors a and b. Then we have the inequalities

$$a_i \leq x_i^m = z_i^m - y_i^m \leq z_i^m - b_i \quad \text{and} \quad b_i \leq y_i^m \leq z_i^m - a_i.$$

Since the sequence z^m converges, it is bounded: There exists a vector $c \in \mathbb{R}^n$ such that $z_i^m \leq c_i$ for all i. Consequently, we have for all i and for all m $a_i \leq x_i^m \leq c_i - b_i$ and $b_i \leq y_i^m \leq c_i - a_i$. Thus the sequences x^m and y^m remain within hypercubes of $\mathbb{R}^n$, which are compact. Hence they have subsequences x^{m_k} and y^{m_k} that converge, respectively, to x and y in $\mathbb{R}^n$. Since X and Y are closed, we conclude that $x \in X$ and $y \in Y$. Then $z = x + y$ and addition is a proper mapping. ∎

We can generalize this proposition in the following fashion.

PROPOSITION 8. Consider a closed subset X of $\mathbb{R}^n$ that is bounded below and a subset Y of $\mathbb{R}^n$ such that

$$(10) \quad \begin{cases} \textbf{i.} & Y \text{ is closed and convex and} \\ \textbf{ii.} & \exists w \in Y \text{ such that } (Y - w) \cap \mathbb{R}_-^n = \{0\} \end{cases}$$

where $\mathbb{R}_-^n$ is the cone of vectors of $\mathbb{R}^n$ whose components are negative. Then the mapping $\{x, y\} \in X \times Y \mapsto x + y$ is proper.

Proof. Let $z^m = x^m + y^m$ be a sequence that converges to $z \in \mathbb{R}^n$ where $x^m \in X$ and $y^m \in Y$. Let us begin by showing, by contradiction, that a subsequence of y^m converges to y: If not, there would exist a subsequence y^{m_k} of y^m for which $\|y^{m_k}\| \to \infty$. Then $u^k = y^{m_k}/\|y^{m_k}\|$ belongs to the unit sphere $S(1) \subset \mathbb{R}^n$ which is compact. We can, therefore, extract from the subsequence u^k a subsequence u^{k_j}, which converges to an element u of $S(1)$. Moreover, the convergent sequence z^m is bounded: There exists a vector $c \in \mathbb{R}^n$ such that for all $i = 1, \ldots, n$, $z_i^m \leq c_i$. Since X is bounded below by a vector $a \in \mathbb{R}^n$, we obtain that $y_i^m = z_i^m - x_i^m \leq c_i - a_i$. Hence for all $i = 1, \ldots, n$ the inequalities

$$u_i^{k_j} = \frac{z_i^{k_j} - x_i^{k_j}}{\|x^{k_j}\|} \leq \frac{c_i - a_i}{\|x^{k_j}\|}$$

imply, taking limits, that $u_i \leq 0$ for all i. Thus the limit u belongs to $\mathbb{R}_-^n$. Moreover, since $\|y^m\| \geq 1$ after a certain rank and since $w \in Y$, we deduce from the fact that Y is convex that

$$\frac{1}{\|y^m\|} y^m + \left(1 - \frac{1}{\|y^m\|}\right) w \in Y.$$

Letting k_j go to infinity, we obtain that $u + w \in Y$, since Y is closed. We have, therefore, established that $u \in (Y - w) \cap \mathbb{R}^n_-$, hence that $u = 0$ according to condition (10)ii. This implies a contradiction since $u = 0$ cannot belong to the sphere $S(1)$.

Thus a subsequence y^{m_p} converges to an element y, which belongs to Y since it is closed. Then the subsequence $x^{m_p} = z^{m_p} - y^{m_p}$ converges to $z - y$, which belongs to X since it is closed. ∎

COROLLARY 1. Suppose that the mapping $\{x, y\} \in X \times Y \mapsto x + y \in \mathbb{R}^n$ is proper. Then if $A \subset X$ and $B \subset Y$ are closed, the sum $A + B$ is also closed. The intersection $X \cap -Y$ is compact.

These last three results are useful in economics.

5. Theorems of Extension by Density

Let E and F be two metric spaces and $A \subset E$ be a *dense subset* of E. This section is devoted to studying the properties of a function f from E to F knowing the properties of the function f from A to F. These theorems play a fundamental role.

Definition 1. Let $A \subset E$ be a subset of E, and let f be a mapping from E to F. We denote by $f_{|A}$ the mapping from A to F defined by

$$(1) \qquad f_{|A}(x) = f(x) \qquad \forall x \in A,$$

and we say that "$f_{|A}$ is the restriction of f to A." If f is a mapping from A to F, we say that a mapping g from E to F "extends" f (or is "an extension" of f) if

$$(2) \qquad f(x) = g(x) \qquad \text{for all} \qquad x \in A.$$

The existence and/or uniqueness of an extension of a function f satisfying particular conditions is very useful.

PROPOSITION 1. *Extension by identities.* Let f and g be two continuous functions from E to F, and let $A \subset E$ be a dense subset of E. If $f(x) = g(x)$ for all $x \in A$, then $f(x) = g(x)$ for all $x \in E$.

Proof. Let $x \in E$. Since A is dense in E, there exists a sequence $\{x_n\}$ of elements $x_n \in A$ that converges to x. Since f and g are continuous, $f(x_n)$ and $g(x_n)$ converge to $f(x)$ and $g(x)$, respectively. Then, since $f(x_n) = g(x_n)$, we obtain that

$$(3) \qquad \begin{cases} d_F(f(x), g(x)) \le d_F(f(x), f(x_n)) + d_F(f(x_n), g(x_n)) + d_F(g(x_n), g(x)) \\ \qquad = d_F(f(x), f(x_n)) + d_F(g(x_n), g(x)). \end{cases}$$

Letting n approach infinity, we conclude that $d_F(f(x), g(x)) \leq 0$, which implies that $f(x) = g(x)$.   ∎

Remark 1.   When A *is dense* in E, this proposition implies the *uniqueness* of an extension of a *continuous* function f from A to F by a *continuous* function from E to F: If g and h are two continuous functions from E to F extending f, then

$$g(x) = h(x) = f(x) \qquad \forall x \in A,$$

which implies that $g(x) = h(x)$ for all $x \in E$, that is, $g = h$.   ∎

Before studying the existence of a continuous function extending a continuous function from $A \subset E$ to F, we shall prove the following proposition.

PROPOSITION 2.   Let f and g be two real-valued continuous functions defined on a metric space E, and let $A \subset E$ be a dense subset of E. If $f(x) \leq g(x)$ for all $x \in A$, then $f(x) \leq g(x)$ for all $x \in E$.

Proof.   Let $x \in E$. There exists a sequence $\{x_n\}$ of elements $x_n \in A$, which converges to x since A is dense in E. As f and g are continuous, $f(x_n)$ and $g(x_n)$ converge to $f(x)$ and $g(x)$, respectively. Since $f(x_n) \leq g(x_n)$, we obtain that

(4)
$$\begin{cases} f(x) - g(x) \leq f(x) - f(x_n) + f(x_n) - g(x_n) + g(x_n) - g(x) \\ \qquad \leq (f(x) - f(x_n)) + (g_n(x) - g(x)). \end{cases}$$

Letting n approach infinity, we conclude that $f(x) - g(x) \leq 0$.   ∎

If A is dense in E and if f is a continuous mapping from A to F, there does not necessarily exist a continuous mapping g from E to F extending f. (See Proposition 6.3.) However, such an extension exists under the following hypotheses.

THEOREM 1. *Theorem of extension by density.*   Let E be a metric space, $A \subset E$ a dense subset of E and F a complete metric space. If f is a uniformly continuous mapping from A to F, there exists a unique extension g of f that is a uniformly continuous mapping from E to F.

Proof.   Consider $x \in E$. There exists a sequence $\{x_n\}$ of elements $x_n \in A$ converging to x since A is dense in E. The function f being uniformly continuous implies that $f(x_n)$ is a Cauchy sequence in F, which converges to a point $g(x)$ of F since F is complete. This point does not depend upon the sequence $\{x_n\}$ which converges to x: If $\{y_n\}$ is another sequence of elements

$y_n \in A$ that also converges to x and if $g(y)$ denotes the limit of the Cauchy sequence $g(y_n)$, we obtain

(5) $d_F(g(x), g(y)) \le d_F(g(x), f(x_n)) + d_F(f(x_n), f(y_n)) + d_F(f(y_n), g(y))$.

But since f is uniformly continuous, for every $\varepsilon > 0$, there exists $\eta > 0$ such that $d_F(f(x_n), f(y_n)) \le \varepsilon$ when $d_E(x_n, y_n) \le \eta$.

Since x_n and y_n converge to x, there exists $n_0(\eta)$ such that $d_E(x_n, y_n) \le \eta$ when $n \ge n_0(\eta)$. Consequently, for $n \ge n_0(\eta(\varepsilon))$

(6) $d_F(g(x), g(y)) \le d_F(g(x), f(x_n)) + d_F(f(y_n), g(y)) + \varepsilon$.

This implies, letting n approach infinity, that $d_F(g(x), g(y)) \le \varepsilon$ for all $\varepsilon > 0$. Thus $d_F(g(x), g(y)) = 0$, and hence $g(x) = g(y)$. We have shown the existence of a function g extending f (for if $x \in A, g(x) = f(x) = \lim f(x_n)$). The function g is uniformly continuous: Given $\varepsilon > 0$, since f is uniformly continuous, there exists $\eta = \eta(\varepsilon)$ such that $d_F(f(x), f(y)) \le \varepsilon$ when $d_F(x, y) \le \eta, x, y \in A$. Moreover, there exist a sequence $\{x_n\}$ of elements $x_n \in A$ and a sequence $\{y_n\}$ of elements $y_n \in A$, that converge to x and y, respectively. Thus there exists $n_0 = n_0(\eta)$ such that

(7) $d_E(x_n, y_n) \le d_E(x_n, x) + d_E(x, y) + d_E(y_n, y) \le 3\dfrac{\eta}{3}$

when $n \ge n_0(\eta)$ and $d_E(x, y) \le \eta/3$. Under these conditions we deduce that

(8) $d_F(g(x), g(y)) \le d_F(g(x), f(x_n)) + \varepsilon + d_F(f(y_n), g(y))$.

Letting n go to infinity, we conclude that

(9) $d_F(g(x), g(y)) \le \varepsilon$ when $d_E(x, y) \le \dfrac{\eta}{3}$.   ∎

Theorem 1 implies the following two theorems, which we shall use constantly in [AFA].

THEOREM 2. Let E be a normed space, $E_0 \subset E$ a dense subspace of E, and F a Banach space. Every continuous linear mapping from E_0 to F has a unique extension to a continuous linear mapping from E to F:

(10) $\mathscr{L}(E_0, F) = \mathscr{L}(E, F)$.

Proof. Let $A \in \mathscr{L}(E_0, F)$. Since A is a continuous linear mapping, it is uniformly continuous from E_0 to F and, according to Theorem 1, has a unique extension to a continuous mapping $\bar{A}$ from E to F. Proposition 1 shows that $\bar{A}$ is linear: Since $A(\lambda x_0 + \mu y_0) = \lambda A(x_0) + \mu A(y_0)$ for all

$x_0, y_0 \in E_0$, and since A is continuous, we deduce that $\bar{A}(\lambda x + \mu y) = \lambda \bar{A}(x) + \mu \bar{A}(y)$ for all $x, y \in E$. Moreover, Proposition 2 shows that the inequalities $\|\bar{A}x_0\| = \|Ax_0\| \le \|A\| \|x_0\|$ for all $x_0 \in E_0$ have extensions to inequalities $\|\bar{A}x\| \le \|A\| \|x\|$ for all $x \in E$, since the norms are continuous. This shows that

$$\|\bar{A}\| = \sup_{x \in E} \frac{\|\bar{A}x\|_F}{\|x\|_E} \le \|A\| = \sup_{x_0 \in E_0} \frac{\|\bar{A}x_0\|_F}{\|x_0\|_E} \le \|\bar{A}\|,$$

that is, that $\|A\| = \|\bar{A}\|$. Hence the mapping $A \in \mathscr{L}(E_0, F) \mapsto \bar{A} \in \mathscr{L}(E, F)$ is an isometry, which is surjective, since every $\bar{A} \in \mathscr{L}(E, F)$ is the unique extension of its restriction $A \in \mathscr{L}(E_0, F)$ to E_0. ∎

THEOREM 3. Let E and F be two normed spaces, $E_0 \subset E$ and $F_0 \subset F$ be two dense subspaces and G be a Banach space. Every continuous bilinear mapping from $E_0 \times F_0$ to G has a unique extension to a continuous bilinear mapping from $E \times F$ to G:

(11) $\mathscr{L}(E_0, F_0; G) = \mathscr{L}(E, F; G)$.

Proof. By using Proposition 3.3 we can identify $\mathscr{L}(E_0, F_0; G)$ with $\mathscr{L}(E_0, \mathscr{L}(F_0, G))$ by associating to a bilinear mapping $A \in \mathscr{L}(E_0, F_0; G)$ the linear mapping $A_1 : x_1 \in E_0 \mapsto A_1 x_1 \in \mathscr{L}(F_0, G)$. Since G is a Banach space by Theorem 2 we can identify $\mathscr{L}(F_0, G)$ and $\mathscr{L}(F, G)$ by identifying $A_1 x_1$ with its unique extension $\overline{A_1 x_1} \in \mathscr{L}(F, G)$. It is clear that the mapping $\bar{A}_1 : x_1 \in E_0 \mapsto \overline{A_1 x_1} \in \mathscr{L}(F, G)$ is linear (since the extension $\overline{\lambda A_1 x_1 + \mu A_1 y_1}$ of a linear combination of linear mappings is the linear combination $\lambda \overline{A_1 x_1} + \mu \overline{A_1 y_1}$ of the extensions). The operator $\bar{A}_1$ is continuous since

$$\|\overline{A_1 x_1}\|_{\mathscr{L}(F, G)} = \|A_1 x_1\|_{\mathscr{L}(F_0, G)} \le \|A_1\|_{\mathscr{L}(E_0, \mathscr{L}(F_0, G))} \|x_1\|_E$$
$$= \|A\|_{\mathscr{L}(E_0, F_0; G)} \|x_1\|_E.$$

Moreover, Proposition 1.10.2 implies that the space $\mathscr{L}(F_0, G) = \mathscr{L}(F, G)$ is a Banach space. Hence Theorem 2 shows that the operator $\bar{A}_1 \in \mathscr{L}(E_0, \mathscr{L}(F, G))$ has a unique extension to a continuous linear operator $\bar{\bar{A}}_1 \in \mathscr{L}(E, \mathscr{L}(F, G))$; the operator $\bar{\bar{A}}_1$ defines a continuous bilinear mapping $\bar{\bar{A}} \in \mathscr{L}(E, F; G)$, which extends A. The mapping $A \in \mathscr{L}(E_0, F_0; G) \mapsto \bar{\bar{A}} \in \mathscr{L}(E, F; G)$ is a surjective isometry, since it is the result of the composition of the surjective isometries

$$A \in \mathscr{L}(E_0, F_0; G) \mapsto A_1 \in \mathscr{L}(E_0, \mathscr{L}(F_0, G)) \mapsto \bar{A}_1 \in \mathscr{L}(E_0, \mathscr{L}(F, G))$$
$$\mapsto \bar{\bar{A}}_1 \in \mathscr{L}(E, \mathscr{L}(F, G)) \mapsto \bar{\bar{A}} \in \mathscr{L}(E, F; G). \qquad ∎$$

6.   Limits

We shall now define the notion of *lower limit* of a function. Let f be a function defined on a metric space E with values in the extended reals $\overline{\mathbb{R}}$. To every $\varepsilon \in {]}0, 1{[}$, we associate the function

$$(1) \qquad\qquad \alpha(\varepsilon) = \inf_{y \in B(x, \varepsilon)} f(y).$$

This is a decreasing function of ε. Therefore,

$$(2) \qquad\qquad \lim_{\substack{\varepsilon \to 0 \\ \varepsilon > 0}} \alpha(\varepsilon) = \sup_{\varepsilon > 0} \alpha(\varepsilon) \qquad \text{exists in} \qquad \overline{\mathbb{R}}.$$

Definition 1.   Let f be a function defined on a metric space E with values in $\overline{\mathbb{R}}$. "The lower limit of f at x" is the expression

$$(3) \qquad \liminf_{y \to x} f(y) = \lim_{\varepsilon \to 0} \inf_{y \in B(x, \varepsilon)} f(y) = \sup_{\varepsilon > 0} \inf_{y \in B(x, \varepsilon)} f(y)$$

and the upper limit of f at x is the expression

$$(4) \qquad \limsup_{y \to x} f(y) = \lim_{\varepsilon \to 0} \sup_{y \in B(x, \varepsilon)} f(y) = -\liminf_{y \to x}[-f(y)].$$

Similarly, if $\{x_n\}$ is a sequence of elements x_n of E, we define the upper and lower limits of $f(x_n)$ by

$$(5) \qquad \liminf_{n \to \infty} f(x_n) = \lim_{n \to \infty} \inf_{m \geq n} f(x_n) = \sup_{n} \inf_{m \geq n} f(x_m)$$

and

$$(6) \qquad \limsup_{n \to \infty} f(x_n) = \lim_{n \to \infty} \sup_{m \geq n} f(x_m) = -\liminf_{n \to \infty}[-f(x_m)].$$

PROPOSITION 1.   Let f be a function defined on E with values in $\overline{\mathbb{R}}$. Then for every $x \in E$,

$$(7) \qquad \inf_{y \in E} f(y) \leq \liminf_{y \to x} f(y) \leq f(x) \leq \limsup_{y \to x} f(y) \leq \sup_{y \in Y} f(y).$$

Proof.   Indeed, we always have

$$(8) \qquad \begin{cases} \inf_{y \in E} f(y) \leq \alpha(\varepsilon) = \inf_{y \in B(x, \varepsilon)} f(y) \leq f(x) \leq \beta(\varepsilon) \\[2mm] \text{where} \quad \beta(\varepsilon) = \sup_{y \in B(x, \varepsilon)} f(y) \leq \sup_{y \in E} f(y). \end{cases}$$

Taking the limit as ε approaches 0, we obtain (7).   ∎

PROPOSITION 2. If $\{x_{n_k}\}$ is a subsequence of a sequence $\{x_n\}$, we obtain

(9)
$$\begin{cases} \liminf_{n \to \infty} f(x_n) \leq \liminf_{k \to \infty} f(x_{n_k}) \leq \limsup_{k \to \infty} f(x_{n_k}) \\ \qquad \leq \limsup_{n \to \infty} f(x_n). \end{cases}$$

Proof. Since $k \leq n_k$, we deduce that

(10)
$$\inf_{m \geq k} f(x_m) \leq \inf_{m \geq n_k} f(x_m) \leq \sup_{m \geq n_k} f(x_m) \leq \sup_{m \geq k} f(x_m).$$

We obtain (9) by letting k approach infinity.   ∎

PROPOSITION 3. Let f and g be two functions defined on a metric space with values in $\mathbb{R}$. Then we have the following properties:

(11)
$$\begin{cases} \text{If } f \leq g, \quad \text{then} \quad \limsup_{y \to x} f(y) \leq \limsup_{y \to x} g(y) \quad \text{and} \\ \liminf_{y \to x} f(y) \leq \liminf_{y \to x} g(y); \end{cases}$$

(12)
$$\limsup_{y \to x}(f(y) + g(y)) \leq \limsup_{y \to x} f(y) + \limsup_{y \to x} g(y);$$

(13)
$$\limsup_{y \to x} f(y) + \liminf_{y \to x} g(y) \leq \limsup_{y \to x}(f(y) + g(y)).$$

Proof.

a. Since $f \leq g$, we deduce that

(14)
$$\sup_{y \in B(x, \varepsilon)} f(y) \leq \sup_{y \in B(x, \varepsilon)} g(y) \quad \text{and} \quad \inf_{y \in B(x, \varepsilon)} f(y) \leq \inf_{y \in B(x, \varepsilon)} g(y),$$

which implies the first assertion x by letting ε approach 0.

b. Since

(15)
$$\sup_{y \in B(x, \varepsilon)} (f(y) + g(y)) \leq \sup_{y \in B(x, \varepsilon)} f(y) + \sup_{y \in B(x, \varepsilon)} g(y),$$

we obtain (12) by letting ε approach 0.

c. Since

(16)
$$\sup_{y \in B(x, \varepsilon)} f(y) + \inf_{y \in B(x, \varepsilon)} g(y) \leq \sup_{y \in B(x, \varepsilon)} (f(y) + g(y)),$$

we obtain (13) by letting ε approach 0.   ∎

LIMITS WITH RESPECT TO A SUBSET. We shall next define and study the properties of limits with respect to a subset.

Definition 2. Let A be a subset of a metric space E, f be a mapping from E to F, and $x \in \bar{A}$. We say that "f has a limit $f_A(x)$ at x with respect to A" if

(17) $\begin{cases} \forall \varepsilon > 0, & \exists \eta = \eta(\varepsilon, x, A) \quad \text{such that} \quad d_F(f_A(x), f(y)) \leq \varepsilon, \\ \text{when} & y \in A \quad \text{and} \quad d_E(x, y) \leq \eta. \end{cases}$

We set

(18) $$f_A(x) = \lim_{\substack{y \to x \\ y \in A}} f(y).$$

EXAMPLE 1. Let $E = \mathbb{R}$ and $x \in \mathbb{R}$. If f is a function from $\mathbb{R}$ to F, we shall say that $f_+(x)$ is the *limit from the right* of f at x if $f_+(x) = f_{]x, \infty[}(x)$ is the limit of f at x with respect to $A = {]}x, \infty{[}$:

(19) $$f_+(x) = \lim_{\substack{y \to x \\ y \geq x}} f(y).$$

We define in the same fashion the *limit from the left* of f at x by

(20) $$f_-(x) = \lim_{\substack{y \to x \\ y \leq x}} f(y).$$

PROPOSITION 4. Let f be a mapping from $A \subset E$ to a metric space F. In order that f have a limit at a point $x \in \bar{A}$ with respect to A, it is necessary and sufficient that if $x = \lim_{n \to \infty} x_n$, $x_n \in A$, then $f_A(x) = \lim_{n \to \infty} f(x_n)$.

Proof.

a. It is obvious that the condition is necessary.

b. In order to show that the condition is sufficient, let us suppose that $z = \lim_{n \to \infty} f(x_n)$ is not the limit of f with respect to A. Then there exists $\varepsilon > 0$ such that for all $\eta = 1/n$, there exists $x_n \in A$ satisfying $d_E(x, x_n) \leq 1/n$ and $d_F(z, f(x_n)) > \varepsilon$. This shows that the sequence x_n converges to x and that the sequence $f(x_n)$ does not converge to z, which is contrary to our hypothesis. ∎

PROPOSITION 5.

a. The limit of f at $x \in \bar{A}$ with respect to A is unique.

b. Suppose that $f_A(x)$ is the limit of f at $x \in \bar{A}$ with respect to A. If $B \subset A$ and $x \in \bar{B}$, then $f_A(x) = f_B(x)$ is the limit of f at x with respect to B.

c. If $f_A(x) = \lim_{\substack{y \to x \\ y \in A}} f(y)$, then $f_A(x) \in \overline{f(A)}$.

Proof.

a. If u and v were two limits of f at $x \in \bar{A}$ with respect to A, we would have for every $\varepsilon > 0$, a number η such that $d_F(u, f(y)) \leq \varepsilon/2$ and $d_F(v, f(y)) \leq \varepsilon/2$ when $y \in A$ satisfies $d_E(x, y) \leq \eta$. We would then obtain that $d_F(u, v) \leq d_F(u, f(y)) + d_F(f(y), v) \leq \varepsilon$ for all $\varepsilon > 0$, and hence that $d_F(u, v) = 0$. This implies that $u = v$.

b. The second assertion is obvious.

c. The third assertion is an immediate result of Proposition 5. ∎

■

Remark 1. We now characterize those continuous functions on a dense subset that has a continuous extension. ∎

■

*PROPOSITION 6. Let $A \subset E$ be a dense subset of a metric space E, and let f be a continuous mapping from A to F. In order that f have a continuous extension g from E to F, it is necessary and sufficient that for each $x \in E$ the limit $\lim_{\substack{y \to x \\ y \in A}} f(y)$ exist. The function g is then defined by

$$(21) \qquad\qquad g(x) = \lim_{\substack{y \to x \\ y \in A}} f(y).$$

Proof.

a. If g is a continuous function from E to F, which is an extension of f, we obtain that $g(x) = \lim_{\substack{y \to x \\ y \in A}} f(y)$ since every point x belongs to the closure of A.

b. *Conversely,* suppose that the limit $\lim_{\substack{y \to x \\ y \in A}} f(y)$ exists. Consider the function g defined by (21). If $x \in A$, then $g(x) = f(x)$. Hence g is an extension of f. It remains to show that g is continuous.

Let $x \in E$ and $\varepsilon > 0$ be given. Since $g(x)$ is the limit of f at x with respect to A, there exists $\eta = \eta(\varepsilon, x)$ such that $f(B(x, \eta) \cap A) \subset B(g(x), \varepsilon)$. We are going to show that if $y \in B(x, \eta)$, $g(y) \in B(g(x), \varepsilon)$): From the fact that $g(y)$ is the limit of f with respect to A, hence with respect to $B(x, \eta) \cap A = B$ since this set is dense in $B(x, \eta)$, we conclude that $g(y)$ belongs to the closure of $f(B(x, \eta) \cap A)$, thus to $B(g(x), \varepsilon)$. Consequently, we have established that $g(y) \in B(g(x), \varepsilon)$ when $y \in B(x, \eta)$. ■

Definition 3. Let $A \subset \mathbb{R}$. A function f from A to $\mathbb{R}$ is said to be increasing (respectively, strictly increasing) if $x < y$ implies $f(x) \leq f(y)$ (respectively, $x < y$ implies $f(x) < f(y)$). A function is decreasing (strictly decreasing) if $-f$ is increasing (strictly increasing). A function is monotone (strictly monotone) if f or $-f$ is increasing (strictly increasing).

Remark 2. It is clear that every *strictly monotone* function is *injective,* that *the sum $f + g$ and the product λf by a scalar $\lambda \geq 0$ of increasing functions are increasing.* ∎

PROPOSITION 7. Suppose $A \subset \mathbb{R}$ and $a = \sup_{y \in A} y = \sup A$. If f is an increasing function from A to $\mathbb{R}$, then

$$
(22) \qquad \lim_{\substack{y \to a \\ y \in A}} f(y) = \sup_{x \in A} f(x) \qquad \text{if} \qquad \sup_{x \in A} f(x) < +\infty.
$$

Proof. Set $\alpha = \sup_{x \in A} f(x)$. For every $n > 0$, there exists $x_n \in A$ such that $\alpha - 1/n < f(x_n) < \alpha$. Consequently, if $y \in [x_n, a[$, we conclude that $f(x_n) \leq f(y) \leq \alpha$, which shows that $\alpha = \lim_{\substack{y \to a \\ y \in A}} f(y)$. ∎

7. Semicontinuous Functions

Consider a real-valued function defined on a metric space E. To say that this function is continuous at x is to say that for every $\varepsilon > 0$, there exists $\eta = \eta(x, \varepsilon)$ such that $f(x) - \varepsilon \leq f(y)$ and $f(y) \leq f(x) + \varepsilon$ for every $y \in B(x, \eta)$. This suggests considering those functions that satisfy only one of the two previously mentioned inequalities.

PROPOSITION 1. Let f be a function defined on a metric space E with values in $\mathbb{R}$. If $x \in E$, the following conditions are equivalent:

$$
(1) \quad
\begin{cases}
\textbf{i.} \quad \forall \varepsilon > 0, \quad \exists \eta = \eta(\varepsilon, x) \quad \text{such that} \quad \forall y \in B(x, \eta); \\
\quad f(x) - \varepsilon \leq f(y) \\
\textbf{ii.} \quad f(x) \leq \liminf_{y \to x} f(y)
\end{cases}
$$

Proof.

a. Let us show that (1)i implies (1)ii. For every $\varepsilon > 0$, we obtain from (1)i that

$$
(2) \qquad f(x) - \varepsilon \leq \inf_{y \in B(x, \eta)} f(y) = \alpha(\eta) \leq \liminf_{y \to x} f(y)
$$

When ε approaches 0, by taking the limit we obtain

$$
(3) \qquad f(x) \leq \liminf_{y \to x} f(y).
$$

b. *Conversely,* we deduce from (1)ii that for every $\varepsilon > 0$, there exists η such that

$$
(4) \qquad f(x) - \varepsilon \leq \alpha(\eta)
$$

where $\alpha(\eta) = \inf_{y \in B(x, \eta)} f(y)$. This inequality implies (1)i. ∎

Definition 1. Let f be a function defined on a metric space E with values in $\mathbb{R}$. We say that f is lower semicontinuous at x if one of the equivalent conditions (1)i or (1)ii is satisfied. We say that f is upper semicontinuous at x if $-f$ is lower semicontinuous at x. A function f is lower (respectively, upper) semicontinuous on E if f is lower (respectively, upper) semicontinuous at every $x \in E$.

Proposition 1 then becomes Proposition 1′.

PROPOSITION 1′. The following conditions are equivalent:

$$(5) \quad \begin{cases} \textbf{i.} & f \text{ is upper semicontinuous at } x. \\ \textbf{ii.} & \forall \varepsilon > 0, \quad \exists \eta = \eta(\varepsilon, x) \quad \text{such that} \quad f(y) \leq f(x) + \varepsilon \\ & \text{for all} \quad y \in B(x, \eta). \\ \textbf{iii.} & \limsup_{y \to x} f(y) = f(x). \end{cases}$$

Definition 2. Let f be a function from E with values in $\mathbb{R}$. The "epigraph" of f is the subset $\mathscr{E}_{p}(f)$ defined by

$$(6) \qquad \mathscr{E}_{p}(f) = \{(x, \lambda) \in E \times \mathbb{R} \text{ such that } \lambda \geq f(x)\}.$$

The following proposition characterizes lower semicontinuous functions on a metric space E.

PROPOSITION 2. The following properties are equivalent:

$$(7) \quad \begin{cases} \textbf{i.} & f \text{ is lower semicontinuous on } E. \\ \textbf{ii.} & \text{The sets } V_{\lambda} = \{x \in E \text{ such that } f(x) > \lambda\} \text{ are open } \forall \lambda \in \mathbb{R}. \\ \textbf{iii.} & \text{The sets } F_{\lambda} = \{x \in E \text{ such that } f(x) \leq \lambda\} \text{ are closed } \forall \lambda \in \mathbb{R}. \\ \textbf{iv.} & \text{The epigraph } \mathscr{E}_{p}(f) \text{ of } f \text{ is closed.} \end{cases}$$

Proof. We shall show successively that (6)i ⇒ (6)iv ⇒ (6)iii ⇒ (6)ii ⇒ (6)i.

a. Suppose that f is lower semicontinuous. Let $\{(x^{n}, \lambda^{n})\}$ be a sequence of points of $\mathscr{E}_{p}(f)$ that converges to (x, λ) in $E \times \mathbb{R}$. Since f is lower semicontinuous and since $f(x^{n}) \leq \lambda^{n}$, Proposition 1 implies that

$$(8) \qquad f(x) \leq \liminf_{n \to \infty} f(x^{n}) \leq \liminf_{n \to \infty} \lambda^{n} = \lim_{n \to \infty} \lambda^{n} = \lambda.$$

b. Suppose that $\mathscr{E}_{p}(f)$ is closed. Consider a sequence of elements $x_{n} \in F_{\lambda}$ that converges to x. Since $f(x_{n}) \leq \lambda$, the sequence of pairs $\{x_{n}, \lambda\}$ is a sequence of elements of $\mathscr{E}_{p}(f)$, which converges to $\{x, \lambda\}$. Since $\mathscr{E}_{p}(f)$ is closed, $\{x, \lambda\} \in \mathscr{E}_{p}(f)$, that is, $f(x) \leq \lambda$. Hence F_{λ} is closed.

c. Suppose that F_λ is closed. Since V_λ is the complement of F_λ, it is open.

d. Let us show that (6)ii implies (6)i. If $x \in E$ and if $\lambda = f(x) - \varepsilon$, we deduce that since V_λ is an open set containing x, it contains a ball of radius η and center x. Hence $\lambda = f(x) - \varepsilon < f(y)$ for all $y \in B(x, \eta)$. ∎

PROPOSITION 3. A real-valued function f is continuous at x (respectively, on E) if and only if it is both upper and lower semicontinuous at x (respectively, on E).

We shall show that the pointwise supremum f of an infinite family of continuous functions f_i, where $f(x) = \sup_{i \in I} f_i(x)$, is lower semicontinuous (although not necessarily continuous).

THEOREM 1. Let $\{f_i\}_{i \in I}$ be a family of real-valued functions f_i that are lower semicontinuous at a point x of a metric space E. The pointwise supremum defined by $f(x) = \sup_{i \in I} f_i(x)$ is a lower semicontinuous function at x when $f(x) < +\infty$.

Proof. Since $f(x) = \sup_i f_i(x)$, we can associate to every $\varepsilon > 0$ an index $i_0 = i_0(\varepsilon)$ such that

$$(9) \qquad\qquad f(x) - \frac{\varepsilon}{2} \le f_{i_0}(x).$$

Moreover, since f_{i_0} is lower semicontinuous at x, there exists $\eta = \eta(\varepsilon, x)$ such that

$$(10) \qquad f_{i_0}(x) - \frac{\varepsilon}{2} \le f_{i_0}(y) \le f(y) = \sup_{i \in I} f_i(y) \qquad \text{for} \qquad y \in B(x, \eta).$$

We then conclude from (9) and (10) that

$$(11) \qquad\qquad f(x) - \varepsilon \le f(y) \qquad \text{for} \qquad y \in B(x, \eta). \qquad ∎$$

PROPOSITION 4. Let f and g be real-valued functions that are lower semicontinuous at x. Then $f + g$ and $\inf(f, g)$ are lower semicontinuous functions at x.

EXAMPLES OF LOWER SEMICONTINUOUS FUNCTIONS

EXAMPLE 1. The preceding theorem implies in particular that the pointwise supremum $f(x) = \sup_{i \in I} f_i(x)$ of a family $\{f_i\}_{i \in I}$ of continuous functions f_i is *at least lower semicontinuous*, but not necessarily continuous.

EXAMPLE 2. We say that a function f from E to $\mathbb{R}$ has a *relative minimum* at a point $x_0 \in E$ if there exists a neighborhood $V(x_0)$ of x_0 such that

$$(12) \qquad\qquad f(x_0) = \min_{x \in V(x_0)} f(x).$$

It is clear that if f has a relative minimum at a point $x_0 \in E$, f is lower semicontinuous at x_0.

EXAMPLE 3. *Characteristic functions of sets.*

Definition 3. If $A \subset E$ is a subset of E, we denote by χ_A the "characteristic function" of A, which is defined by

$$(13) \qquad\qquad \chi_A(x) = \begin{cases} 0 & \text{if} \quad x \notin A, \\ 1 & \text{if} \quad x \in A. \end{cases}$$

PROPOSITION 5. Characteristic functions have the following properties:

$$(14) \quad \begin{cases} \textbf{i.} & \chi_E = 1; \quad \chi_\varnothing = 0. \\[2mm] \textbf{ii.} & \chi_{\complement A} = 1 - \chi_A. \\[2mm] \textbf{iii.} & \chi_A \chi_B = \chi_{A \cap B}; \quad \chi_A + \chi_B = \chi_{A \cup B} + \chi_{A \cap B}. \\[2mm] \textbf{iv.} & \inf_i \chi_{A_i} = \chi_{\cap_i A_i}; \quad \sup_i \chi_{A_i} = \chi_{\cup_i A_i}. \end{cases}$$

for any subsets A, B, A_i $(i \in I)$ of E.

Proof. (*Left as an exercise.*)

Remark 1. The set of characteristic functions of the subsets of E is, therefore, the set $\{0, 1\}^E$. The use of $\{0, 1\}^E$ allows a *convenient* "functional" representation of the set of subsets of E. We see other ways of characterizing sets by means of functions (see gauge of a convex set in Section 8). ■

In particular, the topological properties of sets are characterized by the properties of semicontinuity of their characteristic functions.

PROPOSITION 6. A subset A of a metric space E is open (respectively, closed) if and only if its characteristic function χ_A is lower semicontinuous (respectively, upper semicontinuous) on E.

Proof. To say that χ_A is lower semicontinuous is the same as saying that the subsets $V_\lambda = \{x \text{ such that } \chi_A(x) > \lambda\}$ are open $\forall \lambda \in \mathbb{R}$. But $V_\lambda = E$

if $\lambda < 0$, $V_\lambda = A$ if $\lambda \in [0, 1[$, and $V_\lambda = \varnothing$ if $\lambda \geq 1$; thus V_λ is open for all λ if and only if A is open.

On the other hand, the set A is closed if and only if $\complement A$ is open, that is if and only if $\chi_{\complement A} = 1 - \chi_A$ is lower semicontinuous, that is if and only if χ_A is upper semicontinuous. ∎

Remark 2. Fuzzy sets. The characterization of subsets by their characteristic functions allows us to extend the concept of subset to that of "fuzzy subset." A fuzzy set is any mapping $\chi \in [0, 1]^E$ from E to $[0, 1]$ that associates to each element $x \in E$ the "rate of membership" $\chi(x) \in [0, 1]$ of x in the fuzzy subset χ. We say that x does not belong to the fuzzy set χ if $\chi(x) = 0$. Representing a subset A of E by its characteristic function, we see that it defines a fuzzy set. The concept of fuzzy set is very useful in constructing "models" for various "fuzzy" concepts in the social sciences. ∎

8. Convex Functions

PROPOSITION 1. Let K be a convex subset of a vector space E and f a real-valued function defined on K. The following properties are equivalent:

(1)
$$\begin{cases} \textbf{i.} \quad \text{For every convex combination } x = \sum \alpha_i x_i \text{ of elements } x_i \text{ of} \\ K, f(\sum \alpha_i x_i) \leq \sum \alpha_i f(x_i). \\ \textbf{ii.} \quad \text{For every} \\ \qquad \alpha \in]0, 1[, x, y \in K, f(\alpha x + (1 - \alpha)y) \leq \alpha f(x) + (1 - \alpha)f(y). \\ \textbf{iii.} \quad \text{The epigraph } \mathscr{E}\!p(f) = \{(x, \lambda) \in K \times \mathbb{R} \text{ such that } f(x) \leq \lambda\} \text{ is} \\ \text{a convex subset of } E \times \mathbb{R}. \end{cases}$$

Moreover, any of the equivalent properties (1) imply that

(2) $\forall \lambda \in \mathbb{R}$, the subsets $F_\lambda = \{x \in K \text{ such that } f(x) \leq \lambda\}$ are convex.

Proof.

a. It is clear that (1)i implies (1)ii. Let us show that (1)ii implies (1)iii. Let (x, λ) and (y, μ) be two points of the epigraph of f and take $\alpha \in [0, 1]$. Let us show that the pair $(\alpha x + (1 - \alpha)y, \alpha \lambda + (1 - \alpha)\mu)$ belongs to $\mathscr{E}\!p(f)$. Since $f(x) \leq \lambda$ and $f(y) \leq \mu$, we obtain that $\alpha f(x) \leq \alpha \lambda$ and $(1 - \alpha)f(y) \leq (1 - \alpha)\mu$ and consequently that

(3) $f(\alpha x + (1 - \alpha)y) \leq \alpha f(x) + (1 - \alpha)f(y) \leq \alpha \lambda + (1 - \alpha)\mu.$

b. Let us show that (1)iii implies (1)i. Since $\mathscr{E}\!p(f)$ is a convex set, every convex combination $(x, \lambda) = (\sum \alpha_i x_i, \sum \alpha_i f(x_i))$ of points $(x_i, f(x_i))$ of $\mathscr{E}\!p(f)$ belongs to $\mathscr{E}\!p(f)$; hence $f(\sum \alpha_i x_i) \leq \sum \alpha_i f(x_i)$.

c. Let us show (2). Let $x = \sum \alpha_i x_i$ be a convex combination of points x_i of F_λ. Since $f(x_i) \leq \lambda$ for every i, $\alpha_i \geq 0$ and $\sum \alpha_i = 1$, we obtain that $\sum \alpha_i f(x_i) \leq \lambda$. Thus (1)i implies that $f(\sum \alpha_i x_i) \leq \lambda$, that is, that $x = \sum \alpha_i x_i \in F_\lambda$. ∎

Definition 1. A real-valued function f defined on a convex subset K of a vector space E is said to be "convex" if it satisfies one of the equivalent conditions (1). Furthermore, f is "strictly convex" if

$$(4) \quad \forall \alpha \in]0, 1[, \qquad \forall x, y \in K, \qquad f(\alpha x + (1 - \alpha)y) < \alpha f(x) + (1 - \alpha)f(y).$$

We say that f is "concave" (respectively, "strictly concave") if $-f$ is convex (respectively, strictly convex). A function f is "affine" if it is convex and concave.

Before studying lower semicontinuous convex functions, let us establish the following result.

PROPOSITION 2. Let f and g be two convex functions defined on a convex subset K. Then

$$(5) \qquad \forall \alpha, \qquad \beta \geq 0, \qquad \alpha f + \beta g \qquad \text{is a convex function.}$$

If the pointwise supremum $f(x) = \sup_{i \in I} f_i(x)$ of a family of convex functions f_i is finite on K, then f is convex.

Proof. (*Left as an exercise.*)

We shall now characterize convex functions that are lower semicontinuous at a point of a normed space.

THEOREM 1. Let K be a nonempty convex subset of a normed space and f, a real-valued convex function on K. Then f is lower semicontinuous at $x_0 \in K$ if and only if for all $\varepsilon > 0$, there exists η such that for all $x \in K$

$$(6) \quad \begin{cases} \textbf{i.} \quad f(x_0) \leq f(x) + \varepsilon & \text{if} \quad \|x - x_0\| \leq \eta. \\ \\ \textbf{ii.} \quad f(x_0) \leq f(x) + \dfrac{\varepsilon}{\eta} \|x - x_0\| & \text{if} \quad \|x - x_0\| > \eta. \end{cases}$$

Proof. Inequality (6) clearly implies that f is lower semicontinuous at x_0. *Conversely*, let us suppose that f is lower semicontinuous at x_0. Then for all $\varepsilon > 0$, there exists η such that

$$(7) \qquad f(x_0) \leq f(x) + \varepsilon \qquad \text{when} \qquad \|x - x_0\| \leq \eta.$$

Now consider the case when $\|x - x_0\| > \eta$. Let $y = \theta x + (1 - \theta)x_0$ where $\theta = \eta/\|x - x_0\| < 1$. Then $\|y - x_0\| = \theta\|x - x_0\| = \eta$, and since f is lower semicontinuous at x_0 and also convex, we obtain that

(8)
$$\begin{cases} f(x_0) \leq f(y) + \varepsilon = f(\theta x + (1 - \theta)x_0) + \varepsilon \\ \qquad \leq \theta f(x) + (1 - \theta)f(x_0) + \varepsilon. \end{cases}$$

Consequently,

(9)
$$\begin{cases} f(x_0) \leq \dfrac{1}{\theta}(\theta f(x) + \varepsilon) = f(x) + \dfrac{\varepsilon}{\eta}\|x - x_0\| \\ \text{when} \qquad \|x - x_0\| > \eta. \end{cases}$$

∎

The following important result characterizes continuous convex functions.

THEOREM 2.   Let K be a convex subset of a normed space E and f a real-valued convex function defined on K. In order that f be continuous at a point x_0 of K, it is necessary and sufficient that f be bounded above on a ball $B(x_0, \eta) \subset K$ with radius $\eta > 0$. In this case, f is continuous on the interior $\mathring{K}$ of K and satisfies the following inequalities:

(10)
$$\begin{cases} \textbf{i.}\quad f(x_0) \leq f(x) + \dfrac{a - f(x_0)}{\eta}\|x - x_0\| \qquad \text{for all} \qquad x \in K. \\[2mm] \textbf{ii.}\quad f(x) \leq f(x_0) + \dfrac{a - f(x_0)}{\eta}\|x - x_0\| \qquad \forall x \in K \\[2mm] \text{such that} \qquad \|x - x_0\| \leq \eta \end{cases}$$

where $a = \sup_{y \in B(x_0, \eta)} f(y)$.

Proof.

a.   If f is continuous at x_0, it is clearly bounded by $f(x_0) + \varepsilon$ on a ball $B(x_0, \eta)$ contained in K.

b.   *Conversely*, suppose that x_0 belongs to the interior of K and that

(11) $\qquad\qquad f(y) \leq a \qquad \text{when} \qquad y \in B(x_0, \eta) \subset K.$

Let x be an arbitrary element of K. Set

(12) $\quad y = \theta^{-1}(x_0 - (1 - \theta)x) \qquad \text{where} \qquad \theta = \dfrac{\|x - x_0\|}{\eta + \|x - x_0\|} < 1.$

Then $y - x_0 = ((1 - \theta)/\theta)(x_0 - x) = (\eta/\|x_0 - x\|)(x_0 - x)$ so that $\|y - x_0\| = \eta$; consequently, $y \in B(x_0, \eta) \subset K$. Thus we derive from (11) that $f(y) \leq a$.

Since f is convex and $x_0 = \theta y + (1 - \theta)x$, we obtain

(13) $\qquad f(x_0) \le \theta f(y) + (1 - \theta)f(x) \le \theta a + (1 - \theta)f(x);$

consequently,

(14) $\qquad (1 - \theta)f(x_0) \le \theta(a - f(x_0)) + (1 - \theta)f(x).$

Dividing by $1 - \theta > 0$, we deduce that

(15) $\quad f(x_0) \le f(x) + \dfrac{\theta}{1 - \theta}(a - f(x_0)) = f(x) + \dfrac{a - f(x_0)}{\eta}\|x - x_0\|,$

namely, (10)i.

c. Consider now an element $x \in B(x_0, \eta)$. Set

(16) $\qquad y = \theta^{-1}(x - (1 - \theta)x_0) \qquad$ where $\qquad \theta = \dfrac{\|x - x_0\|}{\eta} \le 1.$

Then $y - x_0 = \theta^{-1}(x - x_0)$ so that $\|y - x_0\| = \eta$; therefore, $y \in B(x_0, \eta) \subset K$. Then (11) implies that $f(y) \le a$. Since f is convex and $x = \theta y + (1 - \theta)x_0$, we obtain

(17) $\qquad \begin{cases} f(x) \le \theta f(y) + (1 - \theta)f(x_0) \le \theta(a - f(x_0)) + f(x_0) \\[2mm] \qquad\qquad = f(x_0) + \dfrac{a - f(x_0)}{\eta}\|x - x_0\|, \end{cases}$

namely, (10)ii. Moreover, the inequalities (10) imply that

(18) $\qquad |f(x) - f(x_0)| \le \dfrac{a - f(x_0)}{\eta}\|x - x_0\| \qquad$ when $\qquad x \in B(x_0, \eta)$

and, hence, the continuity of f at x_0.

d. Suppose (11) holds, and let us show that f is then continuous on the interior $\mathring{K}$ of K. Let $x_1 \in \mathring{K}$, and let $B(x_1, \alpha)$ be a ball with center x_1 contained in K. We are going to show that

(19) $\qquad \begin{cases} f(y) \le b \qquad \text{when} \qquad y \in B(x_1, \beta) \qquad \text{where} \\[2mm] \beta = \min\left(\alpha, \dfrac{\alpha\eta}{\alpha + \|x_1 - x_0\|}\right) \end{cases}$

and thus, applying the first part of Theorem 2, that f is continuous at x_1. To this end we introduce

(20) $\qquad \begin{cases} x_2 = x_0 + \dfrac{1}{1 - \lambda}(x_1 - x_0) = \dfrac{x_1 - \lambda x_0}{1 - \lambda} \qquad \text{where} \\[3mm] \lambda = \dfrac{\alpha}{\alpha + \|x_1 - x_0\|} < 1. \end{cases}$

Then $x_2 \in B(x_1, \alpha) \subset K$ since $\|x_2 - x_1\| = (\lambda/(1 - \lambda))\|x_1 - x_0\| = \alpha$; therefore, $f(x_2)$ is a finite number. We then associate to every $y \in B(x_1, \beta) \subset B(x_1, \lambda\eta) \cap K$ the point $z = \lambda^{-1}(y + \lambda x_0 - x_1)$. Since $z - x_0 = \lambda^{-1}(y - x_1)$, we obtain that $z \in B(x_0, \eta)$ and, from (11), that $f(z) \leq a$. Since $y = \lambda z + x_1 + \lambda x_0 = \lambda z + (1 - \lambda)x_2$, we obtain the inequality $f(y) \leq \lambda f(z) + (1 - \lambda) f(x_2) \leq \lambda a + (1 - \lambda)f(x_2) = b$ when $y \in B(x_1, \beta)$.   ∎

From this theorem we shall establish the following result.

THEOREM 3.   Every convex function f defined on a nonempty open convex subset K of $\mathbb{R}^n$ is continuous on K.

Proof.   Let x_0 be a point of K and $B(x_0, \eta)$ a ball with center x_0 and radius η contained in K. We can then find n points $x_i \in B(x_0, \eta)$ such that the vectors $x_i - x_0$ are linearly independent. The convex hull S of the $(n + 1)$ points x_i $(0 \leq i \leq n)$ contains the nonempty open set U formed by the convex combinations $\sum_{i=0}^{n} \lambda_i x_i$ where $\lambda_i > 0$ for all i. Consequently, since $f(\sum_{i=0}^{n} \lambda_i x_i) \leq \sum_{i=0}^{n} \lambda_i f(x_i)$, the convex function f is bounded above on U, and thus is continuous on K.   ∎

Remark 1.   We shall deduce from Baire's theorem (Theorem 5.3.1) that in Banach spaces (or more generally in Fréchet spaces) if K is a nonempty open convex set and if f is a lower semicontinuous convex function on K, then f is continuous on K. This result implies, in particular, that every lower semicontinuous seminorm is continuous. This property plays a very important role (see [AFA], Chapter 4, Section 4.1).

*GAUGE OF A CONVEX SET.   We are going to associate with a convex set convex functions that characterize it.

Definition 2.   Let K be a convex subset of a vector space E containing the origin. We set

(21) $j_K(x) = \inf\{\lambda \text{ such that } \lambda > 0 \text{ and } \lambda^{-1}x \in K\} \subset [0, \infty]$

and

(22) $\begin{cases} \textbf{i.} & \mathring{B}_K(\alpha) = \{x \in E \text{ such that } j_K(x) < \alpha\}. \\ \textbf{ii.} & B_K(\alpha) = \{x \in E \text{ such that } j_K(x) \leq \alpha\}. \end{cases}$

We shall call the function $j_K : x \in E \mapsto j_K(x)$ defined on E with values in $[0, \infty]$ the "gauge of K." We say that

(23) $\begin{cases} \textbf{i.} & K_0 = \bigcup_{\lambda \geq 0} \lambda K \text{ is the "cone generated by } K." \\ \textbf{ii.} & K_\infty = \bigcap_{\lambda \geq 0} \lambda K \text{ is the "asymptotic cone of } K." \end{cases}$

PROPOSITION 3. If K is the intersection of n convex sets K_i containing the origin, then

(24)
$$j_K(x) = \sup_{i=1,\ldots,n} j_{K_i}(x).$$

Proof. (*Left as an exercise.*)

THEOREM 4. Let $K \subset E$ be a convex subset of a vector space E containing the origin. The gauge $j_K(x)$ has the following properties:

(25)
$$
\begin{cases}
\textbf{i.} & K_0 = \{x \in E \text{ such that } j_K(x) < +\infty\}. \\
\textbf{ii.} & K_\infty = \{x \in E \text{ such that } j_K(x) = 0\}. \\
\textbf{iii.} & \mathring{K} \subset \mathring{B}_K(1) \subset K \subset B_K(1) \subset \bar{K}.
\end{cases}
$$

and

(26)
$$
\begin{cases}
\textbf{i.} & j_K(\lambda x) = \lambda j_K(x) \quad \text{for all} \quad \lambda \geq 0. \\
\textbf{ii.} & j_K(x + y) \leq j_K(x) + j_K(y).
\end{cases}
$$

Moreover, the sets $B_K(1)$ and $\mathring{B}_K(1)$ are convex.

Proof. Consider the set

(27)
$$I(x) = \{\lambda > 0 \text{ such that } \lambda^{-1}x \in K\}.$$

If $x \notin K_0$, then it is clear that $I(x) = \{+\infty\}$. If $x \in K_0$, *then $I(x)$ is a half line.* Indeed, if $\lambda \in I(x)$ and $\mu > \lambda$, then $\mu \in I(x)$ since

$$\mu^{-1}x = (\mu^{-1}\lambda)\lambda^{-1}x = (1 - \mu^{-1}\lambda)0 + (\mu^{-1}\lambda)(\lambda^{-1}x) \in K.$$

Thus

(28)
$$j_K(x) = \inf_{\lambda \in I(x)} \lambda.$$

a. To say that $j_K(x)$ is finite is to say that there exists $\lambda > 0$ such that $\lambda^{-1}x \in K$, that is, such that $x \in \lambda K \subset K_0$.

b. To say that $j_K(x) = 0$ is to say that $\lambda^{-1}x \in K$ for all $\lambda > 0$, that is, that $x \in \lambda K$ for all $\lambda > 0$.

c. If $x \in \mathring{K}$ assumed nonempty, there exists $\eta > 0$ such that $B(x, \eta) \subset K$. Then $y = x + (\eta/\|x\|)x = (1 + \eta/\|x\|)x$ belongs to $B(x, \eta) \subset K$, which implies that

$$j_K(x) \leq \frac{1}{1 + \eta/\|x\|} < 1.$$

Thus $\mathring{K} \subset \mathring{B}_K(1)$.

d. If $x \in \mathring{B}_K(1)$, then $j_K(x) < 1$, and, therefore, $1 \in I(x)$. Thus $x \in K$.

e. If $x \in K$, then $1 \in I(x)$; hence, $j_K(x) \leq 1$, that is, $x \in B_K(1)$.

f. If $x \in B_K(1)$, then $j_K(x) \le 1$; consequently, $\lambda^{-1}x \in K$ for all $\lambda > 1$. Since $x = \lim_{\substack{\lambda \to 1 \\ \lambda > 1}} \lambda^{-1}x$, we deduce that $x \in \overline{K}$.

g. Property (26)i is obvious since $I(\theta x) = \theta I(x)$ for all $\theta \ge 0$.

h. Let us establish (26)ii. If one of the elements x or y does not belong to K_0, the right-hand side of the inequality is infinite, consequently, the inequality holds. Suppose that x and y belong to K. Let $\lambda = j_K(x)$ and $\mu = j_K(y)$. Since $\overline{K}$ is convex,

$$(29) \qquad (\lambda + \mu)^{-1}(x + y) = \frac{\lambda}{\lambda + \mu}(\lambda^{-1}x) + \frac{\mu}{\lambda + \mu}(\lambda^{-1}y) \in \overline{K}$$

because $\lambda^{-1}x$ and $\mu^{-1}y$ belong to $B_K(1) \subset \overline{K}$. This implies that $(\lambda + \mu)$ belongs to $I(x + y)$; consequently, that $j_K(x + y) \le j_K(x) + j_K(y)$.

We conclude then that the sets $\mathring{B}_K(1)$ and $B_K(1)$ are convex.   ∎

PROPOSITION 4.   Let K be a convex subset of a normed space E containing the origin. If K is symmetric, then $j_K(-x) = j_K(x)$. If K is closed, then j_K is lower semicontinuous. If $0 \in \mathring{K}$, then j_K is a uniformly continuous function on E and satisfies:

$$(30) \qquad |j_K(x_1) - j_K(x_2)| \le M\|x_1 - x_2\| \qquad \text{for all} \qquad x_1, x_2 \in E.$$

Proof.

a. If K is symmetric, then $I(x) = I(-x)$ since if $\lambda^{-1}x \in K$, then $(-\lambda)^{-1}x = \lambda^{-1}(-x) \in K$. Thus $j_K(x) = j_K(-x)$.

b. If K is closed, then $K = \overline{K} = B_K(1)$; therefore, the sets $B_K(\alpha) = \alpha B_K(1) = \{x \in E \text{ such that } j_K(x) \le \alpha\}$ are closed for all α. Thus j_K is lower semicontinuous.

c. If $0 \in \mathring{K}$, then a ball $B(0, \alpha)$ with center 0 and radius α is contained in K. Hence for all $x \in E$, $(\alpha/\|x\|)x \in B(0, \alpha) \subset K$, which implies that $j_K(x) < +\infty$. If x_1 and x_2 are elements of E, then $x = \alpha((x_1 - x_2)/\|x_1 - x_2\|)$ belongs to $B(0, \alpha)$ and, consequently,

$$|j_K(x_1) - j_K(x_2)| \le j_K(x_1 - x_2) = \frac{\|x_1 - x_2\|}{\alpha}j_K(x) \le \frac{\|x_1 - x_2\|}{\alpha}. \qquad ∎$$

We obtain the following as a corollary to the preceding results.

PROPOSITION 5.   The gauge j_K of a symmetric convex set K for which $K_0 = E$ is a seminorm. The gauge is lower semicontinuous if K is closed and uniformly continuous if 0 belongs to the interior of K.

APPLICATIONS: PRODUCTION SETS AND PRODUCTION FUNCTIONS. We interpret the space $E = \mathbb{R}^n$ as a commodity space, and we denote by $K \subset \mathbb{R}^n$ a subset that we interpret as the net "production set" of a firm. We generally suppose that

$$(31) \qquad\qquad\qquad 0 \in K,$$

that is, that it is possible to produce nothing. Moreover, we say that a firm has a nonincreasing return to scale if

$$(32) \qquad\qquad K \text{ is a convex subset of } E = \mathbb{R}^n,$$

that is, if two productions x_1 and x_2 are possible and if $\alpha \in [0, 1]$, then the production $\alpha x_1 + (1 - \alpha)x_2$ is also possible. We further suppose that

$$(33) \qquad\qquad\qquad \text{the set } K \text{ is closed.}$$

This means that every production x, which is the limit of a sequence of possible productions x_n, is also possible.

If the hypotheses (31), (32), and (33) are satisfied, then $x \in K$ if and only if $j_K(x) \leq 1$. Then the *production function* f_K is the function defined by

$$(34) \qquad\qquad\qquad f_K(x) = j_K(x) - 1,$$

which is a convex function. We say that the productions $x \in K$ for which $j_K(x) = 1$ (or $f_K(x) = 0$) are *efficient productions*, for they maximize the production function $f_K(x)$ on K.

9. Applications to Optimization Theory

Consider a function f defined on a metric space U with values in the interval $]-\infty, +\infty]$ of $\bar{\mathbb{R}}$. We shall call its *domain* the set

$$(1) \qquad\qquad E = \{x \in U \text{ such that } f(x) < +\infty\}.$$

Hence a function f with values in $]-\infty, +\infty]$ describes *both* a set E (its domain) and a real-valued function defined on E. [*Conversely*, if $E \subset U$ and if f is a real-valued function defined on E, we can associate to f its extension f_E defined on U with values in $]-\infty, +\infty]$ by

$$(2) \qquad\qquad f_E(x) = \begin{cases} f(x) & \text{if} \quad x \in E \\ +\infty & \text{if} \quad x \notin E. \end{cases}]$$

Consider the lower bound $\alpha \in \bar{\mathbb{R}}$ of f defined by

$$(3) \qquad\qquad \alpha = \inf_{x \in U} f(x) = \inf_{x \in E} f(x).$$

We denote by $M^b(f)$ the minimal set of f, defined by

(4) $$M^b(f) = \{x \in U \text{ such that } f(x) = \alpha\}.$$

PROPOSITION 1.   The minimal set of f is equal to

(5) $$M^b(f) = \bigcap_{\lambda > \alpha} F_\lambda$$

where $F_\lambda = \{x \in U \text{ such that } f(x) \leq \lambda\}$.

Proof.   If $x \in M^b(f)$, then $f(x) = \alpha \leq \lambda$ for all $\lambda > \alpha$ and, consequently, belongs to the intersection of the F_λ's when $\lambda > \alpha$. Conversely, if $x \in \bigcap_{\lambda > \alpha} F_\lambda$, then $f(x) \leq \lambda$ for all $\lambda > \alpha$. Letting λ approach α from above, we deduce that $f(x) = \alpha$. ∎

Consequently, *the minimal set $M^b(f)$ has all the properties stable under intersection (of a decreasing family of subsets) that are satisfied by the sections F_λ. For example, if the function f is lower semicontinuous (respectively, convex), the sections F_λ are closed (respectively, convex); therefore, the minimal set $M^b(f)$ is closed (respectively, convex).* (See Propositions 7.3 and 8.1).

In particular, since every decreasing sequence of closed sets has the finite intersection property, it follows that if f is lower semicontinuous and if E is compact, the intersection of the F_λ's is nonempty (Theorem 2.6.1). Hence we have established the fundamental theorem of optimization theory.

THEOREM 1. *Weierstrass.*   If the domain E of a lower semicontinuous function $f: E \mapsto]-\infty, +\infty[$ is nonempty and compact, the minimal set $M^b(f)$ is nonempty and compact.

Indeed, the proof of Theorem 1 suggests the introduction of the following definition.

Definition 1.   We shall say that a function $f: U \mapsto]-\infty, +\infty]$ is lower semicontinuous if its sections F_λ are closed for all $\lambda \in \mathbb{R}$ and lower semicompact if its sections F_λ are relatively compact for all $\lambda \in \mathbb{R}$.

For example, every function f whose domain E is compact is clearly lower semicompact.

THEOREM 2. *Weierstrass.*   If $f: U \mapsto]-\infty, +\infty]$ is lower semicontinuous and lower semicompact and if its domain E is nonempty, then its minimal set $M^b(f)$ is nonempty and compact.

Proof.   Under the hypotheses of Theorem 2 the sets $F_\lambda (\lambda > \alpha)$ are nonempty, closed, and relatively compact, hence compact. Since the family of the F_λ's is decreasing, $M^b(f) = \bigcap_{\lambda > \alpha} F_\lambda$ is nonempty. ∎

Remark 1.   If f is a function with values in $[-\infty, +\infty[$, we define its domain $E^{\#} = \{y \in U$ such that $f(y) > -\infty\}$ and its sections $F_{\lambda}^{\#} = \{y \in U$ such that $f(y) \geq \lambda\}$. We say that f is "upper semicompact" if its sections $F^{\#}$ are relatively compact, that is if $-f$ is lower semicompact. If $\beta = \sup_{x \in U} f(x) = \sup_{x \in E^{\#}} f(x)$, we define the maximal set $M^{\#}(f) = \{x \in E$ such that $f(x) = \beta\}$. A theorem analogous to Theorem 2 holds.

Remark 2.   If E is complete (instead of compact) and f is lower semicontinuous, we prove in Section 4.9 Ekeland's Theorem 4.9.6 on approximate minimization. We prove in Section 10.2 of [AFA] that if E is a Hilbert space, $x \mapsto f(x) + \frac{1}{2}\|x - u\|^2$ has a unique minimum.

COERCIVE FUNCTIONS.   We shall give an important example of lower semicompact functions.

Definition 2.   Let E be a subset of $\mathbb{R}^n$ and f a real-valued function defined on E. We say that f is "coercive" if

(6)
$$\lim_{\substack{\|x\| \to \infty \\ x \in E}} \frac{f(x)}{\|x\|} = +\infty$$

and that f is "semicoercive" if

(7)
$$\lim_{\substack{\|x\| \to \infty \\ x \in E}} \frac{f(x)}{\|x\|} > 0.$$

Every coercive function is, of course, semicoercive.

PROPOSITION 2.   Every semicoercive function defined on a subset E of $\mathbb{R}^n$ is lower semicompact.

Proof.   We are going to show that the sections F_{λ} of a semicoercive function f are bounded and then use the fact that the bounded sets of $\mathbb{R}^n$ are relatively compact (Proposition 2.6.3).

By hypothesis there exists $a > 0$ such that $\lim_{\substack{\|x\| \to \infty \\ x \in E}}(f(x)/\|x\|) = 2a$. This means that for every $\varepsilon > 0$ there exists $R(\varepsilon)$ such that, for every $\|x\| \geq R(\varepsilon)$, we have $f(x)/\|x\| \geq 2a - \varepsilon$. Taking $\varepsilon = a$, we deduce that $\|x\| \leq f(x)/a$ when $\|x\| \geq R(a)$. Therefore, we obtain the inequality

(8)
$$\|x\| \leq \max\left(R(a), \frac{1}{a} f(x)\right).$$

Therefore, if $x \in F_{\lambda}$, $f(x) \leq \lambda$ and $\|x\| \leq \max(R(a), \lambda/a)$.   ∎

PROPOSITION 3.   The functions $x \mapsto \|x - u\|^p$ are coercive if $p > 1$ and semicoercive if $p = 1$.

Proof.   We can write

$$\frac{f(x)}{\|x\|} = \|x\|^{p-1} \left\| \frac{x}{\|x\|} - \frac{u}{\|x\|} \right\|^p \geq \|x\|^{p-1} \left| 1 - \frac{\|u\|}{\|x\|} \right|^p.$$

But the right-hand side approaches infinity for $p > 1$ and converges to a limit greater than 1 for $p = 1$.   ∎

We mention the following result on uniqueness.

THEOREM 3.   Let E be a convex subset of a vector space. If f is a strictly convex real-valued function defined on E, there exists at most one solution $\bar{x}$ that minimizes f.

Proof.   Since f is convex, the minimal set $M^b(f)$ is convex. Let x_1 and x_2 be two elements of $M^b(f)$. Then $x = (1/2)x_1 + (1/2)x_2$ belongs to $M^b(f)$. Since f is strictly convex, we conclude that

$$\alpha = f(x) < (1/2)f(x_1) + (1/2)f(x_2) = (1/2)\alpha + (1/2)\alpha = \alpha,$$

which is impossible.   ∎

10.   Applications to Game Theory

We denote by E and F the sets of strategies of two players, Emily and Frank, respectively and by $f(x, y)$ Emily's loss associated with the choice of strategies $x \in E$ and $y \in F$ for Emily and Frank. In the situation of noncooperative behavior between the players (that is, without an exchange of information between them), Emily associates to every $x \in E$, which she controls, the worst loss

(1) $$f^{\#}(x) = \sup_{y \in F} f(x, y),$$

which she can sustain in all cases. Consequently, without other hypotheses, Emily is led to choose $\bar{x} \in E$ minimizing the function $f^{\#}$.

PROPOSITION 1.   We suppose that

(2) $$E \text{ is compact}$$

and that

(3) $$\forall y \in F, \quad x \mapsto f(x, y) \quad \text{is lower semicontinuous.}$$

Then there exists $\bar{x} \in E$ such that

(4) $$\sup_{y \in F} f(\bar{x}, y) = \inf_{x \in E} \sup_{y \in F} f(x, y) = v^{\#}.$$

Proof.   Since the functions $x \mapsto f(x, y)$ are lower semicontinuous, so is the function $f^{\#}$, according to Theorem 7.1.

Since E is compact, Theorem 9.1 of Weierstrass implies the existence of $\bar{x} \in E$, which minimizes $f^{\#}$. This is expressed from (1) by

$$\sup_{y \in F} f(\bar{x}, y) = f^{\#}(\bar{x}) = \inf_{x \in E} f^{\#}(x) = \inf_{x \in E} \sup_{y \in F} f(x, y). \qquad \blacksquare$$

In fact, hypotheses (2) and (3) of Proposition 1 yield a stronger conclusion than (4).

We introduce

(5) $\mathscr{S}$, the set of finite sequences $K = \{y_1, \ldots, y_n\}$ of F

and denote by

(6)
$$v^{\natural} = \sup_{K \in \mathscr{S}} \inf_{x \in E} \max_{y \in K} f(x, y).$$

Since $\max_{y \in K} f(x, y) \leq \sup_{y \in F} f(x, y)$, we deduce that

$$\inf_{x \in E} \max_{y \in K} f(x, y) \leq \inf_{x \in E} \sup_{y \in F} f(x, y) = v^{\#}.$$

This implies that

(7)
$$v^{\natural} \leq v^{\#}.$$

We show that under the topological hypotheses of Proposition 1, we have the equality $v^{\#} = v^{\natural}$.

THEOREM 1.   Suppose that hypotheses (2) and (3) are satisfied. Then there exists $\bar{x} \in E$ such that

(8)
$$\sup_{y \in F} f(\bar{x}, y) = v^{\#} = v^{\natural}.$$

Proof.   It is sufficient to show that there exists $\bar{x} \in E$ such that

(9)
$$\sup_{y \in F} f(\bar{x}, y) \leq v^{\natural}.$$

(Since $v^{\#} \leq \sup_{y \in F} f(\bar{x}, y)$ and $v^{\natural} \leq v^{\#}$, we can then conclude that $v^{\natural} = v^{\#}$.)
Set

(10) $S_y = \{x \in E \text{ such that } f(x, y) \leq v^{\natural}\}.$

Inequality (9) is equivalent to

(11)
$$\bar{x} \in \bigcap_{y \in F} S_y.$$

Hence we must establish that this intersection is nonempty. To this end we verify that the S_y's are closed sets with the finite intersection property.

The set S_y is closed since S_y is a lower section of the lower semicontinuous function $x \mapsto f(x, y)$. Let us show that for every finite sequence $K = \{y_1, \ldots, y_n\} \in \mathscr{S}$ of F, the intersection $\bigcap_{i=1,\ldots,n} S_{y_i} \neq \varnothing$. Since E is compact and since $x \mapsto \max_{i=1,\ldots,n} f(x, y_i) = \max_{y \in K} f(x, y)$ is lower semicontinuous, we deduce that there exists $\bar{x} \in E$, which minimizes this function (see Theorem 10.1). Such an $\bar{x} \in E$ satisfies

$$\max_{y \in K} f(\bar{x}, y) = \inf_{x \in E} \max_{y \in K} f(x, y) \leq \sup_{K \in \mathscr{S}} \inf_{x \in E} \max_{y \in K} f(x, y) \leq v^{\flat};$$

consequently, $\bar{x} \in \bigcap_{i=1,\ldots,n} S_{y_i}$. Since E is compact, the intersection of the closed sets S_y is nonempty, and there exists $\bar{x}$ satisfying (11) and hence (9). ∎

The study of two-person games is continued in Section 5.6.

11. Fixed Points of Contractions. Existence of Solutions to Differential Equations

Let us consider a metric space (E, d) and a continuous mapping f from (E, d) to itself.

Definition 1. An element $x \in E$ is called a "fixed point" of f if

$$(1) \qquad x = f(x).$$

We shall establish in this section a "simple" sufficient condition for the existence of a fixed point and a "constructive method" to approximate this fixed point, called the "method of successive approximations due to Banach–Picard."

THEOREM 1. Let (E, d) be a complete metric space and f a contraction from E to itself (that is, such that there exists $\lambda \in \,]0, 1[$ satisfying $d(f(x), f(y)) \leq \lambda d(x, y)$).

Then there exists a unique fixed point of f.

Moreover, for any arbitrary $x_0 \in E$, the sequence x_0^n defined by

$$(2) \qquad x_0^1 = f(x_0), \ldots, x_0^{n+1} = f(x_0^n), \ldots$$

converges to the fixed point x, and the error between x and x_0^n is bounded above by

$$(3) \qquad d(x, x_0^n) \leq \frac{\lambda^n}{1 - \lambda} d(f(x_0), x_0).$$

For $n = 0$, the solution x satisfies the inequality

$$(4) \qquad d(x, x_0) \leq \frac{d(f(x_0), x_0)}{1 - \lambda}.$$

Proof.   First of all, if a fixed point x exists, it is unique. Indeed, if x and y are two fixed points, we have

$$d(x, y) = d(f(x), f(y)) \leq \lambda d(x, y);$$

consequently, $(1 - \lambda)d(x, y) \leq 0$. Since $1 - \lambda > 0$, this implies that $d(x, y) \leq 0$, that is, that $x = y$.

Let us now show the existence of a fixed point x of f and more precisely that the sequence x_0^n defined by (2) from an arbitrary given point x_0 converges to the fixed point x. Indeed, since f is contracting, we obtain that

$$(5) \qquad d(x_0^{n+1}, x_0^n) = d(f(x_0^n), f(x_0^{n-1})) \leq \lambda d(x_0^n, x_0^{n-1})$$

and by recursion that

$$(6) \qquad d(x_0^{n+1}, x_0^n) \leq \lambda^n d(x_0^1, x_0) = \lambda^n d(f(x_0), x_0).$$

Therefore,

$$(7) \qquad d(x_0^{n+p+1}, x_0^n) \leq \sum_{k=n}^{n+p} d(x_0^{k+1}, x_0^k) \leq \left(\sum_{k=n}^{n+p} \lambda^k\right) d(f(x_0), x_0).$$

Since $\lambda < 1$, the series $\sum_{k=0}^{\infty} \lambda^k = 1/(1 - \lambda)$ converges and (7) implies that the sequence x_0^n is a Cauchy sequence. Since E is a complete metric space, this sequence, therefore, converges to an element x of E. We must show that this element x is the fixed point of E. Since f is continuous, the sequence $f(x_0^n)$ converges to $f(x)$; consequently, we deduce from the equalities $x_0^{n+1} = f(x_0^n)$ that $x = f(x)$. It remains to establish the estimate of the error (3). Since for all p,

$$d(x, x_0^n) \leq d(x, x_0^{n+p+1}) + d(x_0^{n+p+1}, x_0^n)$$

$$\leq d(x, x_0^{n+p+1}) + \left(\sum_{k=n}^{\infty} \lambda^k\right) d(f(x_0), x_0)$$

$$\leq d(x, x_0^{n+p+1}) + \frac{\lambda^n}{1 - \lambda} d(f(x_0), x_0),$$

we obtain (3) by letting p approach infinity.

In Section 4.9, we generalize this theorem for "dissipative mappings" in the framework of dynamical systems.   ■

APPLICATION. INTEGRAL EQUATIONS.   Let us now consider an integral operator: let $K(\cdot, \cdot)$ be a *continuous* function defined on the square $[a, b] \times [a, b]$. Consider a *continuous* function g on $[a, b]$ and a real number

k. We seek a continuous function x on $[a, b]$, which is a solution of the integral equation

$$(8) \qquad x(t) = k \int_a^b K(t, s)x(s)ds + g(t).$$

PROPOSITION 1.   Suppose that the functions K and f are continuous. If $M = \sup_{t, s \in [a, b]} |K(t, s)|$ and if $|k| < 1/(M(b - a))$, then there exists a unique solution to the integral equation (8).

Proof.   The proof depends on finding a fixed point in the space $\mathscr{C}_\infty([a, b])$ of continuous functions on $[a, b]$ for the mapping f defined by

$$(9) \qquad (fx)(t) = k \int_a^b K(t, s)x(s)ds + g(t).$$

It is evident that f is a *continuous* mapping from $\mathscr{C}_\infty([a, b])$ to itself. Let us show that f is contracting if $|k| < 1/(M(b - a))$. If x_1 and x_2 are two continuous functions, we have

$$|f(x_1)(t) - f(x_2)(t)| \le |k| \int_a^b |K(t, s)| |x_1(s) - x_2(s)| ds$$

$$\le |k|M\|x_1 - x_2\|_\infty \int_a^b ds = |k|M|b - a|\|x_1 - x_2\|_\infty.$$

Since the constant $\lambda = |k|M|b - a|$ is strictly less than 1, the mapping f is contracting. Moreover, $\mathscr{C}_\infty([a, b])$ is a complete metric space. Hence Theorem 1 implies the existence of a unique fixed point of f.   ■

APPLICATION.   FUNCTIONAL EQUATIONS

PROPOSITION 2.   Let X be a complete metric space and φ a continuous mapping from $X \times \mathbb{R}$ to $\mathbb{R}$ such that

$$(10) \qquad \begin{cases} \forall x \in X, \qquad \forall y_1, \qquad y_2 \in \mathbb{R} \qquad |\varphi(x, y_1) - \varphi(x, y_2)| \le \lambda |y_1 - y_2| \\ \text{where} \qquad \lambda \in \,]0, 1[. \end{cases}$$

There exists a unique continuous mapping u from X to $\mathbb{R}$ such that

$$(11) \qquad u(x) = \varphi(x, u(x)) \qquad \text{for all} \qquad x \in X.$$

Proof.   Proposition 2 is a consequence of Theorem 1 where we take $E = \mathscr{C}_\infty(X)$ (which is a *complete* metric space) and where the mapping f from $\mathscr{C}_\infty(X)$ to itself is defined by

$$(12) \qquad f(u): x \mapsto \varphi(x, u(x)),$$

for since φ is continuous, f is a continuous mapping on X.

We deduce from (10) that f is a contraction from $\mathscr{C}_\infty(X)$ to itself. Indeed for every fixed x in X we obtain

$$|f(u)(x) - f(v)(x)| = |\varphi(x, u(x)) - \varphi(x, v(x))|$$
$$\leq \lambda|u(x) - v(x)|.$$

Taking the sup over X of the two sides of this inequality, we obtain that

(13)
$$\begin{cases} \|f(u) - f(v)\|_\infty = \sup_{x \in X}|f(u)(x) - f(v)(x)| \\ \leq \sup_{x \in X}|u(x) - v(x)| = \lambda\|u - v\|_\infty, \end{cases}$$

that is, that f is a contraction. Consequently, there exists a fixed point $u \in \mathscr{C}_\infty(X)$ for f, that is, a continuous mapping u from X to $\mathbb{R}$ such that

(14) $u(x) = f(u)(x) = \varphi(x, u(x))$ for all $x \in X.$ ∎

In fact we can improve Theorem 1 in the following fashion.

THEOREM 2.   Let (E, d) be a complete metric space, $\mathring{B}(x_0, \beta)$ the open ball with center x_0 and radius β and f a contraction from $\mathring{B}(x_0, \beta)$ to E:

$$\exists \lambda \in \,]0, 1[\quad \text{such that} \quad d(f(x), f(y)) \leq \lambda d(x, y).$$

Suppose that $d(f(x_0), x_0) < \beta(1 - \lambda)$. Then there exists a unique fixed point $x \in \mathring{B}(x_0, \beta)$ of the function f.

Proof.   Beginning with x_0, we construct the sequence x_0^n by recursion as in Theorem 1, letting $x_0^n = f(x_0^{n-1})$. For this we must show (by recursion) that x_0^n remains in $\mathring{B}(x_0, \beta)$. First, $d(x_0^1, x_0) = d(f(x_0), x_0) < \beta(1 - \lambda) < \beta$. Suppose that x_0^k belongs to $\mathring{B}(x_0, \beta)$ for all $k \leq n - 1$. Then inequalities (5) and (6) are satisfied. They imply that

(15)
$$\begin{cases} d(x_0^n, x_0) \leq d(x_0^n, x_0^{n-1}) + \cdots + d(x_0^1, x_0) \\ \leq \left(\sum_{k=0}^{n-1} \lambda^k\right) d(f(x_0), x_0) < \beta(1 - \lambda)\left(\sum_{k=1}^{n-1} \lambda^k\right) \leq \beta. \end{cases}$$

The sequence x_0^n can then be defined. We show as in theorem 1 that it is a Cauchy sequence that converges to a fixed point x. Inequality (4) shows that it belongs to the ball $\mathring{B}(x_0, \beta)$.   ∎

APPLICATION. EXISTENCE OF A SOLUTION OF A DIFFERENTIAL EQUATION.   Let Ω be a nonempty open set in $\mathbb{R}^n$, I a nonempty *open* interval of $\mathbb{R}$, and φ a mapping from $I \times \Omega$ to $\mathbb{R}^n$. Consider a point $t_0 \in I$ and a point $x_0 \in \Omega$.

We are going to show the existence of a solution of the differential equation

(16) $\begin{cases} \text{i.} & x'(t) = \varphi(t, x(t)), \\ \text{ii.} & x(t_0) = x_0, \end{cases}$

when t is in a neighborhood of t_0.

First, there exist $\alpha_0 > 0$ and $\beta > 0$ such that $[t_0 - \alpha_0, t_0 + \alpha_0] \subset I$ and $B(x_0, \beta) \subset \Omega$ since I and Ω are open.

THEOREM 3. *Cauchy–Lipschitz.* We suppose that

(17) φ is continuous from $I \times \Omega$ to $\mathbb{R}^n$.

Then φ is bounded on $[t_0 - \alpha_0, t_0 + \alpha_0] \times B(x_0, \beta)$ by

$$M = \sup_{|t-t_0| \le \alpha_0} \sup_{\|x-x_0\| \le \beta_0} \|\varphi(t, x)\|.$$

We suppose that φ is Lipschitz with respect to x:

(18) $\begin{cases} \exists k \quad \text{such that} \quad \forall t \in [t_0 - \alpha_0, t_0 + \alpha_0], \\ \|\varphi(t, x_1) - \varphi(t, x_2)\| \le k\|x_1 - x_2\|. \end{cases}$

Let $\alpha < \min(\alpha_0, \beta/(M + k\beta), \lambda/k)$ where $\lambda < 1$ is fixed. Then there exists a unique solution $x(\cdot)$ of the differential equation (16) when $|t - t_0| \le \alpha$.

Proof.   First of all, we are going to show that the mapping f defined by

(19) $$f(x)(t) = x_0 + \int_{t_0}^{t} \varphi(s, x(s))ds$$

satisfies the hypotheses of Theorem 2. By identifying x_0 with the constant function $t \mapsto x_0$, we see that f is defined on the open ball $\mathring{B}_\infty(x_0, \beta)$ from the space $\mathscr{C}_\infty([t_0 - \alpha, t_0 + \alpha], E)$ to itself (for if $x \in \mathring{B}_\infty(x_0, \beta)$, then $\|x(s) - x_0\| \le \|x - x_0\|_\infty < \beta$; therefore, $x(s) \in \Omega$). Let us verify that $\|f(x_0) - x_0\|_\infty < \beta(1 - k\alpha)$. Indeed, for all $t \in [t_0 - \alpha, t_0 + \alpha]$, we obtain

(20) $\begin{cases} \|f(x_0)(t) - x_0\| \le \int_{t_0}^{t} \|\varphi(s, x_0)\|ds \le M|t - t_0| \\[2mm] \qquad\qquad \le M\alpha < \beta(1 - k\alpha)\left(\text{since } \alpha < \dfrac{\beta}{M + k\beta}\right). \end{cases}$

Moreover, f is Lipschitz on this ball: Indeed, if x_1 and x_2 are two functions,

(21) $\begin{cases} \|f(x_1) - f(x_2)\|_\infty \le \int_{t_0}^{t} |\varphi(s, x_1(s)) - \varphi(s, x_2(s))|ds \\[2mm] \qquad\qquad \le k|t - t_0|\|x_1 - x_2\|_\infty \le k\alpha\|x_1 - x_2\|_\infty \\[2mm] \qquad\qquad \le \lambda\|x_1 - x_2\|_\infty \end{cases}$

since $\alpha \leq \lambda/k$. According to Theorem 2, there exists a unique solution $x \in \mathring{B}_\infty(x_0, \beta)$ of the equation

(22)
$$x(t) = x_0 + \int_{t_0}^t \varphi(s, x(s))ds.$$

Since the function $t \in [t_0 - \alpha, t_0 + \alpha] \mapsto x(t) \in \Omega$ is continuous and since the function $s \mapsto \varphi(s, x(s))$ is continuous, we conclude that the solution x defined by (22) is differentiable. We obtain then that if $|t - t_0| < \alpha$, $x'(t) = \varphi(t, x(t))$ and that $x(t_0) = x_0$; therefore, x is a (local) solution of the differential equation. ∎

We prove in Section 4.6 the existence of global solutions x such that $x(t) \in \bar{\Omega}$ for all $t \in I$ (Nagumo's Theorem 4.6.3).

Remark 1. Theorems 1 and 2 on fixed points have many other applications in diverse domains of analysis. In particular, we are going to use these results to establish the inverse function theorem in the following section. ∎

However, the hypothesis that f be a contraction often proves to be too restrictive. There are many results that replace this hypothesis by others. The study of these results constitutes "nonlinear analysis," which is in full development at the moment. We state (without proof) the deepest fixed point theorem, known as the Brouwer theorem.

THEOREM 4. *Brouwer.* Let E be a compact convex subset of $\mathbb{R}^n$. Every continuous mapping f from E to itself has a fixed point.

Remark 2. The proof of this theorem is long, but elementary in the sense that it makes use of simple results of combinatorics. If $n = 1$, we shall prove this result in Remark 5.4.1. ∎

Remark 3. We are going to use this theorem (accepted without proof) to establish in the following section the surjectivity of coercive operators. We also use it in Chapter 5 to prove the Ky-Fan inequality (see Theorem 5.6.3) from which can be proved the major part of the deep results of game theory, mathematical economics, and, more generally, nonlinear analysis and its applications (see [AFA], Chapter 15). ∎

12.   The Inverse Function Theorem and the Surjectivity of Coercive Mappings

We shall begin by proving a first result on homeomorphisms.

PROPOSITION 1. Let E be a Banach space. Consider a contraction (with proportionality factor λ) g defined on an open ball $\mathring{B}(0, \beta)$ (with center

0) with values in E, such that $g(0) = 0$. Then the function f defined by $f(x) = x + g(x)$ satisfies

(1)

   i. $\mathring{B}(0, \beta(1 - \lambda)) \subset f(\mathring{B}(0, \beta))$.

   ii. The mapping f is injective.

   iii. The inverse of f is Lipschitz with proportionality factor $1/(1 - \lambda)$.

Proof.

a. Take $y \in \mathring{B}(0, \beta(1 - \lambda))$. We define the function h on $\mathring{B}(0, \beta)$ by $h(x) = y - g(x)$. The hypotheses of Theorem 2 are satisfied with $x_0 = 0$ since $d(h(0), 0) = \|y\| < \beta(1 - \lambda)$. Hence there exists $x \in \mathring{B}(0, \beta)$ such that $x = h(x) = y - g(x)$, that is, a solution x of the equation $f(x) = x + g(x) = y$. We have, therefore, proved the first statement.

b. Let x_1 and x_2 be two solutions of the equations $x_1 + g(x_1) = y_1$ and $x_2 + g(x_2) = y_2$. Since $x_1 - x_2 + g(x_1) - g(x_2) = y_1 - y_2$, we obtain

$$\big| \|x_1 - x_2\| - \|g(x_1) - g(x_2)\| \big| \leq \|x_1 - x_2 + g(x_1) - g(x_2)\| = \|y_1 - y_2\|.$$

This implies that

$$\|x_1 - x_2\| \leq \|y_1 - y_2\| + \|g(x_1) - g(x_2)\| \leq \|y_1 - y_2\| + \lambda\|x_1 - x_2\|.$$

Thus $\|x_1 - x_2\| \leq (1/(1 - \lambda))\|y_1 - y_2\|$. If we choose $y_1 = y_2$, this implies that $\|x_1 - x_2\| = 0$, that is the uniqueness of the solution. Therefore, statements (1)ii and (1)iii are established.   ■

To prove the inverse function theorem, we must recall the notion of the differential of a function (in the sense of Fréchet).

Definition 1.   Let E and F be two Banach spaces, $x_0 \in E$ and f a function defined on a neighborhood of x_0 with values in F. We say the f is "differentiable at x_0" if there exists a continuous linear operator $Df(x_0) \in \mathscr{L}(E, F)$ such that

(2)
$$\lim_{\|x - x_0\| \to 0} \frac{\|f(x) - f(x_0) - Df(x_0)(x - x_0)\|}{\|x - x_0\|} = 0.$$

The linear operator $Df(x_0)$ is called the "differential of f at x_0." In other words, *a function f is differentiable at x_0 if it can be approximated by the affine operator $x \mapsto f(x_0) + Df(x_0)(x - x_0)$.*

We show that if $Df(x_0) \in \mathscr{L}(E, F)$ is a bicontinuous isomorphism, then f is a homeomorphism on a neighborhood of x_0.

THEOREM 1. *Inverse function theorem.*   Let E and F be two Banach spaces, U an open ball with center 0 and f a function from U to F that vanishes at the origin. We suppose that

(3)
$$\begin{cases} \textbf{i.} & f \text{ is differentiable at every point of } U. \\ \textbf{ii.} & \text{The mapping } x \in U \mapsto Df(x) \in \mathscr{L}(E, F) \text{ is continuous.} \\ \textbf{iii.} & Df(0) \text{ is a bicontinuous isomorphism from } E \text{ onto } F. \end{cases}$$

Then there exist neighborhoods V of $0 \in E$ and W of $0 \in F$ such that f is a homeomorphism from V onto W.

Proof.   We introduce the function g from U to E defined by

(4)
$$g(x) = Df(0)^{-1}f(x) - x,$$

which is possible according to (3)iii. Then $Dg(x) = Df(0)^{-1}Df(x) - 1$. Moreover, $x \mapsto Dg(x)$ is continuous from U to $\mathscr{L}(E, E)$ (according to (3)ii) and vanishes at 0. Since

(5)
$$\frac{d}{dt} g(x + t(y - x)) = Dg(x + t(y - x))(y - x),$$

we obtain, integrating both sides from 0 to 1,

(6)
$$g(y) - g(x) = \int_0^1 Dg(x + t(y - x))(y - x)dt.$$

Since $x \mapsto Dg(x)$ is continuous on a neighborhood of the origin, we deduce that there exists a ball $B(0, \beta)$ such that, if x and $y \in B(0, \beta)$,

$$\|Dg(x + t(y - x))\| = \|Dg(x + t(y - x)) - Dg(0)\| \leq \lambda < 1$$

(where λ is chosen in advance). We obtain, therefore, from (6) that g is contracting on $B(0, \beta)$:

(7) $\|g(y) - g(x)\| \leq \displaystyle\int_0^1 \|Dg(x + t(y - x))\| \; \|y - x\|dt \leq \lambda\|x - y\|.$

We then apply proposition 1 to the function g: The mapping

$$x \mapsto Df(0)^{-1}f(x) = Df(0)^{-1}f(x) - x + x$$

is a homeomorphism from $\mathring{B}(0, \beta)$ onto a neighborhood of the origin of E. A fortiori, $f = Df(0)[Df(0)^{-1}f]$ is also a homeomorphism from $\mathring{B}(0, \beta)$ onto a neighborhood of the origin of F.   ∎

There are many improved versions of this theorem (the implicit function theorem, for instance), as well as global "versions" of the theorem, implying that f is a homeomorphism of the entire space E onto the space F. These results go beyond the limits that we have fixed for this book.   ∎

However, we can give simple sufficient conditions that assure that a mapping from $\mathbb{R}^n$ to itself is injective. Let us denote by $\langle x, y \rangle = \sum_{i=1}^{n} x_i y_i$ the Euclidean scalar product and by $\| \cdot \|$ the associated norm.

Definition 2.   We say that a mapping f from $\mathbb{R}^n$ to itself is "strictly monotone" if, for every pair $x, y \in \mathbb{R}^n$ such that $x \neq y$, we have

$$(8) \qquad \langle f(x) - f(y), x - y \rangle > 0.$$

PROPOSITION 2.   Every strictly monotone mapping f from $\mathbb{R}^n$ to itself is injective.

Proof.   Indeed, suppose that x_1 and x_2 are two different solutions of the equation $f(x) = y$. We would obtain then from the fact that f is strictly monotone that

$$0 < \langle f(x_1) - f(x_2), x_1 - x_2 \rangle = \langle y - y, x_1 - x_2 \rangle = 0,$$

which is a contradiction.   ∎

If $f \in \mathscr{L}(\mathbb{R}^n, \mathbb{R}^n)$ is a linear mapping, the condition for f to be monotone is written

$$(9) \qquad \langle f(x), x \rangle > 0 \qquad \text{for all} \qquad x \neq 0.$$

This amounts to saying that f is "positive definite" (see Section 1.12). We know that in this case f is a homeomorphism (see Corollary 3.1).   ∎

We express condition (9) differently. Since the unit sphere $S(1)$ of $\mathbb{R}^n$ is *compact* and since $x \mapsto \langle f(x), x \rangle$ is continuous and strictly positive on $S(1)$, we deduce that

$$(10) \qquad c = \inf_{x \in S(1)} \langle f(x), x \rangle > 0$$

and, consequently, that every positive definite linear mapping defined on $\mathbb{R}^n$ satisfies

$$(11) \qquad \exists c > 0 \qquad \text{such that} \qquad \langle f(x), x \rangle \geq c \|x\|^2 \qquad \text{for all} \qquad x \in \mathbb{R}^n.$$

(We use (10) with $(x/\|x\|) \in S(1)$.) Condition (11) implies that the function j defined by $j(x) = \langle f(x), x \rangle$ is *coercive*.   ∎

We show that if a continuous (nonlinear) mapping f is such that j is coercive, then f is surjective. To this end we shall admit and use Brouwer's fixed point Theorem 11.4.

Definition 3. We say that a mapping f from $\mathbb{R}^n$ to itself is "coercive" if

$$(12) \qquad \lim_{\|x\| \to \infty} \frac{\langle f(x), x \rangle}{\|x\|} = +\infty.$$

THEOREM 2. Every continuous coercive mapping f from $\mathbb{R}^n$ to $\mathbb{R}^n$ is surjective. If f is also strictly monotone, f is bijective.

Proof. Let $y \in \mathbb{R}^n$. Take $K > \|y\|$. Since f is coercive, there exists $R > 0$ such that

$$(13) \qquad \langle f(x), x \rangle \geq K\|x\| \qquad \text{when} \qquad \|x\| \geq R.$$

Consider the mapping g defined on the ball $B(R)$ with radius R by

$$(14) \quad g(x) = \begin{cases} x - f(x) + y & \text{if} \quad \|x - f(x) + y\| \leq R, \\[2mm] R\,\dfrac{x - f(x) - y}{\|x - f(x) + y\|} & \text{if} \quad \|x - f(x) + y\| \geq R. \end{cases}$$

This mapping g is clearly continuous and maps the compact convex set $B(R)$ to itself. Brouwer's theorem implies the existence of a fixed point $x = g(x)$. We cannot have $\|x - f(x) + y\| > R$: If we did, we would have $x = R((x - f(x) + y)/\|x - f(x) + y\|)$ and, therefore, $\|x\| = R$.

Taking the scalar product with x, we deduce from (13) that

$$R^2 = \langle x, x \rangle = R\,\frac{\langle x, x \rangle - \langle f(x), x \rangle + \langle y, x \rangle}{\|x - f(x) + y\|}$$

$$\leq \frac{R(R^2 - K\|x\| + \|y\|\,\|x\|)}{\|x - f(x) + y\|}$$

and, consequently, that

$$\|x - f(x) + y\| \leq R + \|y\| - K < R.$$

Thus we have obtained a contradiction. Therefore, $\|x - f(x) + y\| \leq R$ and, consequently, $x = x - f(x) + y$, that is $f(x) = y$. Hence f is, indeed, surjective. ∎

*13.   Algorithms for Solving Equation $f(x) = y$.

We present now a whole family of algorithms allowing to approximate a solution x of

$$(1) \qquad \text{find} \qquad x \in \mathbb{R}^n \qquad \text{such that} \qquad f(x) = y$$

by the sequence of solutions x_n of the linear problems

(2)   if x_{n-1} is given, find x_n such that
$$Ax_n = Ax_{n-1} + \rho(y - f(x_{n-1}))$$

where x_0 is given, ρ is a positive scalar and A *is any positive definite linear operator.*

We assume that f satisfies

(3)
$$\begin{cases} \textbf{i.} \;\; \exists c > 0 & \text{such that} & \langle f(x) - f(y), x - y \rangle \geq c\|x - y\|^p \\ \textbf{ii.} \;\; \exists d > 0 & \text{such that} & \|f(x) - f(y)\| \leq d^{1/2}\|x - y\|^{p/2} \\ \text{where } p > 1. \end{cases}$$

This implies that f is continuous, strictly monotone, and coercive (since $\langle f(x), x \rangle / \|x\| \geq c\|x\|^{p-1} - \|f(0)\|$). Hence Theorem 12.2 implies that there exists a unique solution of equation (1). We prove that the sequence x_n converges to x.

THEOREM 3. Assume that f satisfies assumption (3), that A is positive definite, and that $\rho \in \,]0, 2c/\|A^{-1}\|d[$. Then for any $y \in \mathbb{R}^n$, for any $x_0 \in \mathbb{R}^n$, the sequence of solutions x_n of (2) converges to the solution x of $f(x) = y$.

Proof. Let us set $\varepsilon_n = x_n - x$, $\delta_{n+1} = x_{n+1} - x_n$ and

(4) $\varphi(x) = \frac{1}{2}\langle Ax, x \rangle.$

We deduce from equations (1) and (2) that

(5) $A\delta_{n+1} = \rho(f(x) - f(x_n)).$

Then, by using the fact that $\varepsilon_{n+1} = \varepsilon_n + \delta_{n+1}$, we obtain

$$\varphi(\varepsilon_{n+1}) - \varphi(\varepsilon_n) = \varphi(\varepsilon_n + \delta_{n+1}) - \varphi(\varepsilon_n) = \langle A\delta_{n+1}, \varepsilon_n \rangle + \tfrac{1}{2}\langle A\delta_{n+1}, \delta_{n+1} \rangle$$

$$= -\rho\langle f(x_n) - f(x), x_n - x \rangle$$

$$+ \frac{\rho^2}{2}\langle f(x) - f(x_n), A^{-1}(f(x) - f(x_n)) \rangle$$

$$\leq -\rho\left[c - \frac{\rho}{2}\|A^{-1}\|d\right]\|x - x_n\|^p = -b(\rho)\|x - x_n\|^p$$

where $b(\rho) = \rho(c - (\rho/2)\|A^{-1}\|d)$ is positive. Hence

(6) $b(\rho)\|x - x_n\|^p \leq \varphi(\varepsilon_n) - \varphi(\varepsilon_{n+1}).$

This implies that the sequence of nonnegative numbers $\varphi(\varepsilon_n)$ is nonincreasing. Hence it converges to some number a. Now, inequality (6) implies that $\|x - x_n\|^p$ converges to $(1/b(\rho)) (a - a) = 0$.   ∎

EXAMPLES.   Let $f = F \in \mathcal{L}(\mathbb{R}^n, \mathbb{R}^n)$ be a positive definite operator. Let $F = G + H$ be a decomposition of F as the sum of two linear operators. Many "classical" algorithms are obtained by taking $A = G$:

(7)   If x_{n-1} is given, find x_n such that
$$Gx_n = (1 - \rho)Gx_{n-1} - \rho Hx_{n-1} + \rho y.$$

Usually, ρ is required to be equal to 1 (and thus the convergence holds whenever we assume that $\|G^{-1}\|\|F\|^2 < 2c$). In this case, the algorithm is

(8)   If x_{n-1} is given, find x_n such that

$$Gx_n = -Hx_{n-1} + y.$$

For instance, let us consider the decomposition $F = L + D + U$ where D is the diagonal, L the lower triangular matrix, and U the upper triangular matrix. If we take $G = D$, $H = L + U$, we obtain the *Jacobi* algorithm: if (a_i^j) is the matrix of F,

(9)
$$x_{i,k+1} = \frac{1}{a_{ii}}\left(y_i - \sum_{j \neq i} a_i^j x_{j,k}\right).$$

If we take $G = L + D$, $H = U$, we obtain the Gauss–Seidel algorithm:

(10)
$$x_{i,k+1} = \frac{1}{a_{ii}}\left(y_i - \sum_{j=i}^{i-1} a_i^j x_{j,k+1} - \sum_{j=i+1}^{n} a_i^j x_{j,k}\right).$$

By taking $G = L + (1/\omega)D$ and $H = U + (1 - (1/\omega)D$, we obtain the Young–Frankel algorithm:

(11) $x_{i,k+1} = (1 - \omega)x_{i,k} + \dfrac{\omega}{a_{ii}}\left(y_i - \sum_{j=1}^{i-1} a_i^j x_{j,k+1} - \sum_{j=i+1}^{n} a_i^j x_{j,k}\right).$

■

CHAPTER 4

Operations on Metric Spaces

Introduction

The aim of this chapter is to compare the various distances on a set (Sections 1 and 2) and to construct new metric spaces from one or more given metric spaces.

If d_1 and d_2 are two distances defined on the same set E, we say that d_1 is stronger than d_2 if the identity mapping from E to E is uniformly continuous from (E, d_1) to (E, d_2) and that the distances are equivalent if d_2 is also stronger than d_1. It is clear that if the two distances are equivalent, the topological properties of the metric spaces (E, d_1) and (E, d_2) are the same.

If F is a subset of E and if (E, d) is a metric space, we say that (F, d) is a metric subspace of (E, d). The topological properties of metric subspaces are studied in Section 3.

If we consider the product $E = \prod_{i=1}^{n} E_i$ of n metric spaces (E_i, d_i), we can construct a distance on E in such a way that a sequence $x^m = (x_1^m, \ldots, x_i^m, \ldots, x_n^m)$ of E converges if and only if the sequences x_i^m converge in E_i for every i. The topological properties of the product of metric spaces are studied in Section 4.

In analysis, we are often led to consider the case of a finite set of n mappings f_i defined on a set E with values in sets E_i. If there are distances d_i defined on the sets E_i, we can construct a distance for the set E such that a sequence of elements x_n of E converges if and only if the sequences $f_i(x_n)$ converge in (E_i, d_i) for every i. We shall say that (E, d) is the domain of the mappings f_i. We study its topological properties in Section 5.

We can associate to every metric space (E, d) the Banach space $\mathscr{C}_\infty(E)$ of continuous bounded real-valued functions on E, introduced in Chapter 1. In order to characterize the compact subsets of $\mathscr{C}_\infty(E)$, we are led to introduce and study in Section 6 the fundamental notion of an equicontinuous set of functions defined on E. The use of this notion and its properties is very important in analysis in all those problems involving taking limits. To illustrate this, we prove the existence of *global* solutions to differential

133

equations. We also use these results in the last two sections of this chapter, to construct the completion $\hat{E}$ of a metric space E in Section 7 and to construct a distance for the set $\mathscr{F}(E)$ of nonempty closed subsets of a metric space E in Section 8. This allows us to study the continuity of correspondences A from a metric space X to a metric space E, that is the mappings from X to the space $\mathscr{F}(E)$ of nonempty closed subsets of E. Finally in Section 9, we prove the existence of fixed points of correspondences in the framework of dissipative dynamical systems.

1.   Comparison of Distances

The same set E can have several distances which make of E *different* metric spaces. Let us recall that we can always consider the discrete distance $d_0(x, y)$ for E where $d_0(x, y)$ equals 0 if $x = y$ and equals 1 if $x \neq y$. It is important to be able to "compare" two distances d_1 and d_2.

Definition 1.   Let d_1 and d_2 be two distances on E. We say that the distance d_1 is "stronger" (or "finer") than d_2 (or that d_2 is "weaker" or "coarser" than d_1) if the identity mapping i from the metric space (E, d_1) to the metric space (E, d_2) is uniformly continuous. We say that the distances d_1 and d_2 are equivalent if d_1 is stronger than d_2 and d_2 is stronger than d_1.

If d_1 is equivalent to d_2, this implies, for example, that (E, d_1) and (E, d_2) have the same convergent sequences (or similarly, the same Cauchy sequences).

PROPOSITION 1.   In order for a distance d_1 to be stronger than a distance d_2, it is necessary and sufficient that

(1) $\qquad \begin{cases} \forall \varepsilon > 0, \qquad \exists \eta = \eta(\varepsilon) \qquad \text{such that} \qquad d_2(x, y) \leq \varepsilon \\ \text{when} \qquad d_1(x, y) \leq \eta. \end{cases}$

Proof.   Assertion (1) is equivalent to saying that i is uniformly continuous from (E, d_1) to (E, d_2). ∎

EXAMPLE 1.   If there exists $M > 0$ such that

(2) $\qquad\qquad\qquad d_2(x, y) \leq M d_1(x, y) \qquad \forall x, y \in E,$

then d_1 is stronger than d_2.

Remark 1.   The relation "d_1 *is stronger than* d_2" defines a partial pre-ordering relation on the set $\mathscr{D}$ of distances d on E. $\mathscr{D}$ *has a largest element, the discrete distance* d_0, which is stronger than every distance d. Indeed, if $\eta < 1$, $d_0(x, y) \leq \eta$ if and only if $x = y$, which implies that $d(x, y) = 0 \leq \varepsilon$ for any $\varepsilon > 0$ and any distance d on E. ∎

The following proposition provides a way to construct distances that are equivalent to a given distance.

PROPOSITION 2.   Let φ be a function from $[0, +\infty[$ to $[0, +\infty[$ satisfying

(3)
$$\begin{cases} \textbf{i.} & \varphi(0) = 0 \text{ and } \varphi \text{ is } increasing. \\ \textbf{ii.} & \varphi(u + v) \leq \varphi(u) + \varphi(v) \qquad \forall u, v \in [0, +\infty[. \\ \textbf{iii.} & \text{There exists } a > 0 \text{ such that } \varphi \text{ is bijective and bicontinuous} \\ & \text{from } [0, a] \text{ onto } [0, \varphi(a)]. \end{cases}$$

If d is a distance on a set E, then the distance d_1 defined by

(4) $$d_1(x, y) = \varphi(d(x, y)) \qquad \forall x, y \in E$$

is equivalent to d.

Proof.   Condition (3)iii implies in particular that $\varphi(u) = 0$ if and only if $u = 0$. Thus $d_1(x, y) = \varphi(d(x, y)) = 0$ if and only if $d(x, y) = 0$, that is if and only if $x = y$. Moreover, $d_1(x, y) = \varphi(d(x, y)) = \varphi(d(y, x)) = d_1(y, x)$, and since φ is increasing, (3)ii implies the triangle inequality:

$$d_1(x, y) = \varphi(d(x, y)) \leq \varphi(d(x, z) + d(y, z)) \leq d_1(x, z) + d_1(y, z).$$

Property (3)iii implies that $\forall \varepsilon > 0$, $\exists \eta = \eta(\varepsilon)$ such that $d(x, y) \leq \eta$ implies that $d_1(x, y) = \varphi(d(x, y)) \leq \varepsilon$. Conversely, if ψ is the inverse function of φ, which is continuous from $[0, \varphi(a)]$ onto $[0, a]$, $\forall \varepsilon > 0$, $\exists \eta_1 = \eta_1(\varepsilon)$ such that if $d_1(x, y) \leq \eta$, $d(x, y) = \psi(d_1(x, y)) \leq \varepsilon$. Then d and d_1 are equivalent. ∎

COROLLARY 1.   Let d be a distance on the set E. Then the distances

(5)
$$\begin{cases} \textbf{i.} & d_1(x, y) = (d(x, y))^{1/2}. \\[2mm] \textbf{ii.} & d_2(x, y) = \log(1 + d(x, y)). \\[2mm] \textbf{iii.} & d_3(x, y) = \inf(1, d(x, y)). \\[2mm] \textbf{iv.} & d_4(x, y) = \dfrac{d(x, y)}{1 + d(x, y)}. \end{cases}$$

are equivalent to $d(x, y)$. In particular, the distance $d_4(x, y)$ is bounded by 1 and

(6) $$d(x, y) = \frac{d_4(x, y)}{1 - d_4(x, y)}.$$

Proof.   Indeed, the functions $\varphi_1(u) = u^{1/2}$, $\varphi_2(u) = \log(1 + u)$, $\varphi_3(u) = \inf(1, u)$ and $\varphi_4(u) = u/(1 + u)$ satisfy the conditions (3). Moreover, the inverse of $\varphi_4(u)$ is the function $\psi_4(u) = u/(1 - u)$, which is continuous from $[0, 1[$ to $[0, +\infty[$. ∎

Remark 2.   It may be useful to replace the distance $d(u, v)$ by the equivalent distance $d_4(u, v)$, which makes E a *bounded metric space*. For example, if $(E, \| \cdot \|)$ is a normed vector space, $d(x, y) = \|x - y\|/(1 + \|x - y\|)$ is a distance on E, bounded by 1 and *equivalent to the distance defined by the norm*. ∎

PROPOSITION 3.   Let $d_1, \ldots, d_i, \ldots, d_n$ be n distances on E. Then

$$(7) \qquad\qquad d(x, y) = \sup_{1 \le i \le n} d_i(x, y)$$

is a distance on E that is stronger than each distance d_i and $B(x, \varepsilon) = \bigcap_{i=1}^{n} B_i(x, \varepsilon)$ $\forall x$ and $\forall \varepsilon > 0$.

Proof.   It is clear that $d(x, y)$ is a distance: $d(x, y) = 0$ if and only if $d_i(x, y) = 0$, that is if and only if $x = y$; $d(x, y) = d(y, x)$ and

$$d(x, y) \le \sup_{1 \le i \le n} (d_i(x, z) + d_i(y, z)) \le \sup_{1 \le i \le n} d_i(x, z) + \sup_{1 \le i \le n} d_i(y, z)$$

$$= d(x, z) + d(y, z).$$

Moreover, since $d_i(x, y) \le d(x, y)$, the distance d is stronger than the distance d_i. Finally, if $y \in B(x, \varepsilon)$, $d_i(x, y) \le d(x, y) \le \varepsilon$ $\forall i$ and $y \in \bigcap_{i=1}^{n} B_i(x, \varepsilon)$. Conversely, if $y \in \bigcap_{i=1}^{n} B_i(x, \varepsilon)$, then $d_i(x, y) \le \varepsilon$ for all i, which implies that $d(x, y) = \sup_{1 \le i \le n} d_i(x, y) \le \varepsilon$, that is, that $y \in B(x, \varepsilon)$. ∎

Remark 3.   We can generalize Proposition 3 to the case of an increasing sequence of distances d_k by associating to this sequence the distance d defined by

$$(8) \qquad\qquad d(x, y) = \sum_{k=1}^{\infty} 2^{-k} \frac{d_k(x, y)}{1 + d_k(x, y)}.$$

Propositions 1.13.1 and 1.13.2 can be transposed without difficulty (it suffices to replace $p_k(x - y)$ by $d_k(x - y)$). In particular, the distance d defined by (8) is stronger than each of the distances d_k. ∎

In the case of normed vector spaces, we translate the comparison between two distances in terms of norms.

Definition 2. Let E be a vector space. We say that a norm $\| \cdot \|_1$ is "stronger" than a norm $\| \cdot \|_2$ if the distance d_1 associated with $\| \cdot \|_1$ is stronger than the distance d_2 associated with $\| \cdot \|_2$ and that the norms $\| \cdot \|_1$ and $\| \cdot \|_2$ are equivalent if the distances d_1 and d_2 are.

PROPOSITION 4. Let E be a vector space. In order for $\| \cdot \|_1$ to be stronger than $\| \cdot \|_2$, it is necessary and sufficient that there exists $M > 0$ such that

$$(9) \qquad \|x\|_2 \le M \|x\|_1 \qquad \forall x \in E.$$

In order for $\| \cdot \|_1$ to be equivalent to $\| \cdot \|_2$, it is necessary and sufficient that there exist $m > 0$ and $M > 0$ such that

$$(10) \qquad m\|x\|_1 \le \|x\|_2 \le M \|x\|_1 \qquad \forall x \in E.$$

Proof. The inequality (9) expresses the fact that i is continuous from $(E, \| \cdot \|_1)$ onto $(E, \| \cdot \|_2)$ and (10) that i is bicontinuous from $(E, \| \cdot \|_1)$ onto $(E, \| \cdot \|_2)$. ∎

Remark 4. We have already shown that all the norms on $\mathbb{R}^n$ are equivalent (see Proposition 3.3.6) and in particular that the norms $\|x\|_p$ ($1 \le p \le \infty$) are equivalent among themselves since for $p \le q$,

$$\|x\|_q \le \|x\|_p \le n^{(q-p)/pq}\|x\|_q \qquad \forall x \in \mathbb{R}^n.$$

In the case of the space of sequences l^1, the norm $\|x\|_p$ is *stronger* than the norm $\|x\|_q$ when $p < q$, but the *norms are not equivalent*.

On the space of continuous functions $\mathscr{C}_\infty(-1, +1)$, on the contrary, the norm $\|f\|_q$ is stronger than the norm $\|f\|_p$ when $p \le q$, since from Proposition 1.12.4, $\|f\|_p \le 2^{(q-p)/pq}\|f\|_q$. The norms $\|f\|_p$ and $\|f\|_q$ are *not comparable* on the space $\mathscr{K}(-\infty, +\infty)$ of continuous functions with compact support. ∎

Remark 5. We can show that if $(E, \| \cdot \|_1)$ and $(E, \| \cdot \|_2)$ are Banach spaces, the fact that $\| \cdot \|_1$ is stronger than $\| \cdot \|_2$, implies that these norms are equivalent (see [AFA], Chapter 4). ∎

2. Properties Associated with Stronger Distances

If a distance d_1 is stronger than a distance d_2, that is if the identity mapping from (E, d_1) to (E, d_2) is uniformly continuous, the following proposition compares the topological properties of (E, d_1) and (E, d_2).

PROPOSITION 1.   Let d_1 be a stronger distance than d_2. Set $E_1 = (E, d_1)$ and $E_2 = (E, d_2)$. Then:

1..   Every open set of E_2 is open in E_1.
2.   Every closed set of E_2 is closed in E_1.
3.   Every neighborhood of E_2 is a neighborhood in E_1.
4.   The interior of A in E_2 is contained in the interior of A in E_1.
5.   Every compact set of E_1 is compact in E_2.
6.   Every convergent sequence of E_1 is convergent in E_2.
7.   The closure of A in E_1 is contained in the closure of A in E_2.
8.   If A is dense in E_1, A is dense in E_2.
9.   If f is a continuous mapping from E_2 to a metric space F, it is also continuous from E_1 to F.
10.   If g is a continuous mapping from a metric space G to E_1, it is also continuous from G to E_2.

Proof.   Since the identity mapping i from E_1 to E_2 is continuous, we conclude that the inverse image of every open set, of every closed set and of every neighborhood of E_2 is, respectively, an open set, a closed set, a neighborhood of E_1.

If x is in the interior of A in E_2, A contains a neighborhood of E_2 (hence of E_1) containing x. Thus x is in the interior of E_1.

Since i is continuous, the image under i of a compact set is compact and of a convergent sequence is convergent; therefore, 5 and 6 are established. Moreover, $i(\bar{A}) \subset \bar{A}$, which establishes 7 and 8.

If f is continuous from E_2 to F, then $f = f \circ i$ is continuous from E_1 to F. Similarly if g is continuous from G to E_1, then $g = i \circ g$ is continuous from G to E_2.   ■

PROPOSITION 2.   Let d_1 be a stronger distance than d_2. If the space (E, d_1) is compact, then the distances d_1 and d_2 are equivalent and (E, d_2) is compact.

Proof.   Since (E, d_1) is compact and the identity mapping i is uniformly continuous from (E, d_1) onto (E, d_2), then (E, d_2) is compact and i^{-1} is uniformly continuous from (E, d_2) onto (E, d_1).   ■

*PROPOSITION 3.   Let d_1 be a stronger distance than d_2. If

(1) $\begin{cases} \textbf{i.} & \text{the space } (E, d_2) \text{ is complete and} \\ \textbf{ii.} & \text{every ball } B_1(x, \varepsilon), \text{ which is closed in } (E, d_1), \text{ is closed in } (E, d_2) \end{cases}$

then (E, d_1) is complete.

Proof.   Let $\{x_n\}$ be a Cauchy sequence in $E_1 = (E, d_1)$:

(2)
$$\begin{cases} \forall \varepsilon, & \exists n_0 = n_0(\varepsilon) \quad \text{such that} \quad d_1(x_n, x_m) \leq \varepsilon/2 \\ \text{when} & n, m \geq n_0. \end{cases}$$

Since d_1 is stronger than d_2, $\{x_n\}$ is a Cauchy sequence in $E_2 = (E, d_2)$. Consequently, since E_2 is complete, this sequence converges to an element $x \in E$ in E_2. Let us show that

(3)
$$\begin{cases} \forall \varepsilon, & \exists n_0 = n_0(\varepsilon) \quad \text{such that} \quad d_1(x, x_n) \leq \varepsilon \\ \text{when} & n \geq n_0, \end{cases}$$

that is, that x_n converges to x in E_1.

According to (2), the sequence $\{x_m\}_{m \geq n_0}$ is contained in $B_1(x_{n_0}, \varepsilon/2)$ *which is closed in* E_2. Since x_m converges to x in E_2, we conclude that $x \in B_1(x_{n_0}, \varepsilon/2)$ and that, consequently, for $n \geq n_0$,

$$d_1(x, x_n) \leq d_1(x, x_{n_0}) + d_1(x_{n_0}, x_n) \leq \frac{2\varepsilon}{2} = \varepsilon,$$

which establishes (3).   ■

PROPOSITION 4.   Consider two metric spaces F and G such that

(4) F is contained in G and F has a stronger distance.

If E is dense in F and if F is dense in G, then E is dense in G.

Proof.   Let us fix $\varepsilon > 0$ and $z \in G$. Since F is dense in G, there exists $y \in F$ such that $d_G(y, z) \leq \varepsilon/2$. Since the distance on F is stronger than the distance on G, there exists η such that $d_G(x, y) \leq \varepsilon/2$ when $d_F(x, y) \leq \eta$. But since E is dense in F, we can associate to each $y \in F$ an element $x \in E$ such that $d_F(x, y) \leq \eta$ and, therefore, such that $d_G(x, y) \leq \varepsilon/2$. Thus $d_G(x, z) \leq d_G(x, y) + d_G(y, z) \leq \varepsilon$, which shows that E is dense in G.   ■

Remark 1. E can be dense in G without being dense in F.   ■

3.   Metric Subspaces

Definition 1.   Let (E, d) be a metric space and $F \subset E$ a subset of E. We shall say that (F, d) is a metric subspace of (E, d) with the distance that is induced by that on E.

Then the canonical injection i from F to E is continuous and the inverse image of a subset A of E is equal to $i^{-1}(A) = A \cap F$. This leads to the following proposition.

PROPOSITION 1.   Let (F, d) be a metric subspace of (E, d).

1. $A \subset F$ is open in the subspace (F, d) if and only if there exists an open set B of the space (E, d) such that $A = B \cap F$.

2. $A \subset F$ is closed in the subspace (F, d) if and only if there exists a closed set B of the space (E, d) such that $A = B \cap F$.

3.   A set $A \subset F$ is a neighborhood of $x \in F$ in the subspace (F, d) if and only if there exists a neighborhood B of x in (E, d) such that $A = B \cap F$.

4.   The closure of a subset $A \subset F$ in the space (F, d) is equal to $\bar{A} \cap F$ where $\bar{A}$ is the closure of A in the space (E, d).

5.   If $A \subset F$ is compact in (F, d), it is compact in (E, d).

6.   If f is continuous from (E, d) to (G, δ), its restriction to F is continuous from (F, d) to (G, δ).

Proof.

1.   If B is open in (E, d), then $A = B \cap F = i^{-1}(B)$ is open in (F, d) since i is continuous. *Conversely,* if A is open in (F, d), we can associate to $x \in A$ an $\varepsilon(x)$ such that $F \cap B(x, \varepsilon(x)) \subset A$. If we set $B = \bigcup_{x \in A} \mathring{B}(x, \varepsilon(x))$, which is an open set in (E, d), we see that

$$F \cap B = F \cap \bigcup_{x \in A} \mathring{B}(x, \varepsilon(x)) = \bigcup_{x \in A} (F \cap \mathring{B}(x, \varepsilon(x))) = A.$$

2.   If A is closed in F, then $F \cap \complement A$ is open in (F, d) and there exists an open set $\complement B$ of (E, d) such that $F \cap \complement A = F \cap \complement B$. Since $A \subset F$, then $A = F \cap A = F \cap B$, where B is closed in (E, d).

3.   If $A \subset F$ is a neighborhood of $x \in F$ in (F, d), there exists an open set B_0 of (E, d) such that $A \supset (F \cap B_0) \ni x$. Thus $B = A \cup B_0$ is a neighborhood of x in (E, d) such that $A = B \cap F$. *Conversely,* if B is a neighborhood of x in (E, d), $i^{-1}(B) = B \cap F = A$ is a neighborhood of x in (F, d).

4.   If $x \in F$ is in the closure of $A \subset F$ in (F, d), this means that for every neighborhood B of x in (E, d), $(B \cap F) \cap A = B \cap A \neq \varnothing$, that is that $x \in F$ belongs to the closure of A in (E, d).

5.   If $A \subset F$ is compact, $i(A) \subset F$ is compact since i is continuous.

6.   Indeed, $f|_F$ is the composition $f \circ i$ of two continuous mappings.   ■

Remark 1.   If $B \subset F$ is open (respectively, closed) in the metric space (F, d), it is not necessarily open (respectively, closed) in the metric space (E, d). For example, if F is a subset of E, which is neither open nor closed in (E, d), F is both open and closed in the metric space (F, d).   ■

PROPOSITION 2.   In order that every open set A in a metric subspace (F, d) be open in the metric space (E, d), it is necessary and sufficient that F be open in E. In order that every closed set A in a metric subspace (F, d)

be closed in the metric space (E, d), it is necessary and sufficient that F be closed in E.

Proof. The condition is necessary: F being open in (F, d), it is open in (E, d). The condition is sufficient: If F is open in (E, d), every open set A in (F, d) is the intersection of an open set B of (E, d) and the open set F; hence A is open in (E, d). The arguments are similar for closed sets. ∎

4. The Product of a Finite Number of Metric Spaces

Let us consider a *finite number* of metric spaces (E_i, d_i).

PROPOSITION 1.   Let $E = \prod_{i=1}^{n} E_i$ be the product of the sets E_i. Then

(1)
$$\begin{cases} d_\infty(x, y) = \max_{1 \le i \le n} d_i(x_i, y_i) & \text{where} \\ x = (x_1, \ldots, x_n), & y = (y_1, \ldots, y_n) \in E \end{cases}$$

is a distance on E.

Proof.   Indeed, $d_\infty(x, y) = 0$ if and only if $d_i(x_i, y_i) = 0$, that is, if and only if $x_i = y_i$ for all $i = 1, \ldots, n$. Furthermore, $d_\infty(x, y) = d_\infty(y, x)$ and since $d_i(x_i, y_i) \le d_i(x_i, z_i) + d_i(y_i, z_i)$, we conclude that $d_\infty(x, y) \le d_\infty(x, z) + d_\infty(y, z)$. ∎

Remark 1.   We observe that $d_\infty(x, y) = \|\mathbf{d}(x, y)\|_\infty$ where $\mathbf{d}$ is the mapping from $E \times E$ to $\mathbb{R}^n$ that associates to (x, y) the vector $\mathbf{d}(x, y) = (d_i(x_i, y_i))_{1 \le i \le n}$. More generally, if $\|\cdot\|$ is a norm on $\mathbb{R}^n$ (equivalent to $\|\cdot\|_\infty$), the function $d(x, y) = \|\mathbf{d}(x, y)\|$ is a distance on E equivalent to $d_\infty(x, y)$. In particular, if $1 \le p \le q \le \infty$, the distances

(2)
$$d_p(x, y) = \left(\sum_{i=1}^{n} d_i(x_i, y_i)^p \right)^{1/p}$$

satisfy the inequalities

(3)
$$d_q(x, y) \le d_p(x, y) \le n^{(q-p)/pq} \, d_q(x, y). \qquad ∎$$

Definition 1.   We say that the set $E = \prod_{i=1}^{n} E_i$ with the distance $d_\infty(x, y)$ defined by (1) is the product of the metric spaces (E_i, d_i).

In particular, when the spaces E_i are vector spaces, the product $E = \prod_{i=1}^{n} E_i$ is also a vector space.

PROPOSITION 2.   Let $E = \prod_{i=1}^{n} E_i$ be the product of n normed vector spaces $(E_i, \| \cdot \|_i)$. Then the norms

$$(4) \quad \begin{cases} \textbf{i.} \quad \|x\|_\infty = \max_{1 \le i \le n} \|x_i\|_i \\[2em] \textbf{ii.} \quad \|x\|_p = \left(\sum_{i=1}^{n} \|x_i\|_i^p \right)^{1/p} \quad \text{where} \quad 1 \le p \le +\infty \end{cases}$$

are equivalent norms on E.

Proof.   (*Left as an exercise.*)

PROPOSITION 3.   If the spaces $(E_i, ((\cdot, \cdot))_i)$ are prehilbert spaces, then $E = \prod_{i=1}^{n} E_i$ is a prehilbert space for the scalar product

$$(5) \qquad\qquad ((x, y)) = \sum_{i=1}^{n} ((x_i, y_i))_i .$$

Proof.   (*Left as an exercise.*)

The following proposition motivates the choice of the distances d_∞ on the product $E = \prod_{i=1}^{n} E_i$ of n metric spaces.

PROPOSITION 4.   A sequence $\{x^m\}$ of elements $x^m = (x_1^m, \ldots, x_n^m)$ of $E = \prod_{i=1}^{n} E_i$ converges to an element $x = (x_1, \ldots, x_n)$ of E if and only if for every $i = 1, \ldots, n$, the sequence $\{x_i^m\}$ converges to x_i in E_i.
   A function $f : x \mapsto f(x) = (f_1(x), \ldots, f_n(x))$ from a metric space F to the product $E = \prod_{i=1}^{n} E_i$ of n metric spaces E_i is continuous if and only if for every $i = 1, \ldots, n$, the functions $x \mapsto f_i(x)$ are continuous from F to E_i.

Proof.   Indeed, $d(x^m, x) = \max_{1 \le i \le n} d_i(x_i^m, x_i)$ converges to 0 if and only if for every $i = 1, \ldots, n$, $d_i(x_i^m, x_i)$ converges to 0. Similarly, $d(f(x), f(y)) = \max_{1 \le i \le n} d_i(f_i(x), f_i(y)) \le \varepsilon$ when $d_F(x, y) \le \eta$ if and only if for every $i = 1, \ldots, n$, $d_i(f_i(x), f_i(y)) \le \varepsilon$ when $d_F(x, y) \le \eta$.   ∎

From this we derive the following results.

PROPOSITION 5.   Let $A_i \subset E_i$ be a subset of E_i $(i = 1, \ldots, n)$. The product $A = \prod_{i=1}^{n} A_i$ of the subsets A_i is open (or closed, complete, compact, or bounded) if and only if the subsets A_i are open (or closed, complete, compact, or bounded).

Proof. Indeed, the ball $B(x, \varepsilon)$ with center $x \in E = \prod_{i=1}^{n} E_i$ and radius ε is of the form

(6) $$B(x, \varepsilon) = \prod_{i=1}^{n} B(x_i, \varepsilon) \qquad \text{where} \qquad x = (x_1, \ldots, x_n)$$

since $d_\infty(x, y) = \max_{i=1,\ldots,n} d_i(x_i, y_i) \le \varepsilon$ if and only if $d_i(x_i, y_i) \le \varepsilon$ for every $i = 1, \ldots, n$. Consequently, $A = \prod_{i=1}^{n} A_i$ is open if and only if for every $x \in A$, there exists $\varepsilon > 0$ such that

$$A = \prod_{i=1}^{n} A_i \supset B(x, \varepsilon) = \prod_{i=1}^{n} B(x_i, \varepsilon)$$

that is if and only if $\forall i = 1, \ldots, n,\ A_i \supset B(x_i, \varepsilon)$.

Similarly, according to the preceding proposition, x belongs to the closure of $A = \prod_{i=1}^{n} A_i$ if and only if $x = (x_1, \ldots, x_n)$ is the limit of a sequence of elements $x^m = (x_1^m, \ldots, x_n^m)$ of $\prod_{i=1}^{n} A_i$, that is, if and only if for every $i = 1, \ldots, n$, x_i is the limit of a sequence of elements x_i^m of A_i, namely, if and only if for every $i = 1, \ldots, n$, x_i belongs to the closure of A_i. Thus

(7) $$\overline{\prod_{i=1}^{n} A_i} = \prod_{i=1}^{n} \bar{A}_i,$$

and A is closed if and only if the A_i's are closed. Moreover, a sequence of elements x^m of $A = \prod_{i=1}^{n} A_i$ is a Cauchy sequence if and only if the sequences of elements x_i^m of A_i are Cauchy sequences. Therefore, A is complete if and only if the A_i's are complete.

$A = \prod_{i=1}^{n} A_i$ is compact if and only if A is complete and can be covered by a finite number of open balls $B^0(x, \varepsilon) = \prod_{i=1}^{n} B_i^0(x_i, \varepsilon)$ with radius $\varepsilon > 0$ arbitrarily small, that is if and only if each of the A_i's is complete and can be covered by a finite number of open balls $B_i^0(x_i, \varepsilon)$. Hence A is compact if and only if the A_i's are compact. Finally, $A = \prod_{i=1}^{n} A_i$ is bounded if and only if

(8) $$\sup_{x \in A} \sup_{y \in A} d(x, y) = \max_{1 \le i \le n} \sup_{x_i \in A_i} \sup_{y_i \in A_i} d_i(x_i, y_i)$$

is finite, that is, if and only if the A_i's are bounded. ∎

PROPOSITION 6.

a. The projections π_i of E onto E_i are uniformly continuous.

b. If $A \subset E$ is open (or compact or bounded), its projections $\pi_i(A)$ are open (or compact or bounded).

Proof.

a. Indeed, π_i is Lipschitz since

$$(9) \qquad d_i(\pi_i x, \pi_i y) \leq \max_{1 \leq j \leq n} d_j(\pi_j x, \pi_j y) = d_\infty(x, y).$$

b. If $A \subset E$ is compact, then its projections $\pi_i(A)$ are compact.

If $A \subset E$ is bounded, its ith projection $\pi_i(A)$ is bounded since, according to (9),

$$(10) \qquad \sup_{x \in A} \sup_{y \in A} d_i(\pi_i x, \pi_i y) \leq \sup_{x \in A} \sup_{y \in A} d_\infty(x, y) = \delta(A) < +\infty.$$

c. To show that the projection $\pi_i(A)$ of A is open when A is open, we shall show that

$$(11) \qquad \pi_i(A) = \bigcup_{\substack{a_j \in E_j \\ j \neq i}} A(a_1, \ldots, a_{i-1}, a_{i+1}, \ldots, a_n)$$

is the union of the sets

$$(12) \qquad A(a_1, \ldots, a_{i-1}, a_{i+1}, \ldots, a_n)$$
$$= \{x_i \in E_i \text{ such that } \{a_1, \ldots, a_{i-1}, x_i, a_{i+1}, \ldots, a_n\} \in A\},$$

and we shall show that these are open. Indeed, to say that $x_i \in \pi_i(A)$ is the same as saying that there exist $a_j \in E_j, j \neq i$, such that

$$\{a_1, \ldots, a_{i-1}, x_i, a_{i+1}, \ldots, a_n\}$$

belongs to A, which establishes (11). On the other hand,

$$A(a_1, \ldots, a_{i-1}, a_{i+1}, \ldots, a_n)$$

is clearly the inverse image of A under the mapping

$$(13) \qquad x_i \mapsto \{a_1, \ldots, a_{i-1}, x_i, a_{i+1}, \ldots, a_n\}.$$

But this mapping is clearly continuous, which implies, since A is open, that $A(a_1, \ldots, a_{i-1}, a_{i+1}, \ldots, a_n)$ is also open. Thus $\pi_i(A)$ is open by (11). ∎

Remark 2. The projection $\pi_i(A)$ of a closed set A is not necessarily closed: For example, the hyperbola $H = \{(x_1, x_2) \in \mathbb{R}^2 \text{ such that } x_1 x_2 - 1 = 0\}$ is a closed subset of $\mathbb{R}^2$ (for if $(x_1^n, x_2^n) \in H$ converges to (x_1, x_2), then $x_1 x_2 - 1 = 0$ and $(x_1, x_2) \in H$), but its projections onto $\mathbb{R}$ are both equal to the complement of 0 in $\mathbb{R}$, which is not closed. ∎

PROPOSITION 7. Let f be a mapping from $E = \prod_{i=1}^n E_i$ to a metric space F. If f is continuous at the point $a = (a_1, \ldots, a_n) \in E$, the mappings $f(\hat{a}_i, \cdot): x_i \mapsto f(a_1, \ldots, a_{i-1}, x_i, a_{i+1}, \ldots, a_n)$ are continuous at a_i for all $i = 1, \ldots, n$.

Remark 3.   The converse is not in general true. For example, consider the function f from $\mathbb{R}^2$ to $\mathbb{R}$ defined by

$$(14) \qquad f(x_1, x_2) = \begin{cases} \text{i.} & \dfrac{x_1 x_2}{x_1^2 + x_2^2} & \text{if} \quad (x_1, x_2) \neq (0, 0), \\[3mm] \text{ii.} & 0 & \text{if} \quad (x_1, x_2) = (0, 0). \end{cases}$$

Since $f(x, x) = 1/2$ for every $x \neq 0$, we see that f cannot be continuous at the point $(0, 0)$.

Definition 2.   We say that a mapping f from $E = \prod_{i=1}^{n} E_i$ to a metric space F is "separately continuous" at a point $a = (a_1, \ldots, a_n) \in E$ if for every i, the mappings $f(\hat{a}_i, \cdot)$ are continuous at a_i.

The function f defined by (14) is separately continuous at the point $(0, 0)$, for $f(x_1, 0) = f(0, x_2) = 0$.

Remark 4.   We can construct a metric for a *countable product* $E = \prod_{i=1}^{\infty} E_i$ of metric spaces (E_i, d_i). It can be verified as in Section 1.13 that

$$(15) \qquad \mathbf{d}(x, y) = \sum_{i=1}^{\infty} 2^{-i} \, \frac{d_i(x_i, y_i)}{1 + d_i(x_i, y_i)}$$

is a distance on E such that a sequence of elements $x^n = \{x_i^n\}_{i=1, \ldots}$ converges to $x = \{x_i\}_{i=1, \ldots}$ if and only if each sequence x_i^n converges to x_i.   ∎

GRAPHS

Definition 3.   Let f be a mapping from E to F. We call the graph of f, $G(f)$, the subset of $E \times F$ defined by

$$(16) \qquad G(f) = \{(x, f(x))\}_{x \in E}.$$

PROPOSITION 8.   If f is a continuous mapping from a metric space E to a metric space F, then the graph of f, $G(f)$, is closed in $E \times F$.

Proof.   Let $\{(x^m, f(x^m))\}$ be a sequence of elements of the graph $G(f)$ that converges to (x, y) in $E \times F$. Then the sequence x^m converges to x and $f(x^m)$ converges to y. Since f is continuous, $f(x^m)$ converges to $f(x)$; consequently, $y = f(x)$, and $(x, y) = (x, f(x))$ belongs to $G(f)$.   ∎

Remark 5.   The graph of a function f can be closed without f being continuous.   ∎

However, we obtain the following result.

PROPOSITION 9.   Let f be a mapping from a metric space E to a metric space F. If F is compact and if the graph $G(f)$ is closed in $E \times F$, then f is continuous.

Proof.   Suppose that a sequence $\{x^m\}$ of elements x^m of E converges to x. Let us show that $f(x^m)$ converges to $f(x)$. Since F is compact, $\{f(x^m)\}$ has a subsequence $\{f(x^{m_k})\}$ that converges to y in F. Since the graph $G(f)$ is closed, the subsequence $\{(x^{m_k}, f(x^{m_k}))\}$ converges to $(x, y) = (x, f(x))$ and every subsequence of $\{(x^m, f(x^m))\}$ converges to $(x, f(x))$. Thus $f(x)$ is the unique cluster point of $\{f(x^m)\}$ and since F is compact, $f(x)$ is the limit of $\{f(x^m)\}$. ∎

Remark 6.   We prove an analogous theorem for linear mappings from a *Banach* space to another: A linear mapping is continuous if and only if its graph is closed (see [AFA], Chapter 4). ∎

Remark 7.   Let us recall that a function that is real valued (or which has values in $\mathbb{R}$) can also be *characterized* by its epigraph: A function is lower semicontinuous if and only if its epigraph is closed, and convex if and only if its epigraph is convex (see Sections 3.7 and 3.8). We can also verify that if $f = \sup_{i \in I} f_i$ is the pointwise supremum of the functions f_i, then

$$(17) \qquad \mathscr{E}\!\not{\!\!\rho}\!\left(\sup_{i \in I} f_i \right) = \bigcap_{i \in I} \mathscr{E}\!\not{\!\!\rho}(f_i).$$

5.   Domains of Mappings

Let us consider:

$$(1) \qquad \begin{cases} \textbf{i.} & \text{A set } E. \\ \textbf{ii.} & n \text{ metric spaces } (F_i, d_i). \\ \textbf{iii.} & n \text{ mappings } f_i \text{ from } E \text{ to } F_i (i = 1, \ldots, n). \end{cases}$$

We are going to construct a distance on E for which the mappings f_i will be continuous.

Definition 1.   We say that the family of the f_i's is "collectively injective" if

$$(2) \qquad \forall x, y \in E, x \neq y, \exists i \in [1, \ldots, n] \qquad \text{for which} \qquad f_i(x) \neq f_i(y),$$

and we say that the "family of the f_i's is closed" if

$$(3) \qquad \begin{cases} \text{for every sequence } \{f_i(x^n)\} \text{ that converges to } y_i \in F_i, \text{ for every} \\ i = 1, \ldots, n, \text{ there exists } x \in E \text{ such that } y_i = f_i(x) \text{ for all } i = 1, \ldots, n. \end{cases}$$

We say that the subset

(4) $G(\{f_i\}) = \{(f_1(x), \ldots, f_n(x))\}_{x \in E}.$

of $F = \prod_{i=1}^{n} F_i$ is the "graph of the mappings f_i" and that the distance on E defined by

(5) $d(x, y) = \max_{1 \leq i \leq n} d_i(f_i(x), f_i(y))$

is the "distance of the graph" of E.

Remark 1.   Consider the mapping θ from E to $F = \prod_{i=1}^{n} F_i$ defined by

(6) $\theta x = (f_1(x), \ldots, f_n(x)).$

If the f_i's are collectively injective, this means that θ is injective. Moreover, $\theta(E) = G(\{f_i\})$ is the graph of the f_i's.

If the "family of the f_i's is closed," this means that the graph $G(\{f_i\})$ is closed in $F = \prod_{i=1}^{n} F_i$. One notes that θ is an isometry from the domain E with the distance of the graph to the product F of the spaces F_i.   ■

PROPOSITION 1.   Suppose that the f_i's are collectively injective. Let E have the distance of the graph. Then:

a.   The mappings f_i are uniformly continuous.

b.   A mapping g from a metric space G to E is continuous if and only if for every $i = 1, \ldots, n$, the mappings $f_i \circ g$ from G to F_i are continuous.

c.   If the spaces F_i are complete and if the family of the f_i's is closed, then E is a complete metric space.

Proof.

a.   It is clear that f_i is uniformly continuous since

(7) $d_i(f_i(x), f_i(y)) \leq \max_{1 \leq j \leq n} d_j(f_j(x), f_j(y)) = d(x, y).$

b.   To say that g is continuous at $u \in G$ is to say that $\forall \varepsilon > 0$, there exists η such that $d_G(u, v) \leq \eta$ implies

$$d(g(u), g(v)) = \max_{1 \leq j \leq n} d_j(f_j(g(u)), f_j(g(v))) \leq \varepsilon.$$

This means that the mappings $f_j \circ g$ are continuous at u.

c.   Since θ is an isometry from E onto the graph $G(\{f_i\})$ of the mappings f_i, E is complete if and only if $G(\{f_i\})$ is a complete subspace of $F = \prod_{i=1}^{n} F_i$. If the F_i's are complete and if $G(\{f_i\})$ is closed, then $F = \prod_{i=1}^{n} F_i$ is complete and so is $G(\{f_i\})$.   ■

Remark 2. In particular, if f is an injective mapping from E to a metric space (F, d), f is an isometry from E with the distance $\delta(x, y) = d(f(x), f(y))$ to (F, d). If F is complete and if the image $f(E)$ is closed in F, (E, δ) is a complete space.   ∎

EXAMPLE 1. Suppose that

(8) $\begin{cases} \textbf{i.} & E \text{ is a subset of a metric space } (F, d), \\ \textbf{ii.} & f \text{ is a mapping from } E \text{ to } F. \end{cases}$

Then we can consider E as the domain of the mappings $\{i, f\}$ where i is the canonical injection from E to F. These two mappings are continuous when E has the distance of the graph

(9) $$\delta(x, y) = \max(d(x, y), d(f(x), f(y))).$$

The "graph of the mappings i and f" coincides with the graph of the function f since

(10) $$G(i, f) = G(f) = \{x, f(x)\}_{x \in E} \subset E \times F.$$

This graph is closed if and only if

(11) $\begin{cases} \text{for every sequence } \{x^n\} \text{ of elements } x^n \text{ of } E \text{ such that } x^n \text{ converges} \\ \text{to } x \text{ in } E \text{ and } f(x^n) \text{ converges to } y \text{ in } F, \text{ then } y = f(x). \end{cases}$

In this case, if F is complete, so is E.

6. The Space $\mathscr{C}_\infty(E)$ of Continuous Bounded Functions (II)—Existence of Global Solutions to Differential Equations

Let us recall that $\mathscr{C}_\infty(E)$ denotes the vector space of continuous bounded real-valued functions on a metric space E with norm

(1) $$\|f\|_\infty = \sup_{x \in E} |f(x)|.$$

We saw that $\mathscr{C}_\infty(E)$ is a *closed subspace* of the vector space $\mathscr{U}(E)$ of bounded real-valued functions on E with norm $\|f\|_\infty$ and, consequently, that $\mathscr{C}_\infty(E)$ is a *Banach space* (that is complete and normed) since we have established that $\mathscr{U}(E)$ is a Banach space (see Sections 1.8 and 1.9).

We now characterize the compact subsets of $\mathscr{C}_\infty(E)$ and to this end introduce the notion of an equicontinuous set of continuous functions.

Let us recall that a real-valued function f on E is continuous at $x_0 \in E$ if for every $\varepsilon > 0$, there exists $\eta = \eta(\varepsilon, x_0, f)$ depending on ε, x_0 and *the function f* such that $|f(x) - f(x_0)| \leq \varepsilon$ when $d(x, x_0) \leq \eta = \eta(\varepsilon, x_0, f)$.

Definition 1. We say that a subset $\mathscr{H} \subset \mathscr{C}(E)$ of the space of continuous function on E is "equicontinuous at x_0" if for every $\varepsilon > 0$, there exists $\eta = \eta(\varepsilon, x_0, \mathscr{H})$ (depending on $\mathscr{H}$ but not on any particular function f of $\mathscr{H}$) such that

(2) $\qquad \begin{cases} \forall f \in \mathscr{H}, & |f(x) - f(x_0)| \le \varepsilon \\ \text{when} & d(x, x_0) \le \eta = \eta(\varepsilon, x_0, \mathscr{H}). \end{cases}$

We say that a subset $\mathscr{H} \subset \mathscr{C}(E)$ is "equicontinuous" if it is equicontinuous at every point of E and that it is "uniformly equicontinuous" if for every ε, there exists $\eta = \eta(\varepsilon, \mathscr{H})$ independent of $x \in E$ such that

(3) $\qquad \forall f \in \mathscr{H}, \qquad |f(x) - f(y)| \le \varepsilon \qquad \text{when} \qquad d(x, y) \le \eta = \eta(\varepsilon, \mathscr{H}).$

EXAMPLE 1. Suppose that there exist two constants c and $\alpha > 0$ such that

(4) $\qquad |f(x) - f(y)| \le cd(x, y)^\alpha \qquad \text{for all} \qquad f \in \mathscr{H}, x, y \in E.$

Then $\mathscr{H}$ is uniformly equicontinuous.

EXAMPLE 2. Every *finite* set of functions continuous at x_0 is equicontinuous at x_0, and every *finite* set of uniformly continuous functions is uniformly equicontinuous. Indeed, if $\mathscr{H} = \{f_1, \ldots, f_n\}$, it suffices to take in the first case $\eta(\varepsilon, x_0, \mathscr{H}) = \min_{1 \le i \le n} \eta(\varepsilon, x_0, f_i)$ and in the second case $\eta(\varepsilon, \mathscr{H}) = \min_{1 \le i \le n} \eta(\varepsilon, f_i)$.

EXAMPLE 3. Let $\mathscr{H} \subset E^* = \mathscr{L}(E, \mathbb{R})$ be a set of continuous linear forms on a normed vector space E. Let us recall that if $f \in E^*$, its dual norm $\|f\|_*$ is defined by

$$\|f\|_* = \sup_{x \in E} \frac{|f(x)|}{\|x\|}.$$

Then $\mathscr{H}$ is uniformly equicontinuous if and only if

$$\sup_{f \in \mathscr{H}} \|f\|_* = M < +\infty,$$

since in this case

(5) $\qquad |f(x) - f(y)| = |f(x - y)| \le \|f\|_* \|x - y\| \le M\|x - y\|.$ ∎

In establishing that $\mathscr{C}_\infty(E)$ is a closed subspace of $\mathscr{U}(E)$, we have shown that every uniform limit of a sequence of continuous functions is continuous (see Section 1.9). We are now going to consider weaker notions of convergence.

Definition 2. Let $\{f_n\}$ be a sequence of real-valued functions f_n defined on a set E. We say that $\{f_n\}$ converges "pointwise" to a real-valued function f if for every $x \in E$, $f_n(x)$ converges to $f(x)$.

Remark 1. Of course, if $\{f_n\}$ converges uniformly to f, it converges pointwise to f. *The converse is false*: Take $E = [0, 1]$ and $f_n(x) = x^n$. Then $\{f_n\}$ converges pointwise to the function f defined by

$$(6) \qquad\qquad f(x) = \begin{cases} 0 & \text{if} & 0 \le x < 1 \\ 1 & \text{if} & x = 1 \end{cases}$$

The functions f_n are continuous on $[0, 1]$, but the pointwise limit f is not continuous at the point $x = 1$. *This shows that the pointwise limit of a sequence of continuous functions is not necessarily continuous and, consequently, that pointwise convergence does not imply uniform convergence.* ∎

Nevertheless, we obtain the following results.

THEOREM 1. Let $\mathcal{H} \subset \mathcal{C}(E)$ be an equicontinuous set of continuous real-valued functions on E. Consider a sequence $\{f_n\}$ of functions f_n of $\mathcal{H}$ and a dense subset A of E. The following properties are equivalent:

$$(7) \qquad \begin{cases} \textbf{i.} & f_n(x) \text{ converges to a number } f(x) \text{ for every } x \in A. \\ \textbf{ii.} & f_n(x) \text{ converges to a number } f(x) \text{ for every } x \in E. \end{cases}$$

Under these conditions f is continuous and satisfies:

$$(8) \qquad \begin{cases} \forall \varepsilon > 0, \qquad \forall x \in E, \qquad \exists \eta = \eta(\varepsilon, x, \mathcal{H}) \qquad \text{such that} \\ |f(x) - f(y)| \le \varepsilon \qquad \text{when} \qquad d(x, y) \le \eta(\varepsilon, x, \mathcal{H}). \end{cases}$$

If we suppose further that E is compact, either of the two properties (7) implies that

$$(9) \qquad\qquad \text{the sequence } \{f_n\} \text{ converges uniformly to } f.$$

Proof.

a. Since (7)ii obviously implies (7)i, let us show that (7)i implies the pointwise convergence of f_n to a function f. We shall show that if $x \in E$, the sequence $\{f_n(x)\}$ is a Cauchy sequence: $\forall \varepsilon > 0$, $\exists n_0(\varepsilon)$ such that if $p, q \ge n_0(\varepsilon)$.

$$(10) \qquad \begin{cases} |f_p(x) - f_q(x)| \le |f_p(x) - f_p(y)| + |f_p(y) - f_q(y)| + |f_q(y) - f_q(x)| \\ \qquad \le \varepsilon \end{cases}$$

by finding upper estimates for the three terms of the middle member of inequality (10). Since $\mathscr{H}$ is equicontinuous at $x \in E$, there exists $\eta = \eta(\varepsilon, x, \mathscr{H})$ such that, for all n,

$$(11) \qquad |f_n(x) - f_n(y)| \leq \frac{\varepsilon}{3} \qquad \text{when} \qquad d(x, y) \leq \eta.$$

Since A is dense in E, there exists $y \in A$ such that $d(x, y) \leq \eta$. Since $f_n(y)$ converges to $f(y)$, because $y \in A$, there exists $n_0(\varepsilon)$ such that

$$(12) \qquad |f_p(y) - f_q(y)| \leq \frac{\varepsilon}{3} \qquad \text{when} \qquad p, q \geq n_0(\varepsilon).$$

Thus with this choice of $y \in A$ and of $n_0(\varepsilon)$, we can estimate each of the three terms of (10) by $\varepsilon/3$ according to (11) and (12).

The sequence $\{f_n(x)\}$, being a Cauchy sequence in $\mathbb{R}$, converges to a limit which we shall denote by $f(x)$.

b. Let us show (8). To every $\varepsilon > 0$ and to every $x \in E$, we associate $\eta = \eta(\varepsilon, x, \mathscr{H})$, and we take $y \in E$ such that $d(x, y) \leq \eta$. Since $|f_n(x) - f_n(y)| \leq \varepsilon$ for every n and since $|f_n(x) - f_n(y)|$ converges to $|f(x) - f(y)|$, we conclude that

$$|f(x) - f(y)| \leq \varepsilon \qquad \text{when} \qquad d(x, y) \leq \eta(\varepsilon, x, \mathscr{H}).$$

c. Suppose now that E is compact and that f_n converges pointwise to a (continuous) function f. To every $\varepsilon > 0$, we shall associate $n_0(\varepsilon)$ such that for every $x \in E$,

$$(13) \qquad |f(x) - f_n(x)| \leq \varepsilon \qquad \text{when} \qquad n \geq n_0(\varepsilon)$$

for establishing (9). Indeed, since E is compact, there exists a finite sequence x_i of elements of E such that

$$\sup_{x \in E} \; \min_{1 \leq i \leq n} d(x, x_i) \leq \eta\left(\frac{\varepsilon}{2}, x, \mathscr{H}\right).$$

We can associate to each x_i an integer $n(\varepsilon, x_i)$ such that

$$(14) \qquad |f(x_i) - f_n(x_i)| \leq \frac{\varepsilon}{2} \qquad \text{when} \qquad n \geq n(\varepsilon, x_i).$$

Take $n_0(\varepsilon) = \max_{i \leq i \leq n} n(\varepsilon, x_i)$ and associate to x an element x_i such that $d(x, x_i) \leq \eta(\varepsilon, x, \mathscr{H})$. Then for every $x \in E$,

$$|f(x) - f_n(x)| \leq |f(x) - f(x_i)| + |f(x_i) - f_n(x_i)| \leq \varepsilon$$

according to (8) and (14). Hence (13) holds.   ∎

Let us cite the following useful property of equicontinuous sets.

PROPOSITION 1. Let $\mathcal{H}$ be an equicontinuous subset of $\mathcal{C}(E)$, $\{f_n\}$ a sequence of functions f_n of $\mathcal{H}$ converging pointwise to f and $\{x_p\}$ a sequence of elements of E converging to x in E. Then

$$\lim_{\substack{n \to \infty \\ p \to \infty}} f_n(x_p) = f(x).$$

Proof. We find upper estimates for the two terms of the right-hand side of the inequality:

$$|f_n(x_p) - f(x)| \leq |f_n(x_p) - f_n(x)| + |f_n(x) - f(x)|.$$

Since $\mathcal{H}$ is equicontinuous, the first term is less than $\varepsilon/2$ when $d(x_p, x) \leq \eta$; this is possible when $p \geq p(\eta)$. The second term is less than $\varepsilon/2$ when $n \geq n(\varepsilon)$ since $f_n(x)$ converges to $f(x)$. ∎

Remark 2. There are other situations where the pointwise convergence of a sequence of continuous functions to a continuous function implies uniform convergence. For example, we obtain the following result.

PROPOSITION 2. Suppose that E is compact and that a sequence $\{f_n\}$ of continuous functions f_n converges pointwise to a continuous function f. If the sequence $\{f_n\}$ has the following property:

$$(15) \quad \begin{cases} \text{there exists a constant } k \geq 1 \text{ such that } \forall n \geq 0, \forall m \geq n, \\ |f(x) - f_m(x)| \leq k|f(x) - f_n(x)| \end{cases}$$

then f_n converges uniformly to f.

Proof. Let ε be fixed. For every $x \in E$ we denote by $n(x)$ the integer such that

$$(16) \qquad |f(x) - f_n(x)| \leq \frac{\varepsilon}{3k} \qquad \text{when} \qquad n \geq n(x).$$

Since the functions f and $f_{n(x)}$ are continuous at x, there exists $\eta(x)$ such that

$$(17) \quad \begin{cases} \max(|f(x) - f(y)|, \quad |f_{n(x)}(x) - f_{n(x)}(y)|) \leq \frac{\varepsilon}{3k} \\ \text{when} \qquad d(x, y) \leq \eta(x). \end{cases}$$

Consider then the open balls $\mathring{B}(x, \eta(x))$, which cover the compact set E. There exists a *finite number* of elements x_i ($1 \leq i \leq n$) such that $E \subset \bigcup_{1=i}^{n} \mathring{B}(x_i, \eta(x_i))$. Set $n_0 = \max_{1 \leq i \leq n} n(x_i)$. Then take $n \geq n_0$, $x \in E$, and consider an element x_i such that $x \in \mathring{B}(x_i, \eta(x_i))$. We conclude from (15) that

(18)
$$\begin{cases} |f(x) - f_n(x)| \le k|f(x) - f_{n(x_i)}(x)| \\ \qquad \le k[|f(x) - f(x_i)| + |f(x_i) - f_{n(x_i)}(x_i)| \\ \qquad + |f_{n(x_i)}(x_i) - f_{n(x_i)}(x)|]. \end{cases}$$

Since $x \in \mathring{B}(x_i, \eta(x_i))$, we deduce from (17) that

(19) $|f(x) - f(x_i)| \le \dfrac{\varepsilon}{3k}$ and $|f_{n(x_i)}(x) - f_{n(x_i)}(x_i)| \le \dfrac{\varepsilon}{3k}.$

Since $n \ge n_0 \ge n(x_i)$, (16) gives us that

(20)
$$|f(x_i) - f_n(x_i)| \le \frac{\varepsilon}{3k}.$$

Inequalities (18), (19), and (20) imply that for all $x \in E$ and for all $n \ge n_0$, $|f(x) - f_n(x)| \le \varepsilon$, that is, that f_n converges uniformly to f. ∎

COROLLARY 1. *Dini.*   Suppose that E is compact and that $\{f_n\}$ is a monotone sequence of continuous functions f_n that converges pointwise to a continuous function g. Then f_n converges uniformly to g.

Proof.   Indeed, if the sequence f_n is, for example, increasing, and if $m \ge n$, then

(21) $|f(x) - f_m(x)| \le |f(x) - f_n(x)|$ for all $x \in E,$

and hypothesis (15) of the preceding proposition is satisfied. ∎

We shall now characterize the compact subsets of $\mathscr{C}_\infty(E)$.

THEOREM 2. *Ascoli.*   Let E be a compact metric space, $\mathscr{H} \subset \mathscr{C}_\infty(E)$ a subset of real-valued functions on E. $\mathscr{H}$ is a compact subset of the Banach space $\mathscr{C}_\infty(E)$ if and only if $\mathscr{H}$ is closed and equicontinuous and for every $x \in E$ the sets $\mathscr{H}(x) = \{f(x)\}_{f \in \mathscr{H}}$ are compact in $\mathbb{R}$.

Proof.

a.   Suppose that $\mathscr{H}$ is compact. Then it is closed. Moreover, if $x \in E$, the mapping $\delta(x): f \in \mathscr{C}_\infty(E) \mapsto f(x) \in \mathbb{R}$ is continuous since it is linear and since $|f(x)| \le \|f\|_\infty$. Then $\mathscr{H}(x) = \delta(x)(\mathscr{H})$, being the image under $\delta(x)$ of the compact set $\mathscr{H}$, is a compact subset of $\mathbb{R}$. Finally, let us show that $\mathscr{H}$

is equicontinuous. Since $\mathcal{H}$ is compact, there exists a finite sequence of functions $f_i \in \mathcal{H}(i = 1, \ldots, n)$ such that

$$(22) \qquad \sup_{f \in \mathcal{H}} \min_{i = 1, \ldots, n} \|f - f_i\|_\alpha \leq \frac{\varepsilon}{3}.$$

Since the f_i's are continuous, there exists $\eta = \eta(\varepsilon, x)$ such that

$$(23) \qquad \max_{1 \leq i \leq n} |f_i(x) - f_i(y)| \leq \frac{\varepsilon}{3} \qquad \text{when} \qquad d(x, y) \leq \eta = \eta(\varepsilon, x).$$

Consequently, if $f \in \mathcal{H}$ and if f_i is such that

$$(24) \qquad \|f - f_i\|_\alpha = \sup_{y \in E} |f(y) - f_i(y)| \leq \frac{\varepsilon}{3},$$

we obtain that if $d(x, y) \leq \eta = \eta(\varepsilon, x)$,

$$(25) \qquad \begin{cases} |f(x) - f(y)| \leq |f(x) - f_i(x)| + |f_i(x) - f_i(y)| + |f_i(y) - f(y)| \\[2mm] \qquad \leq 2\|f - f_i\|_\alpha + |f_i(x) - f_i(y)| \\[2mm] \qquad \leq \dfrac{3\varepsilon}{3} = \varepsilon. \end{cases}$$

Thus $\mathcal{H}$ is equicontinuous.

   b.   *Conversely*, let us show that $\mathcal{H}$ is compact, making use of Theorem 2.6.1. We already know that since $\mathcal{H}$ is closed in the complete space $\mathscr{C}_\infty(E)$, it is complete. Let us show that for every $\varepsilon > 0$, there exists a finite sequence $f_1, \ldots, f_i, \ldots, f_n$ of functions $f_i \in \mathcal{H}$ such that for all $f \in \mathcal{H}$, there exists f_i satisfying

$$(26) \qquad \|f - f_i\|_\alpha = \sup_x |f(x) - f_i(x)| \leq \varepsilon.$$

Since $\mathcal{H}$ is an equicontinuous set, we can associate to every $x \in E$ a number $\eta(x) = \eta(\varepsilon, x, \mathcal{H})$ such that

$$(27) \qquad |f(x) - f(y)| \leq \frac{\varepsilon}{3} \qquad \text{when} \qquad y \in B(x, \eta(x)) \qquad \text{for all} \qquad f \in \mathcal{H}.$$

Since E is compact, we can cover it by a *finite number* of open balls $\mathring{B}(x_j, \eta(x_j))(1 \leq j \leq p)$ We associate to every $f \in \mathscr{C}_\alpha(E)$ the vector $(f(x_1), \ldots, f(x_j), \ldots, f(x_p))$ of $\mathbb{R}^p$ and we consider the subset $\mathcal{H}(x_1, \ldots, x_j, \ldots, x_p)$ of $\mathbb{R}^p$ formed by these vectors as f runs over $\mathcal{H}$.

   Since $\mathcal{H}(x_j)$ is the jth projection of $\mathcal{H}(x_1, \ldots, x_j, \ldots, x_p)$ onto $\mathbb{R}$, and since it is compact by hypothesis, the subset $\mathcal{H}(x_1, \ldots, x_j, \ldots, x_p)$ is a relatively compact subset of $\mathbb{R}^p$. It can, therefore, be covered by a finite sequence of balls of radius $\varepsilon/3$: There exists a sequence $f_1, \ldots, f_i, \ldots, f_n$ of n elements

f_i of $\mathscr{H}$ such that for every $f \in \mathscr{H}$, there exists f_i satisfying

(28)
$$\max_{1 \le j \le p} |f(x_j) - f_i(x_j)| \le \frac{\varepsilon}{3}.$$

Consequently, if x is an arbitrary element of E, it belongs to a ball $\mathring{B}(x_j, \eta(x_j))$ and, hence,

(29)
$$\begin{cases} |f(x) - f_i(x)| \le |f(x) - f(x_j)| + |f(x_j) - f_i(x_j)| + |f_i(x_j) - f_i(x)| \\ \le \dfrac{3\varepsilon}{3} = \varepsilon, \end{cases}$$

which implies that $\|f - f_i\|_\alpha \le \varepsilon$.   ∎

Remark 3.   Theorems 1 and 2 demonstrate the importance of the notion of an equicontinuous set $\mathscr{H}$ of continuous functions. First of all, this is a property that can be verified by analytic means and that serves, according to Theorem 2, to establish that a set of functions is compact in $\mathscr{C}_\infty(E)$, that is, that every sequence of functions in the set has a uniformly convergent subsequence. Secondly, Theorem 1 states more precise properties *that prove to be fundamental in the study of limits*: To establish that a sequence of functions f_n of $\mathscr{H}$ converges uniformly to a function f, it is sufficient to verify only that *for every element x of a dense subset A of E, the sequence $f_n(x)$ converges to $f(x)$ in $\mathbb{R}$*. It is clear that this sufficient condition is much easier to establish, for one is able *to make use of any supplementary properties that the elements x of the dense subset A may possess*.   ∎

Remark 4.   When E and F are *Banach* spaces, we can characterize the equicontinuous sets of linear operators $\mathscr{H} \subset \mathscr{L}(E, F)$ by the following property:

(30) $\forall x \in E$, $\exists M(x)$ such that $\sup_{f \in \mathscr{H}} \|f(x)\|_F \le M(x)$.

This result is often called the Banach–Steinhauss theorem (see [AFA], Chapter 4).   ∎

Remark 5.   Theorems 1 and 2 obviously hold when $\mathscr{C}_\infty(E) = \mathscr{C}_\infty(E, \mathbb{R})$ is replaced by the space $\mathscr{C}_\infty(E, U)$ of continuous bounded functions with values in a Banach space U. We need this generalization for the following application.

APPLICATION: EXISTENCE OF GLOBAL SOLUTION OF A DIFFERENTIAL EQUATION. We continue the study of differential equations initiated in Section 3-11. We are looking for solutions of a differential equation

(31)
$$\begin{cases} \textbf{i.} \quad x'(t) = \varphi(t, x(t)) \\ \textbf{ii.} \quad x(t_0) = x_0 \end{cases}$$

such that $x(t)$ *remains in the closure* $\overline{\Omega}$ *of* $\Omega \subset \mathbb{R}^n$ *for all* $t \in [t_0, t_1]$ (where t_1 is given in advance).

THEOREM 3. *Nagumo.*   We suppose that Ω is bounded and that

(32) φ is continuous from $[t_0, t_1] \times E$ to $\mathbb{R}^n$

(where E is a neighborhood of $\overline{\Omega}$) and that φ satisfies the boundary condition

(33)

$$\lim_{h \to 0+} \frac{\mathcal{O}(h)}{h} = 0 \quad \text{where} \quad \mathcal{O}(h) = \sup\{d(\overline{\Omega}, x + h\varphi(t, x)) | x \in \overline{\Omega}, t \in [t_0, t_1]\}$$

Then, for any x_0 given in $\overline{\Omega}$, there exists a solution $x(\cdot)$ of the differential equation such that $x(t) \in \overline{\Omega}$ for all $t \in [t_0, t_1]$

Proof.   We construct a family of approximate functions $x_n(t)$ of the differential equation. We show that this family is equicontinuous and bounded. By using Ascoli's Theorem 2, we know that a subsequence converges uniformly to a function $x(t)$; we check that this limit $x(\cdot)$ satisfies

(34) $$x(t) = x_0 + \int_{t_0}^t \varphi(t, x(t))dt$$

a.   *Construction of the approximate solutions.* We associate with any positive integer n the number $h_n = (t_1 - t_0)/2^n$ and the points $t_n^j = t_0 + jk_n$ (where j ranges over the integers from 0 to 2^n). By assumption (33), there exists $\eta_k = \eta(2^{-k})$ that, if $h_n \leq \eta_k$, $\mathcal{O}(h_n) \leq 2^{-k-1}h_n$. Therefore, we can construct by recursion the following sequence $\{x_n^j\}_j$ of points $x_n^j \in \overline{\Omega}$ such that

(35) $\begin{cases} \textbf{i.} & x_n^0 = x_0 \\ \textbf{ii.} & \|x_n^{j+1} - x_n^j - h_n\varphi(t_n^j, x_n^j)\| \leq 2^{-k}h_n \end{cases}$

Indeed, if $x_n^j \in \overline{\Omega}$ is given, we know that

$$d(\overline{\Omega}, x_n^j + h_n\varphi(t_n^j, x_n^j)) \leq \mathcal{O}(h_n) \leq 2^{-k-1}h_n;$$

then, there exists $x_n^{j+1} \in \overline{\Omega}$ such that

$$\|x_n^{j+1} - (x_n^j + h_n\varphi(t_n^j, x_n^j))\| \leq \mathcal{O}(h_n) + 2^{-k-1}h_n \leq 2^{-k}h_n.$$

Now, we introduce the piecewise-linear functions $x_n(t)$ which interpolates the x_n^j's at t_n^j, defined by

(36) $$x_n(t) = x_n^j + (t - t_n^j)\left(\frac{x_n^{j+1} - x_n^j}{h_n}\right) \quad \text{if} \quad t \in [t_n^j, t_n^{j+1}[$$

b.   *Convergence of the sequence of approximate solutions.* Since

$$x_n'(t) = \frac{x_n^{j+1} - x_n^j}{h_n} \quad \text{if} \quad t \in]t_n^j, t_n^{j+1}[$$

then (35) and (36) implies that

$$\|x_n'(t)\| \leq 2^{-k} + \|\varphi(t_n^j, x_n^j)\| \leq 1 + M_1 \text{ if } t \in [t_n^j, t_n^{j+1}[$$

where $M_1 = \sup\{\|\varphi(t, x)\|/x \in \overline{\Omega} \text{ and } t \in [t_0, t_1]\}$.

Hence $\|x_n(t) - x_n(s)\| \leq (M_1 + 1)(t - s)$; therefore, the sequence $\{x_n(\cdot)\}$ is *equicontinuous*. Let $M_2 = \sup_{x \in \overline{\Omega}} \|x\|$. Since, if $t \in [t_n^j, t_n^{j+1}[$,

$$\|x_n(t)\| \leq \|x_n^j\| + h_n \|\varphi(t_n^j, x_n^j)\| + 2^{-k} h_n \leq M_1 + M_2 + 1$$

we deduce that the sequences $\{x_n(t)\}$ are *bounded* for every $t \in [t_0, t_1]$.

Therefore, $x_n(t)$ belongs to the *compact* neighborhood $E = \overline{\Omega} + B(\varepsilon_0)$ of $\overline{\Omega}$ where $B(\varepsilon_0)$ is the ball of radius ε_0. Hence Ascoli's theorem implies that the sequence $x_n(\cdot)$ is relatively compact in $\mathscr{C}_\infty(t_0, t_1; \mathbb{R}^n)$ and thus, that a subsequence (again denoted by) $x_n(\cdot)$ converges uniformly to a continuous function $x(\cdot)$.

c.   Existence of a solution. Let us denote by f_n the piecewise-constant function defined by

$$f_n(t) = \varphi(t_n^j, x_n^j) \qquad \text{if} \qquad t \in [t_n^j, t_n^{j+1}[$$

and by f the function defined by

$$f(t) = \varphi(t, x(t))$$

Since φ is uniformly continuous on $[t_0, t_1] \times E$ and since $x_n(\cdot)$ converges uniformly to $x(\cdot)$, we deduce that $f_n(\cdot)$ converges uniformly to f. By adding inequalities (35), we obtain:

$$(37) \qquad \left\| x_n(t_n^j) - x_0 - \int_{t_0}^{t_n^j} f_n(\tau) d\tau \right\| \leq 2^{-k}$$

Now, since any $t \in [t_0, t_1]$ is the limit of a sequence of points t_n^j, then $x(t)$ is the limit of the sequence $x_n(t_n^j)$ and (37) yields:

$$\left\| x(t) - x_0 - \int_{t_0}^t f(\tau) d\tau \right\| = 0$$

Hence the function $x(\cdot)$ satisfies (34), that is, it is a solution of the differential equation. Since, for any t, $x(t)$ is the limit of $x_n(t_n^j) = x_n^j \in \overline{\Omega}$, we deduce that $x(t) \in \overline{\Omega}$.   ∎

7.   Completion of a Metric Space

Let (E, d) be a metric space that is not complete. We shall construct *completions* $(\hat{E}, \hat{d})$ of (E, d), that is *complete metric spaces* $(\hat{E}, \hat{d})$ *for which there exists a mapping i from E to $\hat{E}$ such that*

$$(1) \qquad \begin{cases} \textbf{i.} & i \text{ is an isometry from } (E, d) \text{ to } (\hat{E}, \hat{d}). \\ \textbf{ii.} & i(E) \text{ is dense in } \hat{E}. \end{cases}$$

First, we shall establish the fundamental property of completions.

THEOREM 1. Let i be an isometry from a metric space (E, d) to a completion $(\hat{E}, \hat{d})$ of E. If f is a uniformly continuous function from E to a complete metric space F, there exists a unique uniformly continuous function $\hat{f}$ from $\hat{E}$ to F such that

$$(2) \qquad\qquad \hat{f}(ix) = f(x) \qquad \text{for all} \qquad x \in E.$$

Proof. We begin by defining $\hat{f}$ on $i(E)$ by setting

$$(3) \qquad\qquad \hat{f}(ix) = f(x) \qquad \text{for all} \qquad x \in E.$$

Since i is an isometry and since f is uniformly continuous from E to F, we conclude that $\hat{f}$ is a uniformly continuous function from $i(E)$ to F.

Since $i(E)$ is dense in $\hat{E}$ and since F is complete, $\hat{f}$ has an extension to a unique uniformly continuous function $\hat{f}$ from $\hat{E}$ to F (see Theorem 3.5.1), which establishes the theorem. ∎

Remark 1. This theorem implies that the set $\mathscr{C}_u(E, F)$ of uniformly continuous mappings from a metric space E to a *complete* metric space F is equal to the space $\mathscr{C}_u(\hat{E}, F)$ in the following sense: The mapping θ, which associates to a function $f \in \mathscr{C}_u(E, F)$ the unique function $\hat{f} \in \mathscr{C}_u(\hat{E}, F)$ defined in the conclusion of the preceding theorem, is a bijection; indeed, if $\hat{f} \in \mathscr{C}_u(\hat{E}, F)$, the function $f \in \mathscr{C}_u(E, F)$ defined by $f(x) = \hat{f}(ix)$ is equal to $\theta^{-1}\hat{f}$. ∎

COROLLARY 1. Let i be an isometry from (E, d) to a completion $(\hat{E}, \hat{d})$. If f is an isometry from E to a complete metric space F, there exists an isometry $\hat{f}$ from $\hat{E}$ to F such that

$$(4) \qquad\qquad \hat{f}(ix) = f(x) \qquad \text{for all} \qquad x \in E.$$

Proof. We already know that there exists a unique mapping $\hat{f}$ from $\hat{E}$ to F satisfying (4) where $\hat{f}$ is defined by

$$(5) \qquad\qquad \hat{f}(\hat{x}) = \lim_{ix_n \to \hat{x}} f(x_n),$$

where $\{ix_n\}$ is a sequence which converges to $\hat{x}$. Since $d_F(f(x_n), f(y_n)) = d(x_n, y_n) = \hat{d}(ix_n, iy_n)$, we conclude that

$$d_F(\hat{f}(\hat{x}), \hat{f}(\hat{y})) = \lim_{\substack{i\hat{x}_n \to \hat{x} \\ i\hat{y}_n \to \hat{y}}} d_F(f(x_n), f(y_n))$$

$$= \lim_{\substack{ix_n \to \hat{x} \\ iy_n \to \hat{y}}} \hat{d}(ix_n, iy_n) = \hat{d}(\hat{x}, \hat{y})$$

which shows that $\hat{f}$ is an isometry from $\hat{E}$ to F. ∎

There can exist several completions $(\hat{E}, \hat{d})$ of a metric space (E, d), but they are all isometric as the following theorem shows.

THEOREM 2.   The completion $(\hat{E}, \hat{d})$ of a metric space is unique up to an isometry: If $(\hat{E}_1, \hat{d}_1)$ and $(\hat{E}_2, \hat{d}_2)$ are two completions of the same metric space (E, d), there exists an isometry θ from $(\hat{E}_1, \hat{d}_1)$ onto $(\hat{E}_2, \hat{d}_2)$.

Proof.   Consider the following diagram:

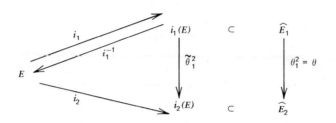

where i_1 and i_2 denote the isometries from (E, d) to $(\hat{E}_1, \hat{d}_1)$ and $(\hat{E}_2, \hat{d}_2)$, respectively. Then i_1^{-1} is an isometry from $i_1(E)$ onto E and $\tilde{\theta}_1^2 = i_2 \circ i_1^{-1}$ is an isometry from $i_1(E)$ onto $i_2(E)$, since it is the composition of two surjective isometries. Similarly, $\tilde{\theta}_2^1 = i_1 \circ i_2^{-1}$ is an isometry from $i_2(E)$ onto $i_1(E)$ and $(\tilde{\theta}_2^1)^{-1} = \tilde{\theta}_1^2$. Moreover, since $i_1(E)$ is dense in $\hat{E}_1$ and since $\hat{E}_2$ is complete, $\tilde{\theta}_1^2$ has an extension to an isometry θ_1^2 from $\hat{E}_1$ to $\hat{E}_2$. Similarly, $\tilde{\theta}_2^1$ has an extension to an isometry θ_2^1 from $\hat{E}_2$ to $\hat{E}_1$. Since $\theta_2^1 \theta_1^2 x = x$ for all $x \in i_1(E)$, we conclude that $\theta_2^1 \theta_1^2 x = x$ for all $x \in \hat{E}_1$ since $i_1(E)$ is dense in $\hat{E}_1$. Similarly, $\theta_1^2 \theta_2^1 x = x$ for all $x \in \hat{E}_2$. This shows that θ_1^2 and θ_2^1 are inverse isometries, and θ_1^2 is the desired isometry θ. ∎

We are now going to construct a completion of the metric space (E, d) and establish thereby the following result.

THEOREM 3.   Every metric space (E, d) has a completion $(\hat{E}, \hat{d})$.

Proof.   Consider the space $\mathscr{C}_\infty(E)$ of continuous and bounded functions on E which is a complete normed space for the norm

$$(6) \qquad \qquad \|f\|_\infty = \sup_{y \in E} |f(y)|.$$

Let

$$(7) \qquad \qquad a \in E \text{ be any element fixed.}$$

The mapping i from E to $\mathscr{C}_\infty(E)$ defined by

(8) $$x \mapsto i(x): y \mapsto i(x)(y) = d(x, y) - d(a, y)$$

is an isometry from (E, d) to $(\mathscr{C}_\infty(E), \|\cdot\|_\infty)$. Indeed, we obtain for any $y \in E$

(9) $$\begin{cases} |i(x)(y)| = |d(x, y) - d(a, y)| \\ \qquad \leq |d(a, y) + d(a, x) - d(a, y)| \leq d(a, x) \end{cases}$$

Therefore,

(10) $$\|i(x)\|_\infty \leq d(a, x)$$

and thus i maps E into $\mathscr{C}_\infty(E)$.

We obtain from

(11) $$\begin{cases} \textbf{i.} & |d(x_1, y) - d(x_2, y)| \leq d(x_1, x_2) \quad \text{for all} \quad y \in E \\ \textbf{ii.} & |d(x_1, x_2) - d(x_2, x_2)| = d(x_1, x_2) \quad \text{for} \quad y = x_2 \end{cases}$$

that

(12) $$\begin{cases} \|i(x_1) - i(x_2)\|_\infty = \sup_{y \in E} |i(x_1)(y) - i(x_2)(y)| \\ \qquad\qquad = \sup_{y \in E} |d(x_1, y) - d(x_2, y)| \\ \qquad\qquad = d(x_1, x_2). \end{cases}$$

We denote by

(13) $$\hat{E} = \overline{i(E)} \text{ the closure of } i(E) \text{ in } \mathscr{C}_\infty(E).$$

Since $\mathscr{C}_\infty(E)$ is complete, $\hat{E}$ is a complete metric space for the distance $\hat{d}(\hat{x}, \hat{y}) = \|\hat{x} - \hat{y}\|_\infty$.

We have, therefore, established that i is an isometry from (E, d) to the complete space $(\hat{E}, \hat{d})$, that is that $(\hat{E}, \hat{d})$ is a completion of (E, d). ∎

Remark 2. In order to show the existence of a completion, we have explicitly constructed an isometry from (E, d) to a complete space $(\hat{E}, \hat{d})$ such that $i(E)$ is dense in $\hat{E}$. *There are other constructions of completions.* Let us note, for example, the following method, analogous to the Cantor construction of the real numbers.

We consider the set $\mathscr{E}$ of Cauchy sequences on E and the equivalence relation $\mathscr{R}$ on $\mathscr{E}$ defined by $\{x_n\}\mathscr{R}\{y_n\}$ if and only if $\lim_{n\to\infty} d(x_n, y_n) = 0$.

We set $\hat{E} = \mathscr{E}/\mathscr{R}$ with the distance $\hat{d}(\{x_n\}, \{y_n\}) = \lim_{n\to\infty} d(x_n, y_n)$. We verify then that the mapping i from E to $\hat{E} = \mathscr{E}/\mathscr{R}$ associating to $x \in E$ the class of the constant sequence $\{x, x, \ldots, x, \ldots\}$ is an isometry from E to $\hat{E}$, that $\hat{E}$ is complete and that $i(E)$ is dense in $\hat{E}$.

This construction of $\hat{E} = \mathcal{E}/\mathcal{R}$ allows us to define $\mathbb{R}$ as the completion of $\mathbb{Q}$, although this is unnecessary in our framework for we have defined $\mathbb{R}$ by a system of axioms.

Let us remark that the construction of the completion $(\hat{E}, \hat{d})$ as a subset of $\mathscr{C}_\infty(E) = \mathscr{C}_\infty(E, \mathbb{R})$ uses the fact that $\mathbb{R}$ is complete and, consequently, *cannot serve to construct $\mathbb{R}$ as the completion of $\mathbb{Q}$.*   ∎

Remark 3. *The choice of a completion of a metric space.* When in a particular problem it is sufficient to replace a metric space (E, d) by a complete space $(\hat{E}, \hat{d})$, we can choose a completion arbitrarily (up to an isomorphism). Once we have made this choice, it is convenient to identify *the isometry i from E to Ê with a canonical injection from E to Ê, that is, we identify E with a dense subspace of Ê by setting i(x) = x for all x ∈ E.*

This allows us to define for example *the spaces $L^p(0, 1)$ as the completions of the normed spaces* $(\mathscr{C}(0, 1), \|x\|_p)$ of continuous functions on $[0, 1]$ with the norm

$$(14) \qquad \|x\|_p = \left(\int_0^1 |x(t)|^p \right)^{1/p} \qquad \text{for} \quad 1 \le p.$$

But it is often necessary in order to satisfy the requirements of a *more precise* problem to choose a completion by constructing it explicitly. We have already indicated two examples of explicit construction. *In the case of the spaces $L^p(0, 1)$, integration theory serves to construct explicitly the completion $L^p(0, 1)$ of $(\mathscr{C}(0, 1), \|x\|_p)$ by identifying it with the space of classes of measurable functions integrable of order p. This method allows the use of the theorems of integration theory in addition to the theorems on normed vector spaces.*   ∎

Completion shall be used as a fundamental tool for constructing Hilbert spaces of analysis. See [AFA], Chapters 5, 6, and 7.

Remark 4. *Injection of a bounded metric space in a Banach space.* The construction of the completion carried out in the proof of the existence of a completion shows that every bounded metric space can be identified with a metric subspace of a Banach space, namely the space $\mathscr{C}_\infty(E)$. Also, it is clear that $i(E)$ is an equicontinuous set of the set $\mathscr{C}_\infty(E)$. We shall generalize this fact in the following section.   ∎

8.   The Space $\mathscr{F}(E)$ of Nonempty Closed Subsets of a Metric Space E

We shall denote by $\mathscr{F}(E)$ the set of nonempty closed subsets of a metric space E. It is useful in many problems to define a distance on $\mathscr{F}(E)$. Such a situation, for example, is that of the stability of minimization problems when the set on which we minimize undergoes perturbations.

EXAMPLE 1.   Let X and E be two metric spaces, f a real-valued function defined on E, $A(x) \subset E$ a closed subset of E depending on $x \in X$. We propose then to study the function $\alpha(x)$ defined by

$$(1) \qquad\qquad \alpha(x) = \inf_{y \in A(x)} f(y),$$

which plays *an important role in optimization and in mathematical economics.* We shall see that if f is continuous on E and if the mapping $x \mapsto A(x)$ from X to $\mathscr{F}(E)$ is continuous, the function $\alpha(x)$ depends continuously on the parameter x.   ∎

Let $a \in E$ be fixed. We define the application j from $\mathscr{F}(E)$ to the Banach space $\mathscr{C}_\infty(E)$ of continuous bounded functions on the metric space E by

$$(2) \qquad\qquad \forall A \in \mathscr{F}(E) \qquad j(A)(x) = d(x,\ A) - d(x,\ a)$$

where $d(x,\ A) = \inf_{y \in A} d(x,\ y)$.

PROPOSITION 1.

a.   The mapping j from $\mathscr{F}(E)$ to $\mathscr{C}_\infty(E)$ is injective.
b.   The image $j(\mathscr{F}(E))$ is a (uniformly) equicontinuous set of $\mathscr{C}_\infty(E)$.
c.   We obtain

$$(3) \qquad \begin{cases} \textbf{i.} & \|j(A) - j(B)\|_\infty = \sup_{x \in E} |d(x,\ A) - d(x,\ B)|) \\[2mm] \textbf{ii.} & \|j(A)\|_\infty \le d(a,\ A) \end{cases}$$

Proof.

  a.   Suppose that $j(A) = j(B)$, that is, that $d(x,\ A) = d(x,\ B)$ for all $x \in E$. If $x \in A$, then $d(x,\ B) = 0$, which implies that $x \in \bar{B} = B$ since B is closed. Thus $A \subset B$. Similarly, if $x \in B$, $d(x,\ A) = 0$, and, consequently, $B \subset \bar{A} = A$. Therefore, $A = B$, and j is injective.
  b.   The set $j(\mathscr{F}(E))$ of functions $j(A)$ as A runs over $\mathscr{F}(E)$ is uniformly equicontinuous since

$$|j(A)(x) - j(A)(y)| = |d(x,\ A) - d(y,\ A) - (d(x,\ a) - d(y,\ a)| \le 2d(x,\ y)$$

for all $x,\ y \in E$ and for all $A \in \mathscr{F}(E)$.
  c.   The proof of (3) is obvious.   ∎

Since j is injective and since $\mathscr{C}_\infty(E)$ is a metric space, we can define a distance on $\mathscr{F}(E)$ making use of j as an isometry.

Definition 1. If (E, d) is a bounded metric space, the distance $h(A, B)$ on $\mathscr{F}(E)$ defined by

(4) $$h(A, B) = \sup_{x \in E} |d(x, A) - d(x, B)| = \|j(A) - j(B)\|_{\infty}$$

is called the "Hausdorff distance" on $\mathscr{F}(E)$. We shall set

(5)
$$\begin{cases}
\textbf{i.} & h_+(A, B) = \sup_{x \in E} (d(x, A) - d(x, B)) \\
\textbf{ii.} & h_-(A, B) = \sup_{x \in E} (d(x, B) - d(x, A)) = h_+(B, A).
\end{cases}$$

PROPOSITION 2. The Hausdorff distance

$$h(A, B) = \max(h_+(A, B), h_-(A, B))$$

satisfies:

(6)
$$\begin{cases}
\textbf{i.} & h_+(A, B) = \sup_{x \in B} d(x, A) = \sup_{x \in B} \inf_{y \in A} d(x, y) \\
\textbf{ii.} & h_-(A, B) = \sup_{x \in A} d(x, B) = \sup_{x \in A} \inf_{y \in B} d(x, y)
\end{cases}$$

and

(7) $$d(a, b) = h_+(\{a\}, \{b\}) = h_-(\{a\}, \{b\}).$$

Proof. Indeed since $d(x, B) = 0$ when $x \in B$,

(8)
$$\begin{cases}
\sup_{x \in B} d(x, A) = \sup_{x \in B} (d(x, A) - d(x, B)) \leq \sup_{x \in E} (d(x, A) - d(x, B)) \\
\qquad\qquad = h_+(A, B).
\end{cases}$$

Conversely, for all $x \in E$, for all $b \in B$, and for all $a \in A$, we have

(9) $$d(x, a) \leq d(x, b) + d(b, a).$$

Taking the inf with respect to $a \in A$, we obtain

(10) $$d(x, A) \leq d(x, b) + d(b, A) \leq d(x, b) + \sup_{x \in B} d(x, A),$$

and taking the inf with respect to $b \in B$, we obtain

(11) $$d(x, A) \leq d(x, B) + \sup_{x \in B} d(x, A).$$

Taking the sup with respect to $x \in E$ in (11) implies

$$h_+(A, B) = \sup_{x \in E} (d(x, A) - d(x, B)) \leq \sup_{x \in B} d(x, A) \leq h_+(A, B).$$

We obtain (6)ii by interchanging A and B. We have already established (7) since $j(\{b\})(x) = j(b)(x) = d(x, b) - d(x, a)$. ∎

*THEOREM 1.   If E is a complete metric space, $\mathcal{F}(E)$ is also a complete metric space.

Proof.   Let $\{A_n\}$ be a Cauchy sequence of nonempty closed sets $A_n \subset E : \forall \varepsilon, \exists n_0(\varepsilon)$ such that

$$(12) \qquad h(A_n, A_p) = \max(h_+(A_n, A_p), h_-(A_n, A_p)) \leq \varepsilon$$

when $n, p \geq n_0(\varepsilon)$. Let us show that A_n converges to the closed set

$$(13) \qquad A = \bigcap_n B_n \qquad \text{where} \qquad B_n = \overline{\bigcup_{m \geq n} A_m}.$$

We observe that B_n is a (decreasing) sequence of nonempty closed sets and that, consequently, A is closed.

a.   Let us show that A is not empty and that

$$(14) \qquad h_-(A_n, A) = \sup_{a_n \in A_n} d(a_n, A) \leq \varepsilon \qquad \text{when} \qquad n \geq n_0(\varepsilon).$$

Let $a_n \in A_n$ be fixed. We shall construct by recursion a subsequence n_j such that $n_0 = n$ and

$$(15) \qquad \begin{cases} n_j \text{ is the smallest integer } n_j > n_{j-1} \text{ such that } h(A_{n_j}, A_p) \leq \varepsilon/(2^{j+1}) \\ \text{when } p \geq n_j. \end{cases}$$

This is possible since $\{A_n\}$ is a Cauchy sequence. Then we construct the sequence $\{a_{n_j}\}$ of elements $a_{n_j} \in A_{n_j}$ in the following fashion:

$$(16) \qquad \begin{cases} \textbf{i.} \quad a_{n_0} = a_n \\[2mm] \textbf{ii.} \quad a_{n_{j+1}} \text{ is an element of } A_{n_{j+1}} \text{ such that} \\[2mm] \qquad d(a_{n_j}, a_{n_{j+1}}) \leq d(a_{n_j}, A_{n_{j+1}}) + \dfrac{\varepsilon}{2^{j+1}} \\[4mm] \qquad \leq h_-(A_{n_j}, A_{n_{j+1}}) + \dfrac{\varepsilon}{2^{j+1}} \leq \dfrac{2\varepsilon}{2^{j+1}} = \dfrac{\varepsilon}{2^j}. \end{cases}$$

Hence the sequence $\{a_{n_j}\}$ is a Cauchy sequence since

$$(17) \qquad \begin{cases} d(a_{n_j}, a_{n_k}) \leq d(a_{n_j}, a_{n_{j+1}}) + \cdots + d(a_{n_{k-1}}, a_{n_k}) \\[2mm] \qquad \leq \varepsilon \sum_{l=j}^{k-1} \dfrac{1}{2^l} \leq \varepsilon \sum_{l=j}^{\infty} \dfrac{1}{2^l} \end{cases}$$

is as small as we wish since the series $\sum_{l=0}^{\infty} (1/2^l)$ is convergent. Thus a_{n_j} converges to an element a since E is complete. Since $a_{n_j} \in B_{n_j}$ and since $A = \bigcap_n B_n$, the limit a belongs to A. This implies first that A is not empty. Secondly, since $d(a_n, a_{n_j}) \leq \varepsilon$, we conclude that

$$d(a, a_n) \leq d(a, a_{n_j}) + d(a_{n_j}, a_n) \leq d(a, a_{n_j}) + \varepsilon.$$

Letting j approach infinity, we obtain that

(18) $$d(a, a_n) \leq \varepsilon.$$

Consequently, $d(a_n, A) \leq d(a_n, a) \leq \varepsilon$, which implies that $h_-(A_n, A) \leq \varepsilon$ when $n \geq n_0(\varepsilon)$.

   b.   Let us show that

(19) $$h_+(A_n, A) = \sup_{a \in A} d(a, A_n) \leq \varepsilon \qquad \text{when} \qquad n \geq n_0(\varepsilon).$$

Let $a \in A$. Since $a \in B_n = \overline{\bigcup_{m \geq n} A_m}$, there exist an index $m \geq n$ and $a_m \in A_m$ such that

(20) $$d(a, a_m) \leq \frac{\varepsilon}{3}.$$

Moreover, we can associate to a_m a point $a_n \in A_n$ such that

(21) $$d(a_m, a_n) \leq d(a_m, A_n) + \frac{\varepsilon}{3}.$$

Finally, since $m, n \geq n_0(\varepsilon)$ and since $\{A_m\}$ is a Cauchy sequence, we obtain

(22) $$d(a_m, A_n) \leq d(a_m, A_m) + \frac{\varepsilon}{3} = \frac{\varepsilon}{3} \qquad \text{since} \qquad a_m \in A_m.$$

Thus inequalities (20), (21), and (22) imply

(23) $$d(a, A_n) \leq d(a, a_n) \leq d(a, a_m) + d(a_m, a_n) \leq \frac{3\varepsilon}{3} = \varepsilon. \qquad \blacksquare$$

Remark 1.   The proof of Theorem 1 shows that the limit A of a sequence of closed sets A_n is given by formula (13). ∎ $\blacksquare$

THEOREM 2.   If E is compact, the space $\mathscr{F}(E)$ is compact. In this case, if $D \subset E$ is a dense subset of E, a sequence $\{A_n\}$ of nonempty closed subsets of E converges to A if and only if

(24) $$\forall x \in D, \text{ the sequence } d(x, A_n) \text{ converges to } d(x, A).$$

Proof.   If E is compact, then E is complete and $\mathscr{F}(E)$ is complete. Consequently, since j is an isometry from $\mathscr{F}(E)$ to $\mathscr{C}_\infty(E)$, $j(\mathscr{F}(E))$ is a closed set in $\mathscr{C}_\infty(E)$. Since $j(\mathscr{F}(E))$ is an equicontinuous set (by proposition 1) and since the sets $\{j(A)(z)\}_{A \in \mathscr{F}(E)}$ are bounded for all $x \in E$ (since E is bounded),

Ascoli's Theorem 6.2 implies that $j(\mathscr{F}(E))$ is a compact set in $\mathscr{C}_\infty(E)$. Since j is an isometry, $\mathscr{F}(E)$ is a compact metric space.

Since E is compact and since $j(\mathscr{F}(E))$ is an equicontinuous set, the functions $j(A_n)$ converge uniformly to $j(A)$ if and only if for all $x \in D$, $j(A_n)(x) = d(x, A_n) - d(a, x)$ converges to $j(A)(x) = d(x, A) - d(a, x)$, according to Theorem 6.1. ∎

Definition 2. Let X and E be two metric spaces and $x \mapsto A(x)$ be a mapping from X to $\mathscr{F}(E)$ associating to each x a nonempty closed subset $A(x)$ of E called "correspondence" or "multifunction." We say that "A is continuous" at x_0 if

(25) $\begin{cases} \forall \varepsilon > 0, & \exists \eta = \eta(\varepsilon, x_0) \quad \text{such that} \quad h(A(x_0), A(x)) \le \varepsilon \\ \text{when} & d(x, x_0) \le \eta. \end{cases}$

We say that A is "upper semicontinuous" at x_0 if

(26) $\begin{cases} \forall \varepsilon > 0, & \exists \eta = \eta(\varepsilon, x_0) \quad \text{such that} \quad h_+(A(x_0), A(x)) \le \varepsilon \\ \text{when} & d(x, x_0) \le \eta \end{cases}$

and that A "is lower semicontinuous" at x_0 if

(27) $\begin{cases} \forall \varepsilon > 0, & \exists \eta = \eta(\varepsilon, x_0) \quad \text{such that} \quad h_-(A(x_0), A(x)) \le \varepsilon \\ \text{when} & d(x, x_0) \le \eta. \end{cases}$

It is clear that A is continuous if and only if it is both upper and lower semi-continuous. ∎

We are going to prove some results concerning the stability of minimization problems.

THEOREM 3. Let X and E be two metric spaces, $A : x \mapsto A(x)$ a mapping from X to $\mathscr{F}(E)$, which is lower semicontinuous at x_0 and $f : (x, y) \mapsto f(x, y)$, a real-valued upper semicontinuous function on $X \times E$. Then the function $x \mapsto \alpha(x)$ defined by

(28) $$\alpha(x) = \inf_{y \in A(x)} f(x, y)$$

is upper semicontinuous on X at x_0.

Proof. According to the definition of α, for every $\varepsilon > 0$, there exists $y_0 \in A(x_0)$ such that

(29) $$f(x_0, y_0) \le \alpha(x_0) + \frac{\varepsilon}{2} = \inf_{y \in A(x_0)} f(x_0, y) + \frac{\varepsilon}{2}.$$

Then for all $x \in X$ and for all $y \in A(x)$

(30)
$$\alpha(x) - \alpha(x_0) \le f(x, y) - f(x_0, y_0) + \frac{\varepsilon}{2}.$$

Since f is upper semicontinuous, there exists $\eta = \eta(\varepsilon, x_0, y_0)$ such that

(31)
$$\begin{cases} f(x, y) - f(x_0, y_0) \le \dfrac{\varepsilon}{2} \quad \text{when} \quad d(x, x_0) \le \eta \\ \text{and} \quad d(y, y_0) \le \eta. \end{cases}$$

Since A is lower semicontinuous at x_0, there exists $\alpha = \alpha(\varepsilon, x_0, y_0, \eta)$ such that

(32)
$$\begin{cases} d(y_0, A(x)) - d(y_0, A(x_0)) = d(y_0, A(x)) \le h_-(A(x_0), A(x)) \\ \qquad\qquad\qquad \le \dfrac{\eta}{2} \quad \text{when} \quad d(x, x_0) \le \alpha. \end{cases}$$

Take $\beta = \min(\alpha, \eta)$, x satisfying $d(x, x_0) \le \beta \le \eta$ and $y \in A(x)$ satisfying $d(y_0, y) \le d(y_0, A(x)) + \eta/2$. Thus (32) implies that $d(y_0, y) \le d(y_0, A(x)) + \eta/2 \le \eta$; we obtain (31) and, consequently, according to (30), we also obtain that

(33)
$$\alpha(x) - \alpha(x_0) \le \frac{2\varepsilon}{2} = \varepsilon. \qquad \blacksquare$$

THEOREM 4. Let X and E be two metric spaces, $A: x \mapsto A(x)$ a mapping from X to $\mathscr{F}(E)$, which is upper semicontinuous at x_0, and $f: (x, y) \mapsto f(x, y)$, a real-valued lower semicontinuous function on $X \times E$. If $A(x_0)$ is compact, the function $x \mapsto \alpha(x)$ defined by

(34)
$$\alpha(x) = \inf_{y \in A(x)} f(x, y)$$

is lower semicontinuous on X at x_0.

Proof. Since f is lower semicontinuous, we can associate to every $y \in A(x_0)$ a number $\eta(y)$ such that

(35)
$$f(x_0, y) \le f(x, z) + \varepsilon$$

when $d(x, x_0) \le \eta(y)$ and $d(z, y) \le \eta(y)$. Since $A(x_0)$ is compact, it can be covered by n open balls $\mathring{B}(y_i, \eta(y_i))$ where $y_i \in A(x_0)$. Moreover, $\bigcup_{1 \le i \le n} \mathring{B}(y_i, \eta(y_i))$ is an open set containing $A(x_0)$ and, consequently, containing a ball $B(A(x_0), \alpha)$ where $\alpha > 0$. Since A is upper semicontinuous, there exists η_0 such that

(36) $$h_+(A(x_0), A(x)) \leq \alpha \qquad \text{when} \qquad d(x, x_0) \leq \eta_0.$$

Set $\eta = \min_{1 \leq i \leq n}(\eta_0, \eta(y_i))$ and take x satisfying $d(x, x_0) \leq \eta$. Since $h_+(A(x_0), A(x)) = \sup_{z \in E}(d(z, A(x_0)) - d(z, A(x)) \leq \alpha$ we obtain, taking $z \in A(x)$, that $A(x) \subset B(A(x_0), \alpha) \subset \bigcup_{1 \leq i \leq n} B(y_i, \eta(y_i))$. Consequently, if $z \in A(x)$, there exists $y_i \in A(x_0)$ such that $d(z, y_i) \leq \eta(y_i)$; since $d(x, x_0) \leq \eta(y_i)$, we deduce from (35) that

(37) $$\alpha(x_0) \leq f(x_0, y_i) \leq f(x, z) + \varepsilon.$$

Since this is true for all $z \in A(x)$, we obtain

(38) $$\alpha(x_0) \leq \inf_{z \in A(x)} f(x, z) + \varepsilon = \alpha(x) + \varepsilon \qquad \text{when} \qquad d(x, x_0) \leq \eta. \quad \blacksquare$$

These two theorems imply the following result.

THEOREM 5. Let X and E be two metric spaces, $A: x \mapsto A(x)$ a continuous mapping from E to $\mathscr{F}(E)$ and $f:(x, y) \mapsto f(x, y)$ a continuous real-valued function on $X \times E$. If the $A(x)$'s are compact for all $x \in X$, then the function α defined by

(39) $$\alpha(x) = \min_{y \in A(x)} f(x, y)$$

is a continuous function on X.

*APPLICATION. FIXED POINTS OF A CONTRACTING MAPPING. The fixed point theorem for continuous contracting mappings from E to E can easily be generalized to the case of contracting mappings A from E to $\mathscr{F}(E)$.

Definition 3. Let E be a bounded metric space and A a mapping from E to $\mathscr{F}(E)$. We say that A is contracting if there exists $\lambda \in]0, 1[$ such that

(40) $$h(A(x), A(y)) \leq \lambda d(x, y) \qquad \text{for all} \qquad x, y \in E.$$

We say that $x \in E$ is a fixed point of A if

(41) $$x \in A(x).$$

THEOREM 6. Suppose that E is complete and that A is a contracting mapping from E to $\mathscr{F}(E)$. If the sets $A(x)$ are compact for every $x \in E$, A has a fixed point.

Proof. As in the case of contracting functions, we establish the existence of a fixed point x of A as the limit of a sequence x_n constructed by recursion starting from an arbitrary element $x_0 \in E$. Indeed, since $A(x_0)$ is compact, there exists $x_1 \in E$ such that

(42) $x_1 \in A(x_0)$ and $d(x_1, x_0) = \min\limits_{y \in A(0)} d(y, x_0) = d(x_0, A(x_0))$.

Similarly, since $A(x_n)$ is compact, we define x_{n+1} from x_n by

(43)
$$\begin{cases} x_{n+1} \in A(x_n) \\ \text{and} \\ d(x_{n+1}, x_n) = \min\limits_{y \in A(x_n)} d(y, x_n) \\ \qquad\qquad = d(x_n, A(x_n)). \end{cases}$$

Since $d(x_n, A(x_{n-1})) = 0$, we obtain from (43) and (40) that

(44)
$$\begin{cases} d(x_{n+1}, x_n) = d(x_n, A(x_n)) = d(x_n, A(x_n)) - d(x_n, A(x_{n-1})) \\ \qquad \leq h(A(x_n), A(x_{n-1})) \leq \lambda d(x_n, x_{n-1}) \\ \qquad \leq \lambda^n d(x_0, A(x_0)) \end{cases}$$

and, consequently, that

(45)
$$d(x_{m+1}, x_n) \leq d(x_0, A(x_0))\left(\sum_{k=n}^{m+1} \lambda^k\right).$$

Since $\lambda < 1$, we obtain from (45) that x_n is a Cauchy sequence, which converges to x since E is complete. We must show that $x \in A(x)$, that is, that $d(x, A(x)) = 0$. Indeed, we deduce from the definition of the Hausdorff distance that

$$\begin{aligned} d(x_n, A(x)) - d(x_n, x_{n+1}) &= d(x_n, A(x)) - d(x_n, A(x_n)) \\ &\leq h(A(x), A(x_n)) \leq \lambda d(x, x_n), \end{aligned}$$

that is, that $d(x_n, A(x)) \leq d(x_n, x_{n+1}) + \lambda d(x, x_n)$. Letting n approach infinity, we conclude that $d(x, A(x)) \leq d(x, x) + \lambda d(x, x) = 0$ and, consequently, that $x \in A(x)$. ∎

Remark 2. Similarly, the Brouwer fixed point theorem (Theorem 3.11.4), which we have stated without proof, can be generalized to the case of mappings from E to $\mathscr{F}(E)$. This is the important theorem of Kakutani.

THEOREM 7. Let E be a nonempty compact convex subset of $\mathbb{R}^n$. If A is an upper semicontinuous mapping from E to $\mathscr{F}(E)$ with nonempty compact convex values, A has a fixed point $x \in A(x)$.

We prove it in [AFA], Chapter 15. ∎

In the next section, we shall prove other fixed point theorems for dissipative correspondences.

*9.   Dissipative Dynamical Systems

Let E be a metric space and A be a map from E to the metric space $\mathscr{F}(E)$ of nonempty closed subsets of E.

We regard E as a set of "states"; we consider the *discrete dynamical system* defined by A, which associates to any state x at a given period the set $A(x)$ of states y reached at the next period.

In physical sciences, usually a dynamical system is completely determined by the initial conditions: A is a map from E to E. In situations that involve human decisions or uncertainty, usually there exist more than one "state" obtained from an initial state. Also, control theory supplies dynamical systems defined by maps: $x \in E \mapsto A(x) = \{f(x, u)\}_{u \in U} \subset E$, where U is the set of "controls" u and where, for each control $u \in U$, the map $x \in E \mapsto f(x, u) \in E$ defines a (deterministic) dynamical system.

We address ourselves to the problem of existence of a "fixed point" $\bar{x} \in A(\bar{x})$ (which may be interpreted as a rest point of the dynamical system) and existence of a "stationary point" $x_* = A(x_*)$ (which may be regarded as an end point of the system). In the framework of control theory, $\bar{x}$ is a fixed point if *there exists* a control $\bar{u} \in U$ such that $\bar{x} = f(\bar{x}, \bar{u})$ and x_* is a stationary point if *for all* controls $u \in U$, we have $x_* = f(x_*, u)$.

We present only *basic* results, which have an intrinsic interest.

Definition 1.   A sequence $\{x_n\}_{n \geq 0}$ of elements $x_n \in E$ satisfying

(1) $$x_0 = x, \, x_{n+1} \in A(x_n) \qquad \text{for any} \qquad n \geq 0$$

is called a "trajectory starting at x." We say that $\bigcup_{n \geq 0} x_n$ is the associated "orbit," and we denote by

(2) $$H(x) = \bigcup \{x_n | \text{for all trajectories starting at } x\}$$

the union of orbits starting at x.

We begin by defining weak dissipative systems.

Definition 2.   A function $\varphi : E \to \mathbb{R}_+$ is called a "weak entropy of A" if

(3) $$\forall x \in E, \exists y \in A(x) \qquad \text{such that} \qquad d(x, y) \leq \varphi(x) - \varphi(y)$$

A map $A : E \to \mathscr{F}(E)$ is said "weakly dissipative" if there exists a weak entropy of A.

THEOREM 1.   Let E be complete. Any weakly dissipative upper semi-continuous map A from E to $\mathscr{F}(E)$ has a fixed point.

Proof.   Let φ be a weak entropy. Starting with $x_0 \in E$, inequality (3) allows the construction of a trajectory $\{x_n\}_{n \geq 0}$ satisfying $d(x_n, x_{n+1}) \leq \varphi(x_n) - \varphi(x_{n+1})$. This implies that the sequence of nonnegative numbers $\varphi(x_n)$ is nonincreasing; hence it converges to some $a \geq 0$. Since $d(x_p, x_q) \leq \sum_{n=p}^{q-1} d(x_n, x_{n+1}) \leq \varphi(x_p) - \varphi(x_q)$, $\{x_n\}_{n \geq 0}$ is a Cauchy sequence, which converges to some $\bar{x} \in E$ (E being complete).

Since $x_{n+1} \in A(x_n)$, we obtain $d(x_{n+1}, A(\bar{x})) \leq h_+(A(\bar{x}), A(x_n))$. But $d(x_{n+1}, A(\bar{x}))$ converges to $d(\bar{x}, A(\bar{x}))$ and $h_+(A(\bar{x}), A(x_n))$ converges to 0 since A is upper semicontinuous. Hence $d(\bar{x}, A(\bar{x})) = 0$.   ∎

As a consequence, we obtain:

THEOREM 2 (Caristi–Siegel).   Let f be a continuous map from a complete space E to itself satisfying

$$(4) \qquad \forall x \in E, d(x, f(x)) \leq \varphi(x) - \varphi(f(x))$$

where $\varphi : E \to \mathbb{R}_+$. Then f has a fixed point.

Remark 1.   If f is a contraction (see Section 3.11), it satisfies (4) for φ defined by $\varphi(x) = \sum_{n=0}^{\infty} d(f^n(x), f^{n+1}(x))$, which is bounded above by $(d(x, f(x))/(1 - \lambda)$ where $\lambda < 1$ is the Lipschitz constant. Hence the fixed-point Theorem 3.11.1 for contractions is a particular case of Theorem 2.   ∎

Before proving existence of a stationary point, we need some results. We notice that

$$(5) \qquad \forall x \in E, x \in H(x) \qquad \text{and} \qquad \forall y \in H(x), H(y) \subset H(x).$$

Also, it is clear that the subsets $H(x)$ are "invariant" in the sense that

$$(6) \qquad \forall x \in E, A(H(x)) \subset H(x).$$

Any trajectory $\{x_n\}_{n \geq 0}$ of A is a trajectory of H (in the sense that $x_{n+1} \in H(x_n)$); a trajectory of H may be regarded as a "subtrajectory" of A.

Definition 3.   If $\{x_n\}_{n \geq 0}$ is a trajectory of H, the set

$$(7) \qquad\qquad L(x_n) = \bigcap_{n \geq 0} \overline{H(x_n)}$$

is called the "limit set" of the trajectory.

A set X is "invariant" by a map $B:E \to \mathscr{F}(E)$ if $B(x) \subset X \; \forall x \in X$.

It is obvious that if the subsets $H(x)$ are closed and invariant by a map $B:E \to \mathscr{F}(E)$, then limit sets are invariant by B.

In most instances, the sets $H(x)$ are not closed, but we may replace this assumption by the fact that B is lower semicontinuous.

PROPOSITION 1.   If the subsets $H(x)$ are invariant by a lower semicontinuous map $B:E \to \mathscr{F}(E)$, then the limit sets are also invariant by B.

Proof.   We take $\bar{x} \in \bigcap_{n \geq 0} \overline{H(x_n)}$. Since B is lower semicontinuous, we can associate to any $\varepsilon > 0$ some $\eta > 0$ such that $h_-(B(\bar{x}), B(y_n)) \leq \varepsilon$ when $d(\bar{x}, y_n) \leq \eta$. But, for any $n \geq 0$, we can find $y_n \in H(x_n)$ such that $d(\bar{x}, y_n) \leq \eta$ [since $\bar{x} \in \overline{H(x_n)}$]. Furthermore, $H(x_n)$ being invariant by B, $B(y_n) \subset H(x_n)$ and thus, $h_-(B(\bar{x}), H(x_n)) \leq h_-(B(\bar{x}), B(y_n)) \leq \varepsilon$. By letting ε go to 0, we deduce that $h_-(B(\bar{x}), H(x_n)) = 0$ for all n; this means that $B(\bar{x}) \subset \overline{H(x_n)}$ for any n, that is, that $B(\bar{x}) \subset L(x_n)$.   ∎

Let us also point out the following obvious property.

PROPOSITION 2.   Any cluster point of a trajectory $\{x_n\}_{n \geq 0}$ of H belongs to the limit set $L(x_n)$.

Proof.   If a subsequence $\{x_{n_k}\}$ converges to $\bar{x}$, then for all $p \geq 0$ and $k \geq p$, $x_{n_k} \in H(x_{n_p}) \subset H(x_p)$. Therefore, $\bar{x} \in \overline{H(x_p)}$ for all $p \geq 0$.   ∎

Definition 4.   A function $\varphi:E \to \mathbb{R}_+$ is called an "entropy" of A if

$$\text{(8)} \qquad \forall x \in E, \qquad \forall y \in A(x), \qquad d(x, y) \leq \varphi(x) - \varphi(y).$$

A map $A:E \to \mathscr{F}(E)$ is said "dissipative" if there exists an entropy.

It is clear that any entropy of A is an entropy of H.

PROPOSITION 3.   Assume that E is complete and that $A:E \to \mathscr{F}(E)$ is dissipative. For any $x \in E$, there exists a trajectory of H starting at x converging to $\bar{x}$, whose limit set is equal to $\{\bar{x}\}$.

Proof.   Let φ be an entropy of A; we define

$$\text{(9)} \qquad \rho(x) = \inf\{\varphi(y) | y \in H(x)\}.$$

We *only* use the following consequence of assumption (8)

$$\text{(10)} \qquad \forall x \in E, \; \sup_{y \in H(x)} d(x, y) \leq \varphi(x) - \rho(x).$$

Therefore, the diameter $\delta(H(x)) = \sup\{d(y, z) \mid y, z \in H(x)\}$ satisfies inequality

(11) $$\forall x \in E, \ \delta(H(x)) \leq 2[\varphi(x) - \rho(x)]$$

since $\delta(H(x)) \leq \sup\{d(y, x) + d(x, z) \mid y, z \in H(x)\} = 2 \sup\{d(y, x) \mid y \in H(x)\}$
$\leq 2[\varphi(x) - \rho(x)]$ by (10). *Now*, we associate to φ the trajectory $\{x_n\}_{n \geq 0}$ of H
defined by

(12) $$x_{n+1} \in H(x_n) \qquad \text{where} \qquad \varphi(x_{n+1}) \leq \rho(x_n) + 2^{-n}.$$

Since $H(x_{n+1}) \subset H(x_n)$ and $x_{n+1} \in H(x_{n+1})$, we obtain the relations $\rho(x_n) \leq$
$\rho(x_{n+1})$ and $\rho(x_{n+1}) \leq \varphi(x_{n+1})$. Therefore, the trajectory $\{x_n\}_{n \geq 0}$ defined by
(12) satisfies:

(13) $$\rho(x_{n+1}) \leq \varphi(x_{n+1}) \leq \rho(x_n) + 2^{-n} \leq \rho(x_{n+1}) + 2^{-n}.$$

This implies that $\delta[H(x_{n+1})] \leq 2[\varphi(x_{n+1}) - \rho(x_{n+1})] \leq 2^{-n+1}$ converges
to 0. Since the sequence of closed subset $\overline{H(x_n)}$ is nonincreasing, Theorem 2.2.2
implies that the limit set $L(x_n)$ is made of one point $\{\bar{x}\}$. Furthermore, (13)
implies also that $\{x_n\}_{n \geq 0}$ is a Cauchy sequence, which converges to some limit.
Proposition 2 implies that this limit is equal to $\bar{x}$. ∎

THEOREM 3 (Maschler–Peleg).   Let E be complete. Any dissipative
lower semicontinuous map A from E to $\mathscr{F}(E)$ has a stationary point.

Proof.   By Proposition 3, there exists a trajectory $\{x_n\}$ whose limit set is a
point $\bar{x}$. Since the sets $H(x)$ are invariant by the lower semicontinuous map A,
Proposition 1 implies that $A(\bar{x}) = \{\bar{x}\}$. ∎

In the framework of control theory, we obtain the following.

THEOREM 4 (Siegel).   Let E be complete; let us consider a family of
continuous maps $x \in E \to f(x, u) \in E$ where u ranges over U. If there exists a
common entropy φ:

(14) $$\forall u \in U, \qquad \forall x \in E, \qquad d(x, f(x, u)) \leq \varphi(x) - \varphi(f(x, u))$$

then the maps $f(\cdot, u)$ have a common fixed point $\bar{x}$.

Proof.   By Proposition 3, there exists a limit set $\{\bar{x}\}$; Proposition 1 implies
that $\bar{x}$ is invariant by the continuous maps $f(\cdot, u)$. ∎

Remark 2.   We can obtain existence of a sationary point under weaker
assumptions.

THEOREM 5 (Brézis–Browder).   Let us assume that there exists a lower semicompact continuous function φ satisfying

$$(15) \qquad \forall x \in E, \qquad \forall y \in A(x), \qquad y \neq x, \; \varphi(y) < \varphi(x).$$

For any $x \in E$, there exists a trajectory of H starting at x whose limit set L satisfies $\varphi(L) = a$.

If A is lower semicontinuous, it has a stationary point.

Proof.   We associate to φ the trajectory defined by (12) and satisfying (13). Since $x_n \in H(x_0)$, then $\varphi(x_n) \leq \varphi(x_0)$, and thus the sequence $\{x_n\}$ remains in a compact set; we can extract a subsequence $\{x_{n_k}\}$ converging to $\bar{x} \in E$, which belongs to the limit set $L(x_n) = L$ by Proposition 2. Since the sequence of non-negative scalars $\varphi(x_n)$ is nonincreasing, it converges to some $a \geq 0$; by (13), we get

$$(16) \qquad a = \lim_{n \to \infty} \varphi(x_n) = \lim_{n \to \infty} \rho(x_n).$$

Now, let us choose any $x \in L$. For any n, we can find $y_n \in H(x_n)$ such that $d(x, y_n) \leq 2^{-n}$. Since $\rho(x_n) \leq \varphi(y_n) \leq \varphi(x_n)$ [by (9) and (15)] we deduce that $\varphi(y_n)$ converges to a [by (16)]. Since φ is assumed to be continuous, then $\varphi(x) = a$. *Finally,* any point x of the limit set is a stationary point: indeed, let $y \in A(x)$; since A is lower semicontinuous, Proposition 1 implies that $y \in L$, and thus $\varphi(y) = \varphi(x) = a$. By (15), we conclude that $y = x$.   ∎

APPLICATION: *Ekeland's Variational Principle*

The following variational principle happens to be very useful. It shows how to approximate a minimization problem with no solution by another minimization problem which has a solution.

THEOREM 6 (Ekeland).   Let E be complete and $\varphi : E \to \mathbb{R}_+$ be lower semicontinuous. For any $\varepsilon > 0$, let us choose an "approximate minimum" $y_\varepsilon \in E$ satisfying $\varphi(y_\varepsilon) \leq \inf_{x \in E} \varphi(x) + \varepsilon$. Then there exists $x_\varepsilon \in E$ satisfying

$$(17) \quad d(x_\varepsilon, y_\varepsilon) \leq \sqrt{\varepsilon} \qquad \text{and} \qquad \varphi(x_\varepsilon) = \min_{x \in E} \left[\varphi(x) + \sqrt{\varepsilon}\, d(x_\varepsilon, x) \right]$$

Proof.   We associate to φ the dynamical system defined by the *closed* subsets

$$(18) \qquad H(x) = \{y \in E \text{ such that } \sqrt{\varepsilon}\, d(x, y) \leq \varphi(x) - \varphi(y)\}$$

for which φ is an entropy. By Proposition 3, there exists a trajectory $\{x_n\}$ starting at y_ε and converging to x_ε such that

$$(19) \qquad H(y_\varepsilon) \supset \bigcap_{n \geq 0} H(x_n) = \bigcap_{n \geq 0} \overline{H(x_n)} = \{x_\varepsilon\}.$$

Since $x_\varepsilon \in H(y_\varepsilon)$, we deduce that $\sqrt{\varepsilon}\, d(x_\varepsilon, y_\varepsilon) \leq \varphi(y_\varepsilon) - \varphi(x_\varepsilon) \leq \varepsilon$ by definition of y_ε. Since x_ε is a stationary point of H, then, for all $y \neq x, y \notin H(x_\varepsilon)$; by (18), this implies that $\varphi(x_\varepsilon) < \varphi(y) + \sqrt{\varepsilon}\, d(x_\varepsilon, y)$. ∎

CHAPTER 5

Special Properties
of Metric Spaces

Introduction

We consider in this chapter certain special properties of metric spaces, and we study some applications. The first five sections are independent. The last three deal with applications to game theory and mathematical economics.

We begin in Section 1 with the study of *locally compact* spaces, that is, those spaces for which every point has a compact neighborhood. An infinite discrete metric space is locally compact, as are the spaces $\mathbb{R}^n$. We shall show, moreover, that the only locally compact normed vector spaces are the spaces $\mathbb{R}^n$.

Next we define separable metric spaces, that is, metric spaces in which every point is the limit of elements of a *countable set* D. We know already that compact spaces have this property. We show that the separable locally compact spaces are those locally compact spaces that are a *countable* union of compact sets.

In Section 2 we pursue the study of separable metric space. In particular, we characterize separable normed vector spaces by showing that every point is the limit of linear combinations of a countable family of elements. We then show that the space $\mathscr{C}_\infty(E)$ of continuous functions on a compact set $E \subset \mathbb{R}^n$ is separable; in fact, the Stone–Weierstrass theorem shows that every continuous function of $\mathscr{C}_\infty(E)$ is the uniform limit of polynomials. We also give an explicit construction by means of Bernstein polynomials.

Section 3 is devoted to the proof of *Baire's theorem*, which is one of the fundamental results of analysis. It plays a basic role in the study of linear operators. (See Chapter 4 of [AFA].) For now, we use this result to show that every lower semicontinuous function on an open set U of a complete metric space is bounded on an open subset V of U, that every lower semicontinuous seminorm is continuous and, more generally, that every convex lower semi-

continuous function on a nonempty open set of a Banach space is in fact
continuous.

We study in Section 4 the *connected* subsets of a metric space, that is,
those subsets that cannot be partitioned by open sets (or by closed sets).
We show that a subset A of $\mathbb{R}$ is connected if and only if A is an interval and
that convex sets are connected. We also show that a continuous function
maps connected sets onto connected sets.

The notion of a connected set is often used to establish that the intersection of two nonempty closed subsets is nonempty. We make use of this
property to construct continuous utility functions representing a total
preorder relation. (This resolves a fundamental problem of economic theory
and of decision theory.)

We show in Section 5 that two nonempty closed subsets A and B can be
separated by a continuous bounded function (that is a function which is 0
on A and 1 on B, for example). We use this separation theorem to construct
continuous partitions of unity subordinate to a finite covering of a set
$X = \bigcup_{i=1}^{n} A_i$ by open sets A_i: These are n continuous functions f_i from E
to $[0, 1]$, which vanish on the complement of A_i and for which $\sum_{i=1}^{n} f_i(x) = 1$
for all $x \in E$.

We use partitions of unity in Section 6 to establish the following equality:

$$\inf_{x \in E} \sup_{y \in F} f(x, y) = \sup_{D \in \mathscr{C}(E, F)} \inf_{x \in E} f(x, D(x)),$$

where f is a function defined on $E \times F$ (E compact, F convex), which is lower
semicontinuous with respect to x and concave with respect to y and where
$\mathscr{C}(E, F)$ is the set of continuous mappings from E to F.

This equality can be interpreted in the framework of game theory.
Admitting the Brouwer fixed point theorem (Theorem 3.11.4), it implies
the symmetric equality

$$\inf_{x \in E} \sup_{y \in F} f(x, y) = \inf_{C \in \mathscr{C}(F, E)} \sup_{y \in F} f(C(y), y)$$

under the same hypotheses.

This last equality leads to the important Ky-Fan inequality when $E = F$:

$$\inf_{x \in E} \sup_{y \in E} f(x, y) \leq \sup_{y \in E} f(y, y),$$

which proves to be a useful tool in nonlinear analysis. To illustrate this fact,
we use this result in Section 7 to prove the theorem of Nash on the existence
of a noncooperative equilibrium in an n-person game and in Section 8 to
prove existence of an equilibrium price of a Walras economic model. This
Ky-Fan inequality will be at the origin of the theorems of nonlinear analysis
presented in Chapter 15 of [AFA].

1. Locally Compact Spaces

The real line $\mathbb{R}$ is not a compact space since it is not bounded. However, the closed neighborhoods $[x - \varepsilon, x + \varepsilon]$ of a point x are compact. Similarly, an infinite discrete metric space is not compact since it is not finite, but every point x has a compact neighborhood, namely the set $\{x\}$.

Definition 1.　We say that a metric space E is "locally compact" if every point x of E has at least one compact neighborhood.

In particular, since this compact neighborhood contains a closed ball, which is, therefore, compact, every point x is the center of a compact ball $B(x, \varepsilon(x))$ with radius $\varepsilon(x) > 0$. It is more generally put in the following proposition.

PROPOSITION 1.　Let E be a locally compact space and K be a compact subset of E. Then every neighborhood V of K contains a compact neighborhood.

Proof.　Let V be a neighborhood of K. Every point x of K is the center of a compact ball $B(x, \varepsilon_1(x))$ and of a ball $B(x, \varepsilon_2(x))$ contained in V. Hence the ball $B(x, \varepsilon(x))$, where $\varepsilon(x) = \min(\varepsilon_1(x), \varepsilon_2(x))$ is compact and contained in V, and the open balls $\mathring{B}(x, \varepsilon(x))$ form an open covering of K. Since K is compact, it is contained in the union $\bigcup_{i=1}^{n} \mathring{B}(x_i, \varepsilon(x_i))$ of a finite family of such balls. Consequently, $V = \bigcup_{i=1}^{n} B(x_i, \varepsilon(x_i))$ is a compact neighborhood contained in V.　∎

PROPOSITION 2.　Let E be a locally compact space. Every open subspace and every closed subspace of E is locally compact.

Proof.　Suppose that A is an open subset of E. For every $x \in A$, there exists a compact ball $B(x, \varepsilon_1(x))$ in E and a ball $B(x, \varepsilon_2(x))$ contained in the open set A. Then the ball $B(x, \varepsilon(x))$ where $\varepsilon(x) = \min(\varepsilon_1(x), \varepsilon_2(x))$ is a compact ball in E contained in A, hence a neighborhood of x that is compact in A.

Similarly, if A is closed in E, every point x of A is the center of a compact ball $B(x, \varepsilon(x))$ in E; consequently, $A \cap B(x, \varepsilon(x))$ is compact in A and is a neighborhood of x in A.　∎

We now show that the only locally compact normed vector spaces are the finite dimensional spaces.

THEOREM 1. *F. Riesz.*　A normed vector space E is locally compact if and only if E is of finite dimension.

Proof. If E is of finite dimension, it is isomorphic to a space $\mathbb{R}^n$, which is locally compact since the neighborhoods $\prod_{i=1}^{n} [x_i - \varepsilon, x_i + \varepsilon]$ are compact.

Conversely, suppose that the normed vector space E is locally compact. Then there exists a *compact* ball $B(0, \varepsilon)$ with center 0 and radius ε. We can, therefore, cover this ball by a finite number of balls $\mathring{B}(x_i, \varepsilon/2)$ of radius $\varepsilon/2 (1 \leq i \leq n)$.

Now consider the finite dimensional subspace F generated by the points x_i. We are going to show that $E = F$ and, consequently, that E is of finite dimension.

Indeed, suppose that F is not equal to E. Let $x_0 \in E$ be an element that does not belong to F. Since F is finite dimensional, it is complete and, therefore, closed. Thus

$$(1) \qquad d(x_0, F) = \inf_{y \in F} \|x_0 - y\| = \alpha > 0,$$

and there exists $y_0 \in F$ such that

$$(2) \qquad \alpha = d(x_0, F) \leq \|x_0 - y_0\| \leq d(x_0, F) + \frac{\alpha}{2} = \frac{3\alpha}{2}.$$

Consider the point $z_0 = \varepsilon(x_0 - y_0)/\|x_0 - y_0\|$. This point belongs to the ball $B(0, \varepsilon)$ since $\|z_0\| = \varepsilon$ and, therefore, to one of the balls $B(x_i, \varepsilon/2)$. Hence

$$(3) \qquad \|z_0 - x_i\| = \left\| \varepsilon \frac{x_0 - y_0}{\|x_0 - y_0\|} - x_i \right\| \leq \frac{\varepsilon}{2}.$$

Consequently, the point

$$(4) \qquad y = x_0 - \frac{\|x_0 - y_0\|}{\varepsilon}(z_0 - x_i) = y_0 + \frac{\|x_0 - y_0\|}{\varepsilon} x_i$$

belongs to F (since y_0 and x_i belongs to F). We thus obtain that

$$(5) \qquad \alpha \leq \|x_0 - y\| = \|x_0 - y_0\| \frac{\|z_0 - x_i\|}{\varepsilon} \leq \frac{1}{2} \frac{3\alpha}{2} = \frac{3\alpha}{4},$$

which is impossible. ∎

We know that the countable set $\mathbb{Q}$ is dense in $\mathbb{R}$ and, therefore, that the countable set $\mathbb{Q}^n$ is dense in the locally compact space $\mathbb{R}^n$. More generally, the following proposition characterizes locally compact spaces which contain a countable dense set.

THEOREM 2.   Let E be a locally compact metric space. The following properties are equivalent.

(6) $\left\{\begin{array}{l}\textbf{i.}\quad\text{There exists a countable set } D = \{a_1, \ldots, a_n, \ldots\} \text{ dense in } E.\\[2mm]\textbf{ii.}\quad E \text{ is a countable union of compact sets } K_n.\\[2mm]\textbf{iii.}\quad\text{There exists a sequence of relatively compact open sets } U_n\\ \text{such that } \overline{U}_n \subset U_{n+1} \text{ and } E = \bigcup_n U_n.\end{array}\right.$

Proof.

a.   We shall show that (6)i implies (6)iii. Before constructing the sequence U_n, let us show that E can be covered by a countable family of compact balls B_n. Indeed, every point x is the center of a *compact* ball $B(x, \varepsilon(x))$. Moreover, if $m > 2/\varepsilon(x)$, there exists $a_k \in D$ such that $d(x, a_k) \leq 1/m$ since D is dense in E. Consequently, $x \in B(a_k, 1/m)$ and $B(a_k, 1/m)$ is a compact ball: Indeed, it is contained in $B(x, \varepsilon(x))$ since if $y \in B(a_k, 1/m)$, $d(y, x) \leq d(y, a_k) + d(a_k, x) \leq 2/m \leq \varepsilon(x)$. These compact balls $B(a_k, 1/m)$ form a countable covering $\{B_n\}_{n \in \mathbb{N}}$ of E. We are going to construct the open sets U_k from the B_n's by recursion. For $n = 1$, we take $U_1 = B_1$. Suppose that the U_k's are constructed for $k \leq n$. Since $\overline{U}_n$ is compact, it can be covered by a finite number of compact balls $B(x_i, \varepsilon(x_i))$. We set

$$U_{n+1} = \mathring{B}_{n+1} \cup \bigcup_{\text{finite}} \mathring{B}(x_i, \varepsilon(x_i)).$$

Consequently, $\overline{U}_n \subset U_{n+1}$ and $\overline{U}_{n+1} = B_{n+1} \cup \bigcup_{\text{finite}} B(x_i, \varepsilon(x_i))$ is compact, since it is a finite union of compact sets. Since $\mathring{B}_n \subset U_n$ and $E = \bigcup_n \mathring{B}_n$, we conclude that $E = \bigcup_n U_n$.

b.   It is clear that (6)iii implies (6)ii.

c.   Let us show that (6)ii implies (6)i. Let $E = \bigcup_n K_n$ where the K_n's are compact. According to Proposition 2.6.1, there exists a countable set D_n dense in K_n. Let $D = \bigcup_n D_n$, which is a countable set since it is the countable union of countable sets. Since $K_n = \overline{D}_n$, we obtain that $E = \bigcup_n \overline{D}_n \subset \overline{D}$, that is, that D is dense in E.   ∎

Remark 1. *Compactification of $\mathbb{R}^n$ by a point at infinity.*   We shall show that there exists a bicontinuous bijection from $\mathbb{R}^n$ onto the complement S_0 of a point x_0 of a compact set S. To this end let us consider the sphere $S = S(a, 1/2)$ of $\mathbb{R}^{n+1} = \mathbb{R}^n \times \mathbb{R}$ with radius $1/2$ and center $a = (0, 1/2)$. S is the *compact* set of pairs $(y, t) \in \mathbb{R}^{n+1}$ such that $\|y\|^2 + t^2 - t = 0$, and the only point of S of the form $(y, 0)$ is the point $x_0 = (0, 0)$. It is then clear that the mapping φ from $\mathbb{R}^n$ to $\mathbb{R}^{n+1} = \mathbb{R}^n \times \mathbb{R}$, defined by

$$\varphi(x) = \left(\frac{x}{\|x\|^2 + 1}, \frac{1}{\|x\|^2 + 1}\right),$$

is a bicontinuous bijection of $\mathbb{R}^n$ onto $S_0 = \complement_s x_0$, since its inverse ψ is defined by

$$\psi(y, t) = \frac{y}{t} \quad \text{if} \quad (y, t) \in S, \quad t \neq 0. \qquad \blacksquare$$

Remark 2. More generally, it is possible to prove that every locally compact space E is homeomorphic to the complement $\complement_s\{\omega\}$ of a point ω of a compact space S (called the *Alexandrov compactification*). $\qquad \blacksquare$

2. Separable Metric Spaces—The Stone–Weierstrass Theorem

We shall study the properties of those metric spaces in which every point is the limit of elements of a countable subset.

Definition 1. We say that a metric space E is "separable" if there exists a countable set D dense in E. If E is a normed vector space, we say that a subset B of linearly independent vectors of E is a "topological base" if the vector subset F generated by B is dense in E.

We have already shown that every compact set of a metric space is separable (see Proposition 2.6.1) and have characterized locally compact separable spaces (see Theorem 1.2). $\qquad \blacksquare$

Let us recall that an algebraic base for E consists of linearly independent vectors that generate E. A topological base consists of linearly independent vectors that generate a *dense* subspace of E.

PROPOSITION 1. Let E be a normed space. E is separable if and only if it has a countable topological base.

Proof. If E is separable, every countable set D dense in E generates a vector space dense in E and, consequently, contains a countable base. *Conversely*, let B be a countable base of E and F the vector subspace generated by B, which is dense in E.

Consider the set D of (finite) linear combinations $\sum \lambda_n b_n$ (where $\lambda_n \in \mathbb{Q}$ and $b_n \in B$) with rational coefficients. Since $\mathbb{Q}$ and B are countable, the set D is countable. Moreover, D is dense in E: Indeed, for every $x \in E$ and for every ε, there exists $y = \sum_{n=1}^{k} \mu_n b_n \in F$ such that $\mu_n \in \mathbb{R}$, $b_n \in B$, and $\|x - y\| \leq \varepsilon/2$. Since $\mathbb{Q}$ is dense in $\mathbb{R}$, there exist λ_n's $\in \mathbb{Q}$ such that $|\lambda_n - \mu_n| \leq \varepsilon/2 \sum_{n=1}^{k} \|b_n\|$. Then $z = \sum_{n=1}^{k} \lambda_n b_n \in D$ is such that $\|y - z\| \leq \varepsilon/2$. Consequently, $\|x - z\| \leq \|x - y\| + \|y - z\| \leq \varepsilon$. $\qquad \blacksquare$

PROPOSITION 2.   Let E be a metric space. E is separable if and only if there exists a countable family of open sets U_n such that for every open set U of E and for every element x of U there exists U_n such that $x \in U_n \subset U$.

Proof.

a.   Suppose that E is separable and let $D = \{a_1, \ldots, a_n, \ldots\}$ be a countable dense set in E. The open balls $\mathring{B}(a_n, 1/m)$ form an open family satisfying the necessary condition of the proposition. Indeed, let U be an open set of E and $x \in U$. Let δ be the distance from x to the closed set $\complement U$, and take $m > \delta/2$. Since D is dense in E, there exists $a_n \in D$ such that $x \in \mathring{B}(a_n, 1/m)$. Furthermore, $\mathring{B}(a_n, 1/m) \subset U$. If this were not the case, there would exist $y \in \complement U$ such that $d(y, a_n) < 1/m$. Therefore, we would have $d(x, y) \leq d(x, a_n) + d(y, a_n) \leq 1/2m < \delta = \inf_{y \in \complement U} d(x, y)$, which is a contradiction.
b.   Let us show that the condition is sufficient. Take $a_n \in U_n$ and consider the countable set $D = \{a_1, \ldots, a_n, \ldots\}$. Then D is dense in E since for every $x \in E$ and every open ball $\mathring{B}(x, \varepsilon)$ there exists U_n such that $x \in U_n \subset \mathring{B}(x, \varepsilon)$. Thus $x_n \in \mathring{B}(x, \varepsilon)$.   ∎

We shall show that the Banach space $\mathscr{C}_\infty(E)$ of continuous functions on a compact set E is separable.

THE STONE–WEIERSTRASS APPROXIMATION THEOREM.   The Stone–Weierstrass approximation theorem provides sufficient conditions for a subset $\mathscr{A}$ of the space $\mathscr{C}_\infty(E)$ of continuous functions on a compact set E to be dense in $\mathscr{C}_\infty(E)$.

THEOREM 1.   Let E be a compact metric space and $\mathscr{C}_\infty(E)$ the Banach space of continuous functions on E. Let $\mathscr{A} \subset \mathscr{C}_\infty(E)$ be a vector subset of functions satisfying

(1) $\begin{cases} \textbf{i.} & \text{If } f \text{ and } g \in \mathscr{A}, \text{ then the product } fg \in \mathscr{A}. \\ \textbf{ii.} & \forall x \neq y \text{ there exists } f \in \mathscr{A} \text{ such that } f(x) \neq f(y). \\ \textbf{iii.} & \mathscr{A} \text{ contains the constant functions.} \end{cases}$

Then $\mathscr{A}$ is dense in $\mathscr{C}_\infty(E)$.

Proof.   We shall denote by $\bar{\mathscr{A}}$ the closure of $\mathscr{A}$ in $\mathscr{C}_\infty(E)$. Since $\|fg\|_\infty \leq \|f\|_\infty \|g\|_\infty$, multiplication of functions is continuous. Hence if f and $g \in \bar{\mathscr{A}}$, then $fg \in \bar{\mathscr{A}}$ according to (1)i. Consequently, the closure $\bar{\mathscr{A}}$ of $\mathscr{A}$ also satisfies the hypotheses (1). We can, therefore, suppose henceforth that $\mathscr{A}$ is closed and show that $\mathscr{A} = \mathscr{C}_\infty(E)$. This results from the three following lemmas.

LEMMA 1.   The hypotheses (1) imply that

(2) $\begin{cases} \text{If} \quad x \neq y \quad \text{and if} \quad \alpha \quad \text{and} \quad \beta \in \mathbb{R}, \quad \text{there exists} \\ f \in \mathscr{A} \quad \text{such that} \quad f(x) = \alpha \quad \text{and} \quad f(y) = \beta. \end{cases}$

Proof. According to (1)ii there exists $g \in \mathscr{A}$ such that $g(x) \neq g(y)$. Therefore, $f(z) = \alpha + (\beta - \alpha)(g(z) - g(x))/(g(y) - g(x))$ belongs to $\mathscr{A}$ since $\mathscr{A}$ contains the constants and since the product of two functions of $\mathscr{A}$ belongs to $\mathscr{A}$. Also, f satisfies property (2).   ∎

LEMMA 2.   Suppose that E is compact, that condition (2) is satisfied, and that the following condition also holds:

(3) If f and $g \in \mathscr{A}$, then $\inf(f, g)$ and $\sup(f, g) \in \mathscr{A}$.

Then $\mathscr{A}$ is dense in $\mathscr{C}_\infty(E)$.

Proof. Take $f \in \mathscr{C}_\infty(E)$ and $\varepsilon > 0$. We must show that there exists $g \in \mathscr{A}$ such that

(4) $\qquad\qquad \forall y \in E, \quad f(y) - \varepsilon \leq g(y) \leq f(y) + \varepsilon.$

Let us show, first of all, that we can associate to every $x \in E$ a function g_x such that

(5) $\quad g_x(x) = f(x) \quad$ and $\quad g_x(y) \leq f(y) + \varepsilon \quad$ for all $\quad y \in E.$

Indeed, according to (2), we can associate to every $z \in E$ a function h_z such that

(6) $\qquad\qquad h_z(x) = f(x) \quad$ and $\quad h_z(z) = f(x) + \dfrac{\varepsilon}{2}.$

Since f and h_z are continuous, there exists $\eta(z)$ such that

(7) $\qquad\qquad h_z(y) \leq f(y) + \varepsilon \quad$ when $\quad y \in B(z, \eta(z)).$

But since E is compact, it can be covered by n balls $\overset{\circ}{B}(z_i, \eta(z_i))$. Then set

(8) $\qquad\qquad g_x(y) = \inf_{i = 1, \ldots, n} h_{z_i}(y).$

Since the functions h_{z_i} belong to $\mathscr{A}$, we deduce from hypothesis (3) that $g_x \in \mathscr{A}$. Consequently, (5) results from (6), (7), and (8).

It remains to establish (4). Following (5) we can associate to every x a number $\eta(x)$ such that

(9) $\qquad\qquad f(y) - \varepsilon \leq g_x(y) \quad$ when $\quad y \in B(x, \eta(x))$

since f and g_x are continuous and $f(x) = g_x(x)$. Covering E by q balls $B(x_j, \eta(x_j))$ and setting

(10)
$$g(y) = \sup_{1 \le j \le q} g_{x_j}(y),$$

we deduce from (9) that $f(y) - \varepsilon \le g(y)$ for all $y \in E$. Moreover, (5) implies that $g(y) \le f(y) + \varepsilon$ for all $y \in E$. ∎

It now remains to verify that condition (3) is satisfied under the hypotheses of the theorem. Since

$$\inf(f, g) = (1/2)(f + g - |f - g|.)$$

and

$$\sup(f, g) = (1/2)(f + g + |f - g|),$$

it suffices to prove the following lemma.

LEMMA 3.   Suppose that $\mathscr{A}$ is closed and satisfies (1). Then

(11)   If $f \in \mathscr{A}, |f|$ belongs to $\mathscr{A}$.

Proof. We show that

(12) $\begin{cases} \text{there exists a sequence of polynomials } u_n \text{ converging uniformly} \\ \text{to } t^{1/2} \text{ on the interval } [0, 1]. \end{cases}$

Indeed, if (12) holds we set

$$t(x) = \frac{f(x)^2}{\|f\|^2} \in [0, 1] \quad \text{and} \quad f_n(x) = \|f\| u_n\left(\frac{f(x)^2}{\|f\|^2}\right).$$

Since u_n is a polynomial, we deduce from (1)i and (1)iii that $f_n \in \mathscr{A}$ when $f \in \mathscr{A}$. Then (12) implies that the functions f_n converge uniformly to $(t(x))^{1/2} \|f\| = |f(x)|$.

To establish (12), we can use Bernstein's theorem 4 below or we construct the polynomials u_n by recursion, setting $u_1 = 0$ and

(13) $u_{n+1}(t) = u_n(t) + (1/2)(t - u_n^2(t))$ for $n \ge 1$.

Let us show by recursion that

(14) $u_n(t) \le t^{1/2}$.

Indeed,

(15) $\begin{cases} u_{n+1}(t) - t^{1/2} = u_n(t) - t^{1/2} + (1/2)(t^{1/2} - u_n(t))(t^{1/2} + u_n(t)) \\ \qquad = (u_n(t) - t^{1/2})(1 - (1/2)(t^{1/2} + u_n)). \end{cases}$

But (14) implies that $1 - (1/2)(t^{1/2} + u_n) \geq 1 - (1/2)(2(t^{1/2})) = 1 - t^{1/2} \geq 0$, which together with (14) and (15) implies that $u_{n+1}(t) - t^{1/2} \leq 0$. Moreover, (14) also implies that $u_{n+1}(t) - u_n(t) = (1/2)(t - u_n^2(t)) \geq 0$, that is, that the sequence $u_n(t)$ converges to a number $v(t)$ since it is increasing and bounded above by $t^{1/2}$. According to (13), $v(t)$ satisfies $v(t) = v(t) + (1/2)(t - v^2(t))$, that is $v(t) = t^{1/2}$. Consequently, Dini's theorem (Corollary 4.6.1) implies that u_n converges uniformly to $v(t) = t^{1/2}$.   ∎

APPLICATIONS

THEOREM 2. Let E be a compact subset of $\mathbb{R}^n$. Then the set of polynomial functions on E is dense in $\mathscr{C}_\infty(E)$; consequently, $\mathscr{C}_\infty(E)$ is separable.

Proof. Since the product of two polynomials is a polynomial and since the set of polynomials contains the constants, it remains to show that hypothesis (1)ii is satisfied. If two points x and y of $\mathbb{R}^n$ are different, then $x_i \neq y_i$ for at least one i, which implies that the polynomial $f(z) = z_i$ is such that $f(x) \neq f(y)$. Hence Theorem 2 follows from Theorem 1.   ∎

THEOREM 3. Let E be a compact subset of $\mathbb{R}^n$ and $L^p(E)$ the completion of $\mathscr{C}(E)$ with norm

$$\|f\|_p = \left(\int_E |f(x)|^p \, dx \right)^{1/p} \qquad (1 \leq p < +\infty).$$

Then the set $\mathscr{A}$ of polynomial functions is dense in $L^p(E)$; therefore, $L^p(E)$ is separable.

Proof. Let us recall that $\|f\|_p$ is a norm on $\mathscr{C}(E)$ (see Section 1.12) and that $\forall f \in \mathscr{C}(E)$, $\|f\|_p \leq M\|f\|_\infty$ where $M = (\int_E dx)^{1/p}$. Since $\mathscr{C}(E)$ is dense in $L^p(E)$ (by definition of the completion), and $\mathscr{A}$ is dense in $\mathscr{C}_\infty(E)$, then $\mathscr{A}$ is dense in $L^p(E)$ by Proposition 4.2.4.   ∎

*BERNSTEIN POLYNOMIALS. The Stone–Weierstrass theorem is not constructive in the sense that it does not indicate how to approach a continuous function by a sequence of polynomials. We shall indicate an explicit procedure of approximation in the case where $E = [0, 1]$.

THEOREM 4. *Bernstein.* If f is a continuous function on $[0, 1]$, the polynomials p_n defined by

$$(16) \qquad p_n(x) = \sum_{p=0}^{n} \binom{n}{p} f\left(\frac{p}{n}\right) x^p (1 - x)^{n-p}$$

converge uniformly to f.

Proof. We denote by μ_p the polynomial defined by

$$(17) \qquad \mu_p(x) = \binom{n}{p} x^p (1 - x)^{n-p}.$$

We show that they satisfy the following:

$$(18) \quad \begin{cases} \textbf{i.} & \displaystyle\sum_{p=0}^{n} \mu_p(x) = 1 \\[2ex] \textbf{ii.} & \displaystyle\sum_{p=0}^{n} p\mu_p(x) = nx \\[2ex] \textbf{iii.} & \displaystyle\sum_{p=0}^{n} p(p-1)\mu_p(x) = n(n-1)x^2. \end{cases}$$

Indeed, we consider the formula of Newton

$$(19) \qquad (x + y)^n = \sum_{p=0}^{n} \binom{n}{p} x^p y^{n-p}.$$

Taking $y = 1 - x$, we obtain (18)i. We differentiate (19) with respect to x, and we multiply by x, obtaining

$$nx(x + y)^{n-1} = \sum_{p=0}^{n} p\binom{n}{p} x^p y^{n-p}.$$

Then setting $y = 1 - x$, we obtain (18)ii. Differentiating (19) twice with respect to x and multiplying by x^2 yields

$$n(n-1)x^2(x + y)^{n-2} = \sum_{p=0}^{n} p(p-1)\binom{n}{p} x^p y^{n-p},$$

which, setting $y = 1 - x$, implies (18)iii. Formulas (18) give the formula

$$(20) \qquad \sum_{p=0}^{n} (p - nx)^2 \mu_p(x) = nx(1 - x)$$

$$\left(\text{since } \sum_{p=0}^{n} (p - nx)^2 \mu_p(x) = \sum_{p=0}^{n} n^2 x^2 \mu_p(x) - 2nx \sum_{p=0}^{n} p\mu_p(x) + \sum_{p=0}^{n} p^2 \mu_p(x) \right.$$

$$= n^2 x^2 - 2nx(nx) + (nx + n(n-1)x^2)$$

$$\left. = nx(1 - x) \right).$$

Therefore, the continuous function f on the compact set $[0, 1]$ is bounded by M and uniformly continuous: For every $\varepsilon > 0$, there exists η such that $|f(x) - f(y)| \leq \varepsilon/2$ when $|x - y| \leq \eta$. We can write, according to (18)i, that $f(x) = \sum_{p=0}^{n} f(x)\mu_p(x)$. Thus for every fixed x

$$\left| f(x) - \sum_{p=0}^{n} f\left(\frac{p}{n}\right)\mu_p(x) \right| \leq \sum_{p=0}^{n} \left| f(x) - f\left(\frac{p}{n}\right) \right| |\mu_p(x)|$$

$$= \sum_{|p/n-x| \leq \eta} \left| f(x) - f\left(\frac{p}{n}\right) \right| |\mu_p(x)|$$

$$+ \sum_{|p/n-x| \geq \eta} \left| f(x) - f\left(\frac{p}{n}\right) \right| |\mu_p(x)|.$$

Since $\mu_p(x) \geq 0$ and $\sum \mu_p(x) = 1$, the first term of the right-hand side of this inequality is dominated by $(\varepsilon/2) \sum_{p=0}^{n} \mu_p(x) = \varepsilon/2$. Since f is bounded by M, the second term is dominated by $2M \sum_{|p/n-x| \geq \eta} \mu_p(x)$. But if $|p/n - x| \geq \eta$, $1 \leq (p - nx)^2/n^2\eta^2$ so that

$$2M \sum_{|p/n-x| \geq \eta} \mu_p(x) \leq \frac{2M}{n^2\eta^2} \sum_{p=0}^{n} (p - nx)^2 \mu_p(x)$$

$$= \frac{2M}{n\eta^2} x(1 - x) \leq \frac{2M}{n\eta^2}$$

(according to (20)). This term is less than $\varepsilon/2$ if $n \geq N(\varepsilon)$ (independent of x). Consequently, $\forall \varepsilon > 0$,

$$(21) \qquad \left| f(x) - \sum_{p=0}^{n} f\left(\frac{p}{n}\right)\mu_p(x) \right| \leq \frac{2\varepsilon}{2} = \varepsilon \qquad \text{when} \qquad n \geq N(\varepsilon). \qquad \blacksquare$$

3.   Baire's Theorem

We prove in this section Baire's theorem, which is one of the *fundamental results* in analysis. It is, indeed, the origin of the theorems of Banach and of Banach–Steinhauss. (See Chapter 4, Sections 1 and 3 of [AFA].) We establish some applications dealing with the continuity of convex functions.

   THEOREM 1. *Baire.*   Let E be a complete metric space. Then the following equivalent properties hold:

(1)
i.   For every countable family of open sets U_n dense in E, $\bigcap_n U_n$ is dense in E.

ii.   For every countable family of closed sets F_n with empty interior, $\bigcup_n F_n$ has an empty interior.

Proof.

a. It is clear that properties (1)i and (1)ii are equivalent by taking complements. We establish (1)i.

b. Consider, therefore, a countable family of open sets U_n such that $\overline{U}_n = E$, and let us show that $U_\infty = \bigcap_n U_n$ is dense in E, that is that for every $x \in E$ and for every $\varepsilon > 0$:

$$(2) \qquad B(x, \varepsilon) \cap U_\infty \neq \varnothing.$$

To this end, we show that the limit x_* of a sequence x_n, which we construct by recursion belongs to $B(x, \varepsilon) \cap U_\infty$.

Since U_1 is dense in E, the intersection $\mathring{B}(x, \varepsilon) \cap U_1$ is not empty. Since U_1 is open, there exists a ball $B(x_1, \varepsilon_1)$ such that

$$(3) \qquad B(x_1, \varepsilon_1) \subset \mathring{B}(x, \varepsilon) \cap U_1, \qquad \varepsilon_1 < 1/2.$$

Suppose that we have constructed the $(n - 1)$ first balls $B(x_k, \varepsilon_k)$ such that

$$(4) \qquad B(x_k, \varepsilon_k) \subset \mathring{B}(x_{k-1}, \varepsilon_{k-1}) \cap U_k, \qquad \varepsilon_k < \frac{1}{2^k}.$$

We construct the ball $B(x_n, \varepsilon_n)$ in the following fashion. Since U_n is dense in E, the open set $\mathring{B}(x_{n-1}, \varepsilon_{n-1}) \cap U_n$ is nonempty and contains a ball $B(x_n, \varepsilon_n)$ satisfying

$$(5) \qquad B(x_n, \varepsilon_n) \subset \mathring{B}(x_{n-1}, \varepsilon_{n-1}) \cap U_n, \qquad \varepsilon_n < \frac{1}{2^n}.$$

The sequence of centers x_n of these balls is a Cauchy sequence since if $m \geq n$, then $B(x_m, \varepsilon_m) \subset B(x_n, \varepsilon_n)$; consequently, $d(x_n, x_m) \leq \varepsilon_n \leq 1/2^n$. Therefore, this sequence converges to an element x_* of E *since E is complete.* Moreover, if $m \geq n$, $d(x_n, x_*) \leq d(x_n, x_m) + d(x_m, x_*) \leq \varepsilon_n + d(x_m, x_*)$. Letting m approach infinity, we conclude that $d(x_n, x_*) \leq \varepsilon_n$, that is that $x_* \in B(x_n, \varepsilon_n)$. Consequently, since $B(x_n, \varepsilon_n) \subset \mathring{B}(x, \varepsilon) \cap U_n$, we obtain

$$(6) \qquad x_* \in \bigcap_n B(x_n, \varepsilon_n) \subset \mathring{B}(x, \varepsilon) \cap \bigcap_n U_n = \mathring{B}(x, \varepsilon) \cap U_\infty,$$

which completes the proof of the theorem. ∎

Remark 1. A *Baire space* is any space satisfying one of the two equivalent properties (1)i or (1)ii (called the *Baire properties*). ∎

Remark 2. We can show that if E is locally compact, the Baire properties are satisfied. It suffices to choose a ball $B(x, \varepsilon)$ that is compact. Then the balls $B(x_n, \varepsilon_n)$ defined by (5) are compact, since they are contained in the compact set $B(x, \varepsilon)$ and satisfy the finite intersection property. Hence there exists a point x_* belonging to the intersection of the balls $B(x_n, \varepsilon_n)$ and, consequently, to $\mathring{B}(x, \varepsilon) \cap U_\infty$.   ∎

The following version of Baire's theorem is often used.

THEOREM 2. Let E be a complete metric space. Suppose that $E = \bigcup_{n=1}^\infty F_n$ is a countable union of closed sets F_n. Then at least one of these closed sets F_n has a nonempty interior.

Proof. (*Left as an exercise.*)

We now establish the following property for lower semicontinuous functions using Baire's theorem.

THEOREM 3. Let E be a complete metric space and f a lower semicontinuous function defined on a nonempty open set K with values in $\mathbb{R}$. Then there exists a nonempty open set U contained in K on which f is bounded.

Proof. Consider the sets $F_n = \{x \in K$ such that $f(x) \le n\}$. These sets are closed since f is lower semicontinuous. Moreover, $K = \bigcup_n F_n$. Since K is open, we deduce from (1)ii that one of the F_n's has a nonempty interior U on which the function f is bounded by n.   ∎

COROLLARY 1. Let E be a complete metric space and $\{f_i\}_{i \in I}$ a family of lower semicontinuous functions on E. Suppose that $f(x) = \sup_{i \in I} f_i(x)$ is finite on E. Then every nonempty open set K of E contains a nonempty open set U on which the family $\{f_i\}_{i \in I}$ is uniformly bounded.

Proof. It suffices to apply Theorem 3.3 to the pointwise supremum $f(x) = \sup_{i \in I} f_i(x)$, which is lower semicontinuous.   ∎

Theorem 3 also implies the important sufficient condition below on the continuity of convex functions.

THEOREM 4. Let K be a nonempty open convex subset of a Banach space E and f a convex lower semicontinuous function on K. Then f is continuous on K. In particular, every lower semicontinuous seminorm is continuous.

Proof. Theorem 3 states that f is bounded on a nonempty open set U of K and Theorem 3.8.2 implies that f is continuous on the open set K. In the case of seminorms, we take $K = E$, which is open.   ∎

Remark 3.   These theorems remain true when E is a Fréchet space.   ∎

Remark 4.   The second assertion of Theorem 4 will imply the Banach–Steinhauss theorem (see Remark 4.6.4), which characterizes equicontinuous sets of linear operators. (See Chapter 4, Section 1 of [AFA].) Its importance justifies the following terminology. We call a "*barreled vector space*" any topological vector space such that every lower semicontinuous seminorm is continuous.   ∎

4.   Connected Spaces—Construction of a Continuous Utility Function

It is useful to distinguish the sets E satisfying one of the following equivalent properties.

PROPOSITION 1.   Let E be a metric space. The following assertions are equivalent

(1) $\begin{cases} \textbf{i.} & \text{The only sets both open and closed in } E \text{ are } E \text{ and } \varnothing, \\ \textbf{ii.} & \text{If } E = A \cup B, \text{ where } A \text{ and } B \text{ are nonempty and open (or} \\ & \text{closed), then } A \cap B \neq \varnothing. \end{cases}$

Proof.   (*Left as an exercise.*)

Definition 1.   We say that a metric space E is "connected" if one of two equivalent properties (1) is satisfied. We say that a subset $A \subset E$ is connected if the metric subspace A is connected.

We indicate the fundamental property of connected sets.

THEOREM 1.   Let f be a continuous mapping from a metric space E to a metric space F. If E is connected, $f(E)$ is connected.

Proof.   Suppose that $f(E)$ is not connected: There exist M and N such that

(2) $\begin{cases} \textbf{i.} & f(E) \subset M \cup N, \quad f(E) \cap M \cap N = \varnothing. \\ \textbf{ii.} & M \text{ and } N \text{ are open in } F, \quad f(E) \cap M \neq \varnothing, \\ & \text{and} \quad f(E) \cap N \neq \varnothing. \end{cases}$

Since f is continuous, $f^{-1}(M)$ and $f^{-1}(N)$ are nonempty open sets of E such that

(3) $E = f^{-1}(M) \cup f^{-1}(N) \quad \text{and} \quad f^{-1}(M) \cap f^{-1}(N) = \varnothing.$

This contradicts the hypothesis that E is connected.   ∎

Before giving examples of connected sets, let us establish the three following propositions that enable us to show that a set is connected.

PROPOSITION 2.   If A is a connected set of a metric space E, then every subset B such that

(4) $$A \subset B \subset \bar{A}$$

is connected.

Proof.   Suppose that B is not connected: There exist two nonempty sets M and N, open in B, such that $B = M \cup N$ and $M \cap N = \varnothing$. Since A is dense in B and since M and N are open in B, then $M_0 = M \cap A$ and $N_0 = N \cap A$ are nonempty and are open in the metric subspace A. Moreover, $A = A \cap B = M_0 \cup N_0$ and $M_0 \cap N_0 = \varnothing$. This contradicts the hypothesis that A is connected.   ∎

PROPOSITION 3.   Let E be a metric space, $\{A_i\}_{i \in I}$ a family of connected sets such that $\bigcap_{i \in I} A_i \neq \varnothing$. Then

(5) $$A = \bigcup_{i \in I} A_i \text{ is a connected set.}$$

Proof.   Let $a \in \bigcap_{i \in I} A_i$. Suppose that A is not connected: There exist then nonempty open sets M and N of A such that

(6) $$A = M \cup N \quad \text{and} \quad M \cap N = \varnothing,$$

Suppose that the point a of the intersection of the A_i's belongs to M, for example. Then $M \cap A_i \neq \varnothing$ for every $i \in I$. Moreover, there exists an index i_0 such that $N \cap A_{i_0} \neq \varnothing$. Consequently, the sets $N_0 = N \cap A_{i_0}$ and $M_0 = M \cap A_{i_0}$ are nonempty open sets in A_{i_0} such that

(7) $$A_{i_0} = M_0 \cup N_0 \quad \text{and} \quad M_0 \cap N_0 = \varnothing.$$

This contradicts the hypothesis that A_{i_0} is connected.   ∎

PROPOSITION 4.   Let E be a metric space and $\{A_i\}_{1 \leq i \leq n}$ be a finite sequence of connected sets such that $A_i \cap A_{i+1} \neq \varnothing$ for $1 \leq i \leq n - 1$. Then $\bigcup_{1 \leq i \leq n} A_i$ is connected.

Proof.   The proposition results from the preceding one by recursion on n.   ∎

Let us remark that a discrete set E having at least two elements is not connected, for the set $\{x\}$ is both open and closed and is not equal to E nor to $\varnothing$.   ∎

We show that every interval of the real line $\mathbb{R}$ is connected.

THEOREM 2. The space $\mathbb{R}$ is connected. A set $A \subset \mathbb{R}$ is connected if and only if A is an interval.

Proof. Since $A = \mathbb{R}$ is an interval, the first assertion of the theorem results from the second.

a. Suppose that $A \subset \mathbb{R}$ is a nonempty interval with origin a and extremity b in the extended real line $\bar{\mathbb{R}}$. Suppose that A is not connected: There exist two nonempty sets M and N open in A such that $A = M \cup N$ and $M \cap N = \varnothing$. Suppose, for example, that $x \in M$, $y \in N$, $x < y$. Let z be the least upper bound of $M \cap [x, y]$ in $\bar{\mathbb{R}}$. We show that z does not belong to M nor to N, which is impossible since $z \in [x, y] \subset A$ (since A is an interval) and $A = M \cup N$.

To this end, let us begin by supposing that z belongs to M. Then $z < y$ (since $M \cap N = \varnothing$ and $y \in N$). Then there exists an interval $[z, z + \varepsilon]$ where $\varepsilon > 0$ contained in $[x, y]$ and in M (since M is open), which is impossible since z is the least upper bound of $M \cap [x, y]$. Suppose then that $z \in N$. Thus $x < z$, and there exists an interval $[z - \varepsilon, z]$ contained in $N \cap [x, y]$, which also contradicts the definition of z. Thus z belongs neither to M nor to N.

b. Now suppose that A is connected. If A reduces to a single point, it is clearly an interval. If not, A contains at least two points a and b such that $a < b$. Let us show that

$$(8) \qquad\qquad [a, b] \subset A.$$

If this is not the case, there exists $x \in \,]a, b[$, which does not belong to A. This implies that $A = M \cup N$ where $M = A \cap \,]-\infty, x[$ and $N = A \cap \,]x, +\infty[$. Moreover, M and N are nonempty (since $a \in M$ and $b \in N$), open, and disjoint. This is impossible for we have supposed that A is connected.

It remains to show that (8) implies that A is an interval. For this let us consider the greatest lower bound and least upper bound α and β of A in $\bar{\mathbb{R}}$. If b is an arbitrary element of A, we show that

$$(9) \qquad\qquad]\alpha, b] \subset A \qquad \text{and} \qquad [b, \beta[\,\subset A.$$

It suffices to show that $]\alpha, b] \subset A$. If $\alpha = -\infty$, then for every $x < b$, there exists $a \in A$ such that $a < x$. Thus $x \in [a, b]$ and, consequently, $x \in A$. Hence $]-\infty, b] \subset A$. Otherwise, if $\alpha > -\infty$, for every x such that $\alpha < x < b$, there exists $a \in A$ such that $\alpha < a < x$. Then $x \in [a, b] \subset A$ and, consequently, $]\alpha, b] \subset A$. Clearly (9) implies that A is an interval. ∎

More generally, convex sets are connected.

THEOREM 3. Let E be a normed vector space. Every convex subset K of E is connected.

Proof. Let x_0 be an arbitrary element of K. For every $y \in K$, the set $A(y) = \{\lambda x_0 + (1 - \lambda)y\}_{0 \leq \lambda \leq 1}$ is contained in K, since K is convex. Therefore, $K = \bigcup_{y \in K} A(y)$ and $\bigcap_{y \in K} A(y) \neq \varnothing$. Moreover, the sets $A(y)$ are connected for they are homeomorphic images of the interval $[0, 1]$, which is connected. Thus K is connected according to Proposition 3.   ∎

Theorems 1 and 2 imply the intermediate value theorem.

THEOREM 4. *Bolzano.* Let E be a connected space, f a continuous mapping from E to $\mathbb{R}$. Let a and b be two points of $f(E)$ such that $a < b$. Then for every $c \in]a, b[$, there exists $x \in E$ such that $f(x) = c$.

Proof. Since $f(E)$ is connected in $\mathbb{R}$, $f(E)$ is an interval; consequently, $]a, b[$ is contained in $f(E)$.   ∎

APPLICATION: CONSTRUCTION OF A CONTINUOUS UTILITY FUNCTION.   Consider

(10) $\begin{cases} \text{the set } \mathbb{R}^n_+ \text{ of vectors } x = (x_1, \ldots, x_n) \text{ of } \mathbb{R}^n \text{ whose components} \\ \text{are positive,} \end{cases}$

which can be interpreted as a commodity space, and a given commodity $\xi \in \mathbb{R}^n_+$ all of whose components are strictly positive (ξ can be viewed as a "numéraire").

It is usual for economists to represent the behavior of a consumer by a *utility function* u from $\mathbb{R}^n_+$ to $\mathbb{R}$, which induces on the set of commodities the preference preordering defined by

(11) x is preferred to y if and only if $u(x) \geq u(y)$.

Several utility functions can induce the same preference preordering (verify as an exercise that two utility functions u and v induce the same preordering if and only if there exists a strictly increasing function φ from $\mathbb{R}$ to $\mathbb{R}$ such that $u = \varphi \circ v$). Therefore, only the preference preordering is important to describe the behavior of a consumer in an intrinsic fashion, and not the choice of the utility function. The only way to resolve this situation is to prove that every preference preordering can be represented by a utility function, continuous if possible. The necessary condition for this to be the case is that the sets

$$U_+(m) = \{x \in \mathbb{R}^n_+ \quad \text{such that} \quad u(x) \geq m\}$$

and

$$U_-(m) = \{x \in \mathbb{R}^n_+ \quad \text{such that} \quad u(x) \leq m\}$$

be closed.

Debreu proved that this condition is sufficient (under very weak conditions). We also show the converse under slightly more restrictive conditions. To this end we describe the *preference preordering* of a consumer by

(12)   a total preordering $\leqslant$ on $\mathbb{R}^n_+$,

that is, a binary relation that is *reflexive* ($x \leqslant x, \forall x \in \mathbb{R}^n_+$), *transitive* (if $x \leqslant y$ and if $y \leqslant z$, then $x \leqslant z$), and *total* (if x and $y \in \mathbb{R}^n_+$, then $x \leqslant y$ or $y \leqslant x$). The relation $\simeq$ defined by

(13) $x \simeq y$ if and only if $x \leqslant y$ and $y \leqslant x$

is an equivalence relation called the *indifference relation*.

We denote by

(14) $\begin{cases} \textbf{i.} & U_+(x) = \{y \in \mathbb{R}^n_+ \text{ such that } y \geqslant x\} \\ \textbf{ii.} & U_-(x) = \{y \in \mathbb{R}^n_+ \text{ such that } y \leqslant x\} \end{cases}$

the upper and lower sections of the preordering.

THEOREM 5.   Suppose that the total preordering $\leqslant$ on $\mathbb{R}^n_+$ satisfies

(15) $\forall x \in \mathbb{R}^n_+$ the sets $U_+(x)$ and $U_-(x)$ are closed

and

(16) $\begin{cases} \textbf{i.} & \forall x \in \mathbb{R}^n_+, \quad 0 \leqslant x \text{ (every commodity is preferred to the zero commodity).} \\ \textbf{ii.} & \forall x \in \mathbb{R}^n_+ \text{ there exists } m_x \in \mathbb{R}_+ \text{ such that } x \leqslant m_x \xi \text{ (insatiability with respect to the numéraire).} \\ \textbf{iii.} & \text{If } 0 < m_1 < m_2, \quad \text{then } m_1 \xi \leqslant m_2 \xi \text{ and not } (m_2 \xi \leqslant m_1 \xi) \text{ (the preordering is strictly increasing for the numéraire).} \end{cases}$

Then there exists a continuous function u from $\mathbb{R}^n_+$ to $\mathbb{R}$ such that

(17) $x_1 \leqslant x_2$ if and only if $u(x_1) \leqslant u(x_2)$.

We say that the function u satisfying (17) is a *utility function representing the preference preordering*.

Proof.   Let $x \in \mathbb{R}^n_+$ and $m_x \in \mathbb{R}_+$ be defined by (16)ii. Set:

$$(18) \quad \begin{cases} \textbf{i.} & T_x = [0, m_x]. \\ \textbf{ii.} & T_x^- = \{t \in T_x \text{ such that } t\xi \preccurlyeq x\}. \\ \textbf{iii.} & T_x^+ = \{t \in T_x \text{ such that } t\xi \succcurlyeq x\}. \end{cases}$$

Since the preordering is total, $T_x = T_x^- \cup T_x^+$. Since $0 \preccurlyeq x$ according to (16)i, $0 \in T_x^-$, and since $x \preccurlyeq m_x \xi$ according to (16)ii, $m_x \in T_x^+$. Moreover, (15) implies that the sets T_x^+ and T_x^- are closed.

Since the interval T_x is connected, we deduce that $T_x^+ \cap T_x^- \neq \varnothing$. Therefore, if $m \in T_x^+ \cap T_x^-$, we obtain from (18) and (13) that $m\xi \simeq x$. Hypothesis (16)ii implies in particular that there exists at most one element $m \in \mathbb{R}_+$ such that $m\xi \simeq x$. Thus the intersection $T_x^+ \cap T_x^-$ contains exactly one element $u(x)$ such that

$$(19) \qquad\qquad u(x)\xi \simeq x.$$

This implies that the set of x's such that $u(x) = m$ is the set of elements x equivalent to $m\xi$.

Now let us show (17). According to (16)iii $u(x_1) < u(x_2)$ if and only if $u(x_1)\xi \preccurlyeq u(x_2)\xi$. Since the relation $\preccurlyeq$ is transitive, this is equivalent to saying that $x_1 \preccurlyeq x_2$.

In particular, this implies that for every $m \in \mathbb{R}$,

$$(20) \quad \begin{cases} \textbf{i.} & U_+(m) = \{x \text{ such that } u(x) \geq m\} \quad \text{if} \quad m \geq 0, \\ & U_+(m) = \mathbb{R}^n_+ \quad \text{if} \quad m < 0. \\ \textbf{ii.} & U_-(m) = \{x \text{ such that } u(x) \leq m\} \quad \text{if} \quad m \geq 0, \\ & U_-(m) = \varnothing \quad \text{if} \quad m < 0. \end{cases}$$

Since according to (15) these sets are closed, the function u is upper semicontinuous and lower semicontinuous and, therefore, continuous.   ∎

Remark 1.   We can also use the fact that $[a, b]$ is connected to prove the Brouwer fixed point theorem for continuous mappings f from $[a, b]$ to itself. To this end we show, first of all, that for every integer n there exists x such that $f(x) \in B(x, 1/n)$. [Indeed, if this is not the case, then for every $x \in [a, b]$ we have $f(x) \notin B(x, 1/n)$, that is either $f(x) < x - 1/n$ or $f(x) > x + 1/n$. We denote by A the set of x such that $f(x) > x + 1/n$ and by B the set of x such that $f(x) < x - 1/n$. Since f is continuous, these are open sets. The negation of the conclusion implies then that $[a, b] \subset A \cup B$, that $a \in A$ and $b \in B$ (for n sufficiently large). Since the interval $[a, b]$ is *connected*, the intersection of A and B is nonempty: There exists, therefore, x such that $f(x) < x - 1/n < x + 1/n < f(x)$, which is impossible.]

We denote by K_n the nonempty set of x such that $f(x) \in B(x, 1/n)$. The K_n's form a *decreasing family of closed sets*. Since $[a, b]$ is compact, these closed sets K_n have a nonempty intersection. But to say that $x \in \bigcap_n K_n$ clearly means that $x = f(x)$.   ∎

Remark 2.   The Brouwer theorem in $\mathbb{R}^n$ is proved with the aid of the lemma of "Knaster–Kuratowski–Mazurkiewicz": The intersection of $n + 1$ closed sets F_i of the compact convex set $S^n = \{x \in \mathbb{R}_+^{n+1}$ such that $\sum_{i=0}^n x_i = 1\}$ is nonempty when $x \in \bigcup_{\{i \text{ such that } x_i > 0\}} F_i$ for every $x \in S^n$. If $n = 1$, this lemma is none other than Theorem 2 stating that $[0, 1]$ is connected.   ∎

5.   Partitions of Unity

We show in this section that to every finite covering of a metric space we can associate a partition of unity that is subordinate to it.

Definition 1.   Let f be a real-valued function defined on a metric space E. We call the support of f, denoted supp(f), the smallest closed set S such that $f(x) = 0 \; \forall x \notin S$.

In other words, the support of f is the closure in E of the set of elements $x \in E$ such that $f(x) \neq 0$. Hence it is the set of elements $x \in E$ such that in *every* neighborhood of x there exists a point y such that $f(y) \neq 0$.

Definition 2.   Consider an open covering $\{A_i\}_{i=1,\dots,n}$ of E. A "continuous partition of unity subordinate to this covering" is a family $\{f_i\}_{i=1,\dots,n}$ of continuous functions from E to $[0, 1]$ such that

(1) $\begin{cases} \textbf{i.} \quad \forall x \in E, \quad \sum\limits_{i=1}^n f_i(x) = 1 \\[2mm] \textbf{ii.} \quad \forall i = 1, \dots, n \qquad \text{supp}(f_i) \subset A_i \end{cases}$

Before establishing the existence of a partition of unity subordinate to a finite open covering, we must show the following propositions.

PROPOSITION 1.   *The separation of two closed sets by a continuous function.* Let A and B be two disjoint nonempty closed subsets of a metric space E. Then there exists a continuous function g from E to $[0, 1]$ such that $g(x) = 0$ for all $x \in A$ and $g(x) = 1$ for all $x \in B$.

Proof.   Since A and B are disjoint, $d(x, A) + d(x, B) > 0$ for all $x \in E$. Thus the function g defined by

$$g(x) = \frac{d(x, A)}{d(x, A) + d(x, B)}$$

is a continuous function from E to $[0, 1]$, which vanishes on A and equals 1 on B.   ∎

PROPOSITION 2.   Suppose that $E = A \cup B$ is the union of two open sets. Then there exists an open set W such that

(2) $$\overline{W} \subset A \qquad \text{and} \qquad E = W \cup B.$$

Proof.   First, we shall construct an open set V such that

(3) $$\complement A \subset V \subset \overline{V} \subset B.$$

If $A = E$, we take $V = \varnothing$, and if $B = E$, we take $V = E$. Hence let us consider the case where $A \neq E$ and $B \neq E$. Since $E = A \cup B$, the nonempty closed sets $\complement A$ and $\complement B$ are disjoint. Then according to the preceding proposition, there exists a *continuous* function from E to $[0, 1]$ such that $f(x) = 0$ for all $x \in \complement A$ and $f(x) = 1$ for all $x \in \complement B$. Thus $\complement A$ is contained in the open set $V = \{y \in E \text{ such that } f(y) < 1/2\}$. The closure $\overline{V}$ of V is contained in the closed set $T = \{y \in E \text{ such that } f(y) \leq 1/2\}$, which is clearly contained in B. Thus let us take $W = \complement \overline{V}$, which is open. Since $\complement A \subset V$, we deduce that $\overline{W} = \complement V \subset A$. Since $\overline{V} \subset B$, we conclude that $E = W \cup B$.   ∎

PROPOSITION 3.   Let $\{A_i\}_{i=1,\ldots,n}$ be a finite open covering of E. There exists an open covering $\{W_i\}_{i=1,\ldots,n}$ of E such that

(4) $$\forall i = 1, \ldots, n \qquad \overline{W}_i \subset A_i.$$

Proof.   We construct the covering $\{W_i\}$ by recursion using Proposition 2. Setting $B_1 = \bigcup_{i=2}^{n} A_i$, we obtain that $E = A_1 \cup B_1$. Proposition 2 implies the existence of an open set $W_1 \subset A_1$ such that

(5) $$\overline{W}_1 \subset A_1 \qquad \text{and} \qquad E = W_1 \cup B_1 = W_1 \cup \bigcup_{i=2}^{n} A_i.$$

Now suppose that we have already constructed the open sets

$$W_j \, (1 \leq j \leq k - 1)$$

such that

(6) $$\overline{W}_j \subset A_j \quad \text{if} \quad 1 \leq j \leq k - 1; \qquad E = \bigcup_{i=1}^{k-1} W_i \cup \bigcup_{j=k}^{n} A_j.$$

We define the open set

(7)
$$B_k = \bigcup_{i=1}^{k-1} W_i \cup \bigcup_{j=k+1}^{n} A_j$$

in such a way that, from (6), we obtain $E = A_k \cup B_k$. Proposition 1 implies the existence of an open set $W_k \subset A_k$ such that $\overline{W}_k \subset A_k$ and $E = W_k \cup B_k$. We have, therefore, constructed k open sets W_i such that

(8) $\qquad \overline{W}_i \subset A_i \quad$ if $\quad 1 \le i \le k, \quad E = \bigcup_{i=1}^{k} W_i \cup \bigcup_{i=k+1}^{n} A_j.$ ∎

THEOREM 1. For every finite open covering of a metric space E there exists a continuous partition of unity that is subordinate to it.

Proof. Suppose that $E = \bigcup_{i=1}^{n} A_i$ where the A_i's are open (and non-empty). According to Proposition 3 there exist n open sets $W_i \subset A_i$ such that $\overline{W}_i \subset A_i$ and $E = \bigcup_{i=1}^{n} W_i$. Moreover, according to Proposition 1 on the separation of sets, there exist continuous functions h_i from E to $[0, 1]$ such that

(9)
$$\begin{cases} h_i(x) = 1 & \text{if} \quad x \in \overline{W}_i \\ h_i(x) = 0 & \text{if} \quad x \in \complement A_i \end{cases}$$

since $\overline{W}_i$ and $\complement A_i$ are disjoint nonempty closed sets. Then we set

(10) $\qquad g_i(x) = \max(0, h_i(x) - 1/2).$

The functions g_i are continuous. The support of g_i is contained in the set $\{x \in E \text{ such that } h_i(x) \ge 1/4\}$ which is contained in A_i. Setting $g(x) = \sum_{i=1}^{n} g_i(x)$ (which is positive since $E = \bigcup_{i=1}^{n} W_i$), the functions f_i defined by

(11)
$$f_i(x) = \frac{g_i(x)}{g(x)}$$

form a continuous partition of unity subordinate to the covering of E by the A_i's. ∎

APPLICATION. APPROXIMATION OF CONTINUOUS FUNCTIONS OF SEVERAL VARIABLES. Let E and F be two compact metric spaces and $\mathscr{C}_\infty(E \times F)$ the space of continuous functions on $E \times F$. We are going to show that we can approximate every continuous function $f \in \mathscr{C}_\infty(E \times F)$ by functions of the form

$$\sum_{i,j=1}^{n} f_i(x) g_j(y).$$

THEOREM 2.   The space of finite sums of functions $\{x, y\} \mapsto g_i(x)h_i(y)$ where $g_i \in \mathscr{C}_\infty(E)$ and $h_i \in \mathscr{C}_\infty(F)$ is dense in $\mathscr{C}_\infty(E \times F)$.

Proof.   Let us take $\varepsilon > 0$. We can associate to every pair $\{x, y\} \in E \times F$ open balls $\mathring{B}(x, \eta)$ and $\mathring{B}(y, \delta)$ such that

$$(12) \qquad |f(x', y') - f(x, y)| \leq \varepsilon \quad \text{when} \quad x \in \mathring{B}(x', \eta) \quad \text{and} \quad y \in \mathring{B}(y', \delta)$$

since f is uniformly continuous. Hence we can cover E and F by a finite number of balls $\mathring{B}(x_i, \eta)$ and $\mathring{B}(y_j, \delta)$. Let $\{g_i\}_i$ and $\{h_j\}_j$ be continuous partitions of unity subordinate to the coverings of E and F by the open balls $\mathring{B}(x_i, \eta)$ and $\mathring{B}(y_j, \delta)$. Then the function $f_\varepsilon \in \mathscr{C}_\infty(E \times F)$ defined by

$$(13) \qquad f_\varepsilon(x, y) = \sum_{i,j} f(x_i, y_j)g_i(x)h_j(y)$$

is in a ball with center f and radius ε: Indeed, since the functions g_i and h_j form partitions of unity, we have $\sum_{i,j} g_i(x)h_j(y) = 1$ for every $x \in E$ and $y \in F$. Thus

$$(14) \qquad |f(x, y) - f_\varepsilon(x, y)| \leq \sum_{i,j} |f(x, y) - f(x_i, y_j)|g_i(x)h_j(y).$$

If $g_i(x)h_j(y) > 0$ then $x \in \mathring{B}(x_i, \eta)$ and $y \in \mathring{B}(y_j, \delta)$. Hence (12) and (14) imply that

$$(15) \qquad |f(x, y) - f_\varepsilon(x, y)| \leq \sum_{i,j} \varepsilon g_i(x)h_j(y) = \varepsilon.$$

Since (15) holds for any x and y, we conclude that $\| f - f_\varepsilon \|_\infty \leq \varepsilon$.   ∎

6.   Game Theory—The Lasry Equality, the Ky-Fan Inequality

Consider the players Emily and Frank whose respective sets of strategies are denoted by E and F. We denote by $f(x, y)$ the loss sustained by Emily and the gain of Frank when Emily plays $x \in E$ and her partner Frank plays $y \in F$. Let us recall that

$$(1) \qquad v^{\#} = \inf_{x \in E} \sup_{y \in F} f(x, y)$$

denotes the best loss for Emily when she has no information about the choice made by Frank (see Section 3.10). Suppose that *Frank has information* about the strategies adopted by Emily. Instead of choosing a strategy $y \in F$, Frank is able to choose a "*decision rule*" D, that is, a mapping D from E to F that associates to each strategy played by Emily a strategy $D(x) \in F$. In this case, Frank will associate to such a decision rule the worst gain

$$(2) \qquad f^b(D) = \inf_{x \in E} f(x, D(x))$$

and will seek to maximize $f^b(D)$ on the set of possible decision rule. We always have for any decision rule D

(3) $$f^b(D) = \inf_{x \in E} f(x, D(x)) \le \inf_{x \in E} \sup_{y \in F} f(x, y) = v^{\#}.$$

PROPOSITION 1. If Frank has the right to use all possible decision rules, then

(4) $$\sup_{D \in \mathscr{F}(E, F)} \inf_{x \in E} f(x, D(x)) = \inf_{x \in E} \sup_{y \in F} f(x, y).$$

Proof. Indeed, by definition, we can associate to every $\varepsilon > 0$ and to every $x \in E$ a strategy $D_\varepsilon(x) \in F$ such that

(5) $$\sup_{y \in F} f(x, y) \le f(x, D_\varepsilon(x)) + \varepsilon.$$

From this we deduce that

$$v^{\#} = \inf_{x \in E} \sup_{y \in F} f(x, y) \le f^b(D_\varepsilon) + \varepsilon$$
$$\le \sup_{D \in \mathscr{F}(E, F)} \inf_{x \in E} f(x, D(x)) + \varepsilon.$$

Since this inequality holds for every $\varepsilon > 0$, we obtain the inequality $v^{\#} \le \sup_{D \in \mathscr{F}(E, F)} \inf_{x \in E} f(x, D(x))$, which, together with (3), implies the desired equality (4). ∎

We show that with additional hypotheses equality (4) remains true when we require Frank *to use only continuous decision rules* (in order to describe a "stable" behavior for Frank).

THEOREM 1. *Lasry's equality.* Let us suppose that

(6) $$\begin{cases} \textbf{i.} & E \text{ is compact} \\ \textbf{ii.} & \forall y \in F, \quad x \mapsto f(x, y) \quad \text{is lower semicontinuous} \end{cases}$$

and that

(7) $$\begin{cases} \textbf{i.} & F \text{ is a convex subset of a Banach space} \\ \textbf{ii.} & \forall x \in E, \quad y \mapsto f(x, y) \quad \text{is concave.} \end{cases}$$

Then if $\mathscr{C}(E, F)$ denotes the set of continuous mappings from E to F, we have the following equality

(8) $$\sup_{D \in \mathscr{C}(E, F)} \inf_{x \in E} f(x, D(x)) = \inf_{x \in E} \sup_{y \in F} f(x, y).$$

Proof. We already know from (3) that $\sup_{D \in \mathscr{C}(E, F)} \inf_{x \in E} f(x, D(x)) \leq v^\#$. Thus we must establish the opposite inequality. First of all, we can associate to every $\varepsilon > 0$ a mapping (not necessarily continuous) D_ε from E to F satisfying (5). Moreover, since the functions $x \mapsto f(x, y)$ are lower semi-continuous, there exist neighborhoods $B(x, \eta(x))$ of x such that

$$(9) \qquad \forall z \in B(x, \eta(x)), \qquad f(x, D_\varepsilon(x)) \leq f(z, D_\varepsilon(x)) + \varepsilon.$$

Since E is compact, it can be covered by n balls $\mathring{B}(x_i, \eta(x_i))$. Let $\{g_i\}_{i=1, \ldots, n}$ be a continuous partition of unity subordinate to this covering. We introduce the function D defined by $D(x) = \sum_{i=1}^n g_i(x) D_\varepsilon(x_i)$, which is *continuous*, since the functions g_i are continuous. Finally, since the functions $y \mapsto f(x, y)$ are concave, since $g_i(x) \geq 0$ for all i and $\sum_{i=1}^n g_i(x) = 1$, we obtain

$$(10) \qquad f(x, D(x)) \geq \sum_{i \in I(x)}^n g_i(x) f(x, D_\varepsilon(x_i))$$

where $I(x)$ is the set of integers $i = 1, \ldots, n$ for which $g_i(x) > 0$. This set is not empty since $\sum_{i=0}^n g_i(x) = 1$. Furthermore, if $g_i(x) > 0$, x belongs to the support of g_i, which is contained in the ball $\mathring{B}(x_i, \eta(x_i))$. Hence we deduce from (9) that $f(x, D_\varepsilon(x_i)) \geq f(x_i, D_\varepsilon(x_i)) - \varepsilon$ and from (5) that $f(x_i, D_\varepsilon(x_i)) \geq \sup_{y \in F} f(x_i, y) - \varepsilon \geq v^\# - \varepsilon$. Thus if $i \in I(x)$, we obtain $f(x, D_\varepsilon(x_i)) \geq v^\# - 2\varepsilon$. Therefore, (10) implies that $f(x, D(x)) \geq \sum_{i \in I(x)} g_i(x)(v^\# - 2\varepsilon) = v^\# - 2\varepsilon$. We conclude from this that $\inf_{x \in E} f(x, D(x)) \geq v^\# - 2\varepsilon$ and, consequently, that $\sup_{D \in \mathscr{C}(E, F)} \inf_{x \in E} f(x, D(x)) \geq v^\# - 2\varepsilon$. Letting ε approach 0, we obtain the desired result. ∎

*THE KY-FAN INEQUALITY. We establish the Ky-Fan inequality with the aid of the Lasry equality and the theorem of Brouwer 3.11.4, which we accepted without proof.

First of all, we shall give another expression of $v^\#$. Within the framework of game theory, we now suppose that it is Emily who has information on the choice of strategies by Frank and who has the right to choose *continuous* decision rules $C \in \mathscr{C}(F, E)$ that allow her to associate continuously to every strategy $y \in F$ played by Frank a strategy $C(y) \in E$.

THEOREM 2. We assume the hypotheses (6) and (7) of Theorem 1. Then if $\mathscr{C}(F, E)$ denotes the set of continuous decision rules from F to E, we have the following equality:

$$(11) \qquad \inf_{C \in \mathscr{C}(F, E)} \sup_{y \in F} f(C(y), y) = \inf_{x \in E} \sup_{y \in F} f(x, y).$$

Proof. Theorem 2 results from Theorem 1, Brouwer's fixed point theorem, and Theorem 3.10.1 of game theory. First, inequality

$$\inf_{C \in \mathscr{C}(F, E)} \sup_{y \in F} f(C(y), y) \leq v^{\#}$$

is obviously satisfied. For proving the opposite inequality we make use of the compact convex set

(12) $$S^n = \left\{ \lambda \in \mathbb{R}^n_+ \text{ such that } \sum_{i=1}^n \lambda_i = 1 \right\}.$$

Since E is compact and the functions $x \mapsto f(x, y)$ are lower semicontinuous, Theorem 3.10.1 implies the existence of $\bar{x} \in E$ such that

(13) $$\sup_{y \in F} f(\bar{x}, y) = v^{\#} = \sup_{K = \{y_1, \ldots, y_n\} \in \mathscr{S}} \inf_{x \in E} \max_{i = 1, \ldots, n} f(x, y_i).$$

where $\mathscr{S}$ is the family of finite subsets of F. Thus it suffices to prove that for every finite set $K = \{y_1, \ldots, y_n\}$ and for every continuous mapping $C \in \mathscr{C}(F, E)$ we have

(14) $$\inf_{x \in E} \max_{i = 1, \ldots, n} f(x, y_i) \leq \sup_{y \in F} f(C(y), y).$$

Since F is convex, $\text{co}(K) \subset F$ and $C(\text{co}(K)) \subset E$. Hence

(15) $$\begin{cases} \displaystyle \inf_{x \in E} \max_{i = 1, \ldots, n} f(x, y_i) \leq \inf_{x \in E} \sup_{\lambda \in S^n} \sum_{i=1}^n \lambda^i f(x, y_i) \\[3mm] \displaystyle \qquad\qquad \leq \inf_{\mu \in S^n} \sup_{\lambda \in S^n} \sum_{i=1}^n \lambda^i f\left(C\left(\sum_{j=1}^n \mu^j y_j \right), y_i \right) \\[3mm] \displaystyle \qquad\qquad = \inf_{\mu \in S^n} \sup_{\lambda \in S^n} \varphi(\mu, \lambda) \end{cases}$$

where the function φ is defined on $S^n \times S^n$ by

(16) $$\varphi(\mu, \lambda) = \sum_{i=1}^n \lambda^i f\left(C\left(\sum_{j=1}^n \mu^j y_j \right), y_i \right).$$

Since C is continuous and since the functions $x \mapsto f(x, y_i)$ are lower semicontinuous, we deduce that the functions $\mu \mapsto \varphi(\mu, \lambda)$ are lower semicontinuous. The functions $\lambda \mapsto \varphi(\mu, \lambda)$ are linear, hence, concave. The set S^n is convex and compact. Therefore, Theorem 1 implies that

(17) $$\inf_{\mu \in S^n} \sup_{\lambda \in S^n} \varphi(\mu, \lambda) = \sup_{D \in \mathscr{C}(S^n, S^n)} \inf_{\mu \in S^n} \varphi(\mu, D(\mu)).$$

But from Brouwer's theorem, we can associate to every $D \in \mathcal{C}(S^n, S^n)$ a fixed point $\mu_D \in S^n$. Thus

$$\inf_{\mu \in S^n} \varphi(\mu, D(\mu)) \leq \varphi(\mu_D, D(\mu_D)) = \varphi(\mu_D, \mu_D) \leq \sup_{\mu \in S^n} \varphi(\mu, \mu).$$

This implies then that

(18) $$\sup_{D \in \mathcal{C}(S^n, S^n)} \inf_{\mu \in S^n} \varphi(\mu, D(\mu)) \leq \sup_{\mu \in S^n} \varphi(\mu, \mu).$$

Moreover, since the functions $y \mapsto f(x, y)$ were assumed to be concave, we obtain that

(19)
$$
\begin{cases}
\varphi(\mu, \mu) = \displaystyle\sum_{i=1}^{n} \mu^i f\left(C\left(\sum_{j=1}^{n} \mu^j y_j \right), y_i \right) \\[2ex]
\qquad \leq f\left(C\left(\displaystyle\sum_{j=1}^{n} \mu^j y_j \right), \sum_{j=1}^{n} \mu^j y_j \right) \\[2ex]
\qquad \leq \displaystyle\sup_{y \in F} f(C(y), y).
\end{cases}
$$

Thus inequalities (15), (17), (18), and (19) imply inequality (14), which we needed to show.   ∎

In particular, we derive from this theorem the important Ky-Fan inequality that follows.

THEOREM 3. *Ky-Fan inequality.* Suppose that E is a compact convex subset of a Banach space and that φ is a function from $E \times E$ to $\mathbb{R}$ satisfying

(20) $\begin{cases} \textbf{i.} \quad \forall y \in E, \qquad x \mapsto \varphi(x, y) \qquad \text{is lower semicontinuous.} \\ \textbf{ii.} \quad \forall x \in E, \qquad y \mapsto \varphi(x, y) \qquad \text{is concave.} \end{cases}$

Then there exists $\bar{x} \in E$ such that

(21) $$\sup_{y \in E} \varphi(\bar{x}, y) \leq \sup_{y \in E} \varphi(y, y).$$

Proof. Theorems 3.10.1 and 1 imply the existence of $\bar{x} \in E$ such that

$$\sup_{y \in E} \varphi(\bar{x}, y) = v^{\#} = \inf_{C \in \mathcal{C}(E, E)} \sup_{y \in E} \varphi(C(y), y) \leq \sup_{y \in E} \varphi(y, y)$$

since the identity mapping is continuous from E to E.   ∎

Remark 1. We have obtained the Ky-Fan inequality from the Brouwer fixed point theorem. In fact, *these two results are equivalent.* We derive the Brouwer theorem from the Ky-Fan inequality with the aid of the theorem of projections. (See Chapter 15 of [AFA].)   ∎

7.   Noncooperative Equilibria in n-Person games

We denote by $N = \{1, 2, \ldots, n\}$ the set of n players, and we let $\hat{\imath} = \{j \in N$ such that $j \neq i\}$. A noncooperative game is described when we are given

$$(1) \quad \begin{cases} \textbf{i.} & n \text{ sets of strategies } E^i. \\ \textbf{ii.} & n \text{ loss functions } f_i \colon \prod_{i=1}^{n} E^i \longmapsto \mathbb{R}. \end{cases}$$

The loss function f_i of the ith player associates to each multistrategy $x = \{x^1, \ldots, x^n\} \in \prod_{i=1}^{n} E^i$ the loss $f_i(x) = f_i(x^1, \ldots, x^n)$ that he sustains. Each player i distinguishes the strategy $x^i \in E$ that he controls from the multistrategies $x^{\hat{\imath}} = \{x^1, \ldots, x^{\hat{\imath}}, \ldots, x^n\} \in \prod_{j \neq i} E^j$ employed by the adverse coalition $\hat{\imath}$. From his point of view, the multistrategies decompose as follows:

$$(2) \qquad\qquad x = \{x^i, x^{\hat{\imath}}\} \in E^i \times \prod_{j \neq i} E^j.$$

We describe the noncooperative behavior of the players in the following fashion.

If the adverse coalition employs the multistrategy $x^{\hat{\imath}} \in \prod_{j \neq 1} E^j$, we suppose that the ith player chooses a strategy $x^i \in E^i$ that minimizes the loss function $y^i \longmapsto f_i(y^i, x^{\hat{\imath}})$.

Definition 1.   We say that a multistrategy $x = \{x^1, \ldots, x^n\}$ is a "noncooperative equilibrium" if for every player i the strategy x^i minimizes the loss function $y^i \longmapsto f_i(y^i, x^{\hat{\imath}})$.

In other words, $x = \{x^1, \ldots, x^n\}$ is a noncooperative equilibrium if

$$(3) \qquad \forall i \in N, \qquad f_i(x) = f_i(x^i, x^{\hat{\imath}}) = \min_{y^i \in E^i} f_i(y^i, x^{\hat{\imath}}).$$

Remark 1.   This concept of a noncooperative equilibrium describes a property of *individual stability* for the multistrategy x in the sense that it is not in the interest of any player to modify *alone* his choice x^i when the adverse coalition has adopted $x^{\hat{\imath}}$.

THEOREM 1. *Nash.*   Suppose that

$$(4) \qquad\qquad \forall i \in N, \qquad \text{the sets } E^i \text{ are compact and convex}$$

and that

$$(5) \quad \begin{cases} \forall i \in N, & \text{the functions } f_i \text{ are continuous and the functions} \\ y^i \longmapsto f_i(y^i, x^{\hat{\imath}}) & \text{are convex.} \end{cases}$$

Then there exists a noncooperative equilibrium.

Proof. The theorem is a consequence of Theorem 6.3 of Ky-Fan (and, in fact, of the Brouwer theorem, which we have accepted). We introduce the set E and the function φ defined by:

(6)
$$
\begin{cases}
\textbf{i.} & E = \prod_{i=1}^{n} E^i. \\[2ex]
\textbf{ii.} & \varphi(x, y) = \sum_{i=1}^{n} (f_i(x^i, x^{\hat{i}}) - f_i(y^i, x^{\hat{i}})).
\end{cases}
$$

The set E is compact and convex since it is the product of the compact convex sets E^i (according to hypothesis (4)). Moreover, hypothesis (5) implies clearly that the functions $x \mapsto \varphi(x, y)$ are continuous and that the functions $y \mapsto \varphi(x, y)$ are concave. The Ky-Fan theorem implies, therefore, the existence of $x = \{x^i, x^{\hat{i}}\} \in E$ such that

$$
\sup_{y \in E} \varphi(x, y) \le \sup_{y \in E} \varphi(y, y) = 0,
$$

since $\varphi(y, y) = 0$ for all y. In particular, let us take $y = \{y^i, x^{\hat{i}}\}$. The inequality $\varphi(x, y) \le 0$ can be written

(7)
$$
f_i(x^i, x^{\hat{i}}) - f_i(y^i, x^{\hat{i}}) + \sum_{j \ne i} (f_j(x^j, x^{\hat{j}}) - f_j(y^j, x^{\hat{j}})) \le 0.
$$

But $x = \{x^j, x^{\hat{j}}\} = \{y^j, x^{\hat{j}}\}$ whenever $j \ne i$. Hence (7) implies that

(8)
$$
\forall y^i \in E^i, \qquad f_i(x^i, x^{\hat{i}}) \le f_i(y^i, x^{\hat{i}}),
$$

that is, that x is a noncooperative equilibrium. ∎

Consider, in particular, the case of a two-player zero-sum game, that is, the case where

$$
E^1 = E, \qquad E^2 = F, \qquad f_1(x, y) = f(x, y), \qquad \text{and} \qquad f_2(x, y) = -f(x, y).
$$

PROPOSITION 1. A pair $\{\bar{x}, \bar{y}\}$ is a noncooperative equilibrium of a zero-sum game of two players if and only if

(9)
$$
\forall x \in E, \qquad y \in F, \qquad f(\bar{x}, y) \le f(\bar{x}, \bar{y}) \le f(x, \bar{y}).
$$

In this case, we obtain

(10)
$$
\sup_{y \in F} \inf_{x \in E} f(x, y) = \inf_{x \in E} \sup_{y \in F} f(x, y).
$$

Proof. To say that $\{\bar{x}, \bar{y}\} \in E \times F$ is a noncooperative equilibrium is the same as to say that

(11)
$$
f(\bar{x}, \bar{y}) \le f(x, \bar{y}) \qquad \text{and} \qquad -f(\bar{x}, \bar{y}) \le -f(\bar{x}, y)
$$

for all $x \in E$, $y \in F$. In this case, we obtain inequality (10). ∎

Definition 2.   Every pair $\{\bar{x}, \bar{y}\} \in E \times F$ that satisfies (9) is called a "saddle point" of f.

Theorem 1 of Nash implies the well-known Von Neumann theorem on the existence of a saddle point.

THEOREM 2. *Von Neumann.*   Suppose that

(12) E and F are compact convex sets

and that

(13) $\begin{cases} \textbf{i.} & \forall y \in F, \quad x \mapsto f(x, y) \quad \text{is lower semicontinuous and convex} \\ \textbf{ii.} & \forall x \in E, \quad y \mapsto f(x, y) \quad \text{is upper semicontinuous and concave.} \end{cases}$

Then there exists a saddle point.

Proof.   We apply the Ky-Fan theorem to the function defined on $(E \times F)^2$ by $\varphi(\{\bar{x}, \bar{y}\}, \{x, y\}) = f(\bar{x}, y) - f(x, \bar{y})$, which is lower semicontinuous with respect to $\{\bar{x}, \bar{y}\}$ and concave with respect to $\{x, y\}$. Ky-Fan's theorem then implies the existence of $\{\bar{x}, \bar{y}\}$ such that

(14) $\sup_{\{x, y\}} \varphi(\{\bar{x}, \bar{y}\}, \{x, y\}) \leq \sup_{\{x, y\}}(\varphi(\{x, y\}, \{x, y\}) = 0.$

This inequality can be written as $f(\bar{x}, y) \leq f(x, \bar{y})$ for all $x \in E$, $y \in F$. By taking successively $x = \bar{x}$ and $y = \bar{y}$, we conclude that $f(\bar{x}, y) \leq f(\bar{x}, \bar{y}) \leq f(x, \bar{y})$.   ∎

Remark 2.   In Chapter 2, Section 7 of [AFA], we prove Von Neumann's Theorem 2 without using Brouwer's fixed point theorem.   ∎

Remark 3.   When the loss functions are quadratic we give another proof of Theorem 1 that does not use the Brouwer fixed point theorem but only the theorem of projections. (See Chapter 2, Section 7 of [AFA].)   ∎

Remark 4.   Nash's Theorem 1 is generalized in Chapter 15, Section 7 of [AFA], to the case where strategy sets $E^i(\hat{x}^i)$ depend upon strategies of adverse coalitions.   ∎

8. Walras Equilibrium

We consider a finite dimensional space $U = \mathbb{R}^n$, which represents a commodity space. Its dual represents, therefore, the space of price systems (see Section 1.6) and $S^n = \{p \in \mathbb{R}_+^n \text{ such that } \sum_{k=1}^n p^k = 1\}$ is interpreted as a set of *normalized prices.* We now suppose that the economy is described by an *income function*

(1) $$p \in S^n \mapsto r(p) \in \mathbb{R}$$

that associates to each price system p the *total income* produced by the economy.

Now consider I consumers, $i = 1, \ldots, I$, whose aim is to choose commodities $x^i \in U$, which we call *consumptions.* In this case, we say that I consumptions form an *allocation* if for every price system $q \in S^n$ we have

(2) $$\sum_{i=1}^{I} \langle q, x^i \rangle \leq r(q);$$

that is, for every price system q, the value of the sum of commodities consumed is inferior to the total income produced by the economy.

There is an infinite set of such allocations. We construct a (theoretic) "mechanism" by which we can choose one such allocation by making supplementary hypotheses on the behavior of the consumers and on the distribution of incomes among them.

We suppose that the behavior of consumer i is described by a *demand function* $f_i : S^n \times \mathbb{R} \mapsto U$ associating to every price system $p \in S^n$ and to every income r a consumption $x = f_i(p, r)$ satisfying the property:

(3) $$\langle p, x \rangle = \langle p, f_i(p, r) \rangle \leq r.$$

This property expresses that for each price system p, the value of the demand $f_i(p, r)$ is inferior to the income r. We suppose as well that the income $r(p)$ produced by the economy is shared among the I consumers:

(4) $$r(p) = \sum_{i=1}^{I} r_i(p),$$

where $r_i(p)$ is the part of the total income allocated to consumer i. The problem arises to know if a price system $\bar{p}$ of S^n exists such that

(5)   the consumptions $f_i(\bar{p}, r_i(\bar{p}))$ form an allocation.

Definition 1. We say the price system $\bar{p} \in S^n$, which satisfies (5), is a "Walras equilibrium" associated to the demand functions f_i and the distribution of the total income (4).

THEOREM 1. Suppose that the functions f_i and r_i are continuous and that the function $r(\cdot)$ is convex. Then there exists a Walras equilibrium.

Proof. Consider the function φ defined on the product $S^n \times S^n$ by

(6)
$$\varphi(p, q) = \sum_{i=1}^{I} \langle q, f_i(p, r_i(p)) \rangle - r(q).$$

The set S^n is compact and convex. The function φ is continuous with respect to p and concave with respect to q. Moreover, the inequalities of (3) imply the "Walras law":

(7) $\varphi(q, q) = \sum_{i=1}^{I} (\langle q, f_i(q, r_i(q)) \rangle - r_i(q)) \leq 0$ for all $q \in S^n.$

The Ky-Fan theorem, therefore, implies the existence of $\bar{p} \in S^n$ such that

(8)
$$\sup_{q \in S^n} \varphi(\bar{p}, q) \leq \sup_{q \in S^n} \varphi(q, q) \leq 0.$$

This last inequality expresses the fact that the consumptions $f_i(\bar{p}, r_i(\bar{p}))$ form an allocation.   ■

EXAMPLE. *Exchange economies.* Let us suppose that the I consumers wish to exchange the commodities $w^i \in U$ that they possess. These commodities produce for them the income $r_i(p) = \langle p, w^i \rangle$. An *exchange economy* is an economy in which the incomes $r_i(p)$ come only from the commodities w^i and where $r(p) = \sum_{i=1}^{I} \langle p, w^i \rangle$. In this case, the consumers form an allocation if

(9)
$$\sum_{i=1}^{I} x^i \leq \sum_{i=1}^{I} w^i \qquad \text{(inequality in } U\text{)}.$$

COROLLARY 1. Suppose that the demand functions are continuous. Then there exists a Walras equilibrium for the exchange economy.

Resumé of Results

We group here the principal results established in this book under the following headings: closed sets, open sets, dense sets, compact sets, convex sets and functions, complete spaces, continuity, homeomorphisms, convergence of functions, semicontinuity, optimization theory, and game theory.

1. Closed Sets

1. The family of closed sets of a metric space E has the following properties:
   a. $\emptyset$ and E are closed.
   b. Every intersection of closed sets is closed.
   c. Every finite union of closed sets is closed.

2. In a metric space E, every *closed* set can be written as a *countable intersection of open sets.*

3. If E is *complete* and if F_n is a *decreasing* sequence of closed sets whose diameters $\delta(F_n)$ *tend to* 0, then $\bigcap_{n=1}^{\infty} F_n = \{a\}$.

4. Every *decreasing* sequence of closed sets of a compact space has a nonempty intersection. Every sequence of closed sets F_n of a compact space with the *finite intersection property* has a nonempty intersection.

5. If F_1 and F_2 are two *nonempty closed* sets whose *union is connected*, then $F_1 \cap F_2 \neq \emptyset$.

6. If E is a *complete* metric space and if the interior of a *countable* union of *closed* sets F_n is nonempty, then at least one of these closed sets has a nonempty interior (Baire).

7. The *inverse* image of a closed set under a continuous mapping is closed.

8. The image of a closed set under a *proper* mapping is closed.

9. The product of n closed sets is closed. The projection of a closed set of a product space *is not necessarily closed.*

10. In a normed vector space, if A is a *closed* subset and B a *compact* subset, $A + B$ is *closed*.
11. Every *complete* subset is closed. A closed subset of a complete space is complete.
12. Every compact set is closed.
13. The closure of a convex set is convex and is equal to the closure of its interior when it is nonempty.
14. Every closed set remains closed for any stronger distance. (Every weakly closed set is strongly closed.)
15. A real-valued function is *lower semicontinuous* if and only if its *sections* $f^{-1}(]-\infty, \lambda])$ are *closed* or if and only if its *epigraph is closed*.
16. The *graph* of a *continuous* function f from E to F is *closed*. The converse holds if F is compact (or if E and F are Banach spaces and f is linear).
17. If A and B are two *disjoint closed* sets of a metric space, there exist two *disjoint open* sets U and V such that $A \subset U$ and $B \subset V$.
18. If A and B are two disjoint closed sets of a metric space, there exists a continuous function which is equal to 0 on A and to 1 on B.

2. Open Sets

1. The family of open sets of a metric space E has the following properties:
    a. $\varnothing$ and E are open.
    b. Every union of open sets is open.
    c. Every finite intersection of open sets is open, since A is open if and only if $\complement A$ is closed, or if and only if A is a neighborhood of each of its points.
2. In a metric space, every open set is a countable union of closed sets.
3. Every open covering of a compact space has a finite subcovering.
4. We can associate to every finite open covering $E = \bigcup_{i=1}^{n} U_i$ a covering $E = \bigcup_{i=1}^{n} V_i$ such that for every i, $\overline{V}_i \subset U_i$.
5. We can associate to every finite *open covering* a *continuous partition of unity*.
6. If F is a *complete* metric space, every *intersection* of a *countable* family of *dense open* sets is dense (Baire).
7. A metric space E is separable if there exists a countable family of open sets U_n such that for every nonempty open set U and every $x \in U$, there exists U_n such that $x \in U_n \subset U$.
8. Let E be a locally compact metric space. It is *separable* if and only if there exists a *countable* covering by *relatively compact open* sets U_n such that $\overline{U}_n \subset U_{n+1}$ for all n.
9. The *inverse* image of an *open* set under a *continuous* mapping is *open*.
10. A finite product of open sets is open. The *projection* of an *open* subset of a product space is *open*.

11. If A is an *open* set of a normed space E and if $B \subset E$, then $A + B$ is open.

12. The interior of a *convex* set is convex and is equal to the interior of its closure.

13. Every *convex* function defined on a convex *open* set of $\mathbb{R}^n$ is *continuous*. Every *lower semicontinuous convex* function on a convex *open* set of a *Banach* space is *continuous*.

14. Every open set remains open for any stronger distance. (Every weakly open set is strongly open.)

3. Dense Sets

1. If $A \subset B \subset C$ and if A is dense in C, then B is dense in C, but A is not necessarily dense in B. If A is dense in B and B is dense in C, then A is dense in C.

2. If A is dense in E, it remains dense for any *weaker* distance.

3. Let A be a dense subset of E. If f and g are two continuous mappings on E such that $f = g$ on A, then $f = g$. If f and g are two real-valued continuous functions on E such that $f \leq g$ on A, then $f \leq g$ on E.

4. If A is *dense* in E, if f is a *uniformly continuous* mapping from A to a *complete* metric space F, then f has a *unique extension* to a *uniformly continuous* mapping from E to F (extension by density).

5. We say that a metric space E is *separable* if it contains a *countable dense set*. A normed space is separable if and only if it has a *countable topological base*.

6. Every *compact* space is *separable*.

7. A locally compact metric space is *separable* if and only if it is a *countable union of relatively compact open sets*.

8. If Ω is a compact subset of $\mathbb{R}^n$, the spaces $C_\infty(\Omega)$ and $L^p(\Omega)$ are separable (Stone–Weierstrass).

4. Compact Sets

1. A subset K of E is *compact* if and only if one of the following equivalent conditions is satisfied:

   a. Every sequence of elements $x_n \in K$ has a *cluster point* in K.

   b. K is *closed* and every family of closed sets with the *finite intersection property* has a nonempty intersection.

   c. K is *closed* and every open covering has a finite subcovering (*Borel–Lebesgue*).

   d. K is *complete* and for every $\varepsilon > 0$, K can be covered by a *finite number* of balls of *radius* ε.

2. Hence every *compact* set is *closed* and *complete*. It is also *bounded* and *separable*.

3. If a sequence x_n of a compact K has a *unique* cluster point x, then the sequence x_n *converges* to x.

4. A subset K of $\mathbb{R}^n$ is *compact* if and only if it is *closed* and *bounded*.

5. Let E be a compact metric space. A subset $\mathcal{H} \subset \mathcal{C}_\infty(E)$ is *compact* if and only if it is *equicontinuous, closed*, and its projections $\mathcal{H}(x) = \{f(x)\}_{f \in \mathcal{H}}$ are bounded for all $x \in E$ (Ascoli).

6. Let E be a *compact* metric space. Then the space $\mathcal{F}(E)$ of its closed nonempty subsets is compact.

7. Every *closed* subset of a *compact* set is *compact*.

8. Every intersection of compact sets is compact. Every finite union of compact sets is compact.

9. If E is a normed space and if A and B are closed subsets of E, then $A + B$ is closed if A is compact and is compact if A *and* B are compact.

10. The convex hull of a compact set in $\mathbb{R}^n$ is compact.

11. Every finite product of compact sets is compact. The projection of a compact subset of a product space is compact.

12. A *compact* set remains compact for any *weaker* distance. (A strongly compact set is weakly compact.)

13. If E is compact, every *weaker* distance on E is *equivalent* to the original distance.

14. Every *continuous* function on a compact set is *uniformly continuous*.

15. The image of a *compact* set under a *continuous* mapping is *compact*.

16. The *inverse* image of a *compact* set under a *proper* mapping is *compact*.

17. Every real-valued function on a compact set is bounded and attains its bounds.

18. If E is *compact*, if an *increasing* sequence of *continuous* functions converges *pointwise* to a *continuous* function g, then the convergence is in fact uniform (Dini).

19. If E is *compact*, if $\mathcal{H}$ is an *equicontinuous* set of $\mathcal{C}_\infty(E)$, if $A \subset E$ is a dense subset of E, then every sequence of functions $f_n \in \mathcal{H}$ such that $f_n(x)$ converges to $f(x)$ for every $x \in A$ (only) in fact converges *uniformly*. In this case if a sequence x_p of elements of E converges to x, then $f_n(x_p)$ converges to $f(x)$.

5. Convex Sets and Functions

1. K is *convex* if and only if every convex combination of elements of K belongs to K.

2. The *image* and the *inverse image* of a *convex* set under a *linear* mapping is *convex*.

3. The intersection of a family of convex sets is convex.
4. Every product of convex sets is convex. The projection of a convex subset of a product space is convex.
5. If K and L are convex, $K + L$ is also convex.
6. Every convex set is connected.
7. If K is convex, if $x_0 \in \overset{\circ}{K}$ and if $x_1 \in \overline{K}$, then $\lambda x_0 + (1 - \lambda)x_1 \in \overset{\circ}{K}$ for all $\lambda \in \,]0, 1[$.
8. The closure and the interior of a convex set are convex. If $\overset{\circ}{K} \neq 0$, then $\overset{\circ}{K} = \overline{\overset{\circ}{K}}$ and $\overline{K} = \overset{\circ}{\overline{K}}$.
9. The convex hull of the union of a finite family of *compact convex* sets is compact. The convex hull of a *compact* set of $\mathbb{R}^n$ is *compact*.
10. Let K be a *closed convex* subset containing the origin. Then $K = \{x \in E$ such that $j_K(x) \leq 1\}$, where j_K denotes the *gauge of K*; it is a *convex lower semicontinuous* function which is *positively homogeneous* from E to $]-\infty, +\infty]$. The cone generated by K is the domain of its gauge j_K, and the asymptotic cone is the kernel of j_K. The set K is *symmetric* if and only if its gauge is *even*, and K has a *nonempty interior* if and only if j_K is (uniformly) *continuous*.
11. A real-valued function on a convex set K is *convex* if and only if its *epigraph* is *convex*.
12. The *sections $F_\lambda = \{x \in E$ such that $f(x) \leq \lambda\}$ of a convex* function are convex.
13. Let f be a convex function defined on a convex set K of a normed space E:

   a. f is lower semicontinuous at x_0 if and only if $\forall \varepsilon > 0, \exists \eta > 0$ such that

   $$\begin{cases} \textbf{i.} \quad f(x_0) \leq f(x) + \varepsilon \quad & \text{if} \quad \|x - x_0\| \leq \eta. \\[2mm] \textbf{ii.} \quad f(x_0) \leq f(x) + \dfrac{\varepsilon}{\eta}\|x - x_0\| \quad & \text{if} \quad \|x - x_0\| > \eta. \end{cases}$$

   b. f is *continuous* at $x_0 \in \overset{\circ}{K}$ if and only if it is *bounded above* by $a < +\infty$ on a *ball* $B(x_0, \eta)$, with center x_0, which is contained in K. In this case, f satisfies the inequalities

   $$\begin{cases} \textbf{i.} \quad f(x_0) \leq f(x) + \dfrac{a - f(x_0)}{\eta}\|x - x_0\| \quad & \text{for all} \quad x \in K. \\[2mm] \textbf{ii.} \quad f(x) \leq f(x_0) + \dfrac{a - f(x_0)}{\eta}\|x - x_0\| \quad & \text{if} \quad \|x - x_0\| \leq \eta. \end{cases}$$

   c. f is *continuous* on $\overset{\circ}{K}$ if and only if it is continuous at a point $x_0 \in \overset{\circ}{K}$.

14. Every convex function defined on an open convex set K of $\mathbb{R}^n$ is continuous.

15. Every lower semicontinuous convex function on an *open* convex set K of a *Banach* space is continuous.

16. Every lower semicontinuous seminorm on a Banach space is continuous.

6. Complete Spaces

1. Every metric space *has a completion* that is unique up to an isomorphism.

2. The spaces $\mathbb{R}$, $\mathbb{C}$, $\mathbb{R}^n$, l^p ($1 \leq p \leq \infty$), $\mathscr{U}(X)$, $\mathscr{C}_\infty(X)$ are *complete* spaces.

3. If F is a *Banach* space, $\mathscr{L}(E, F)$ is a *Banach* space. The topological dual space of a normed space is a Banach space.

4. If E is a complete metric space, the space $\mathscr{F}(E)$ of its nonempty closed subsets is complete.

5. Every *compact* metric space is *complete*.

6. Every *closed* subset of a *complete* space is *complete*.

7. The *finite product* of *complete* sets is *complete*.

8. Consider a *collectively injective* and *closed* family of n mappings f_i from a set E to *complete metric spaces* F_i. Then E is *complete* for the *distance of the graph*.

9. If E is *complete* and if f is an *isometry* from E to F, then $f(E)$ is *complete*.

10. If d_1 is a *stronger* distance than d_2, if E is *complete for* d_2, and if every ball that is closed for d_1 is closed for d_2, then E is *complete for* d_1.

11. If A is *dense* in E and if f is a uniformly continuous mapping from A to a *complete* space F, then f has a unique extension to a uniformly continuous mapping from E to F (Extension by density).

7. Continuity

1. Let f be a mapping from E to F, and let $x \in E$. The following conditions are equivalent:

    a. f is continuous at x.

    b. The image under f of a sequence that is convergent to x is a sequence that is convergent to $f(x)$.

    c. $\forall \varepsilon > 0$, $\exists \eta > 0$ such that $d_F(f(x), f(y)) \leq \varepsilon$ when $d_E(x, y) \leq \eta$.

    d. for every neighborhood V of $f(x)$, there exists a neighborhood U of x such that $f(U) \subset V$.

    e. for every neighborhood V of $f(x)$, $f^{-1}(V)$ is a neighborhood of x.

2. Let f be a mapping from E to F. The following conditions are equivalent:

    a. f is continuous on E.

    b. The inverse image of every open set is open.

    c. The inverse image of every closed set is closed.

    d. The image of the closure of any set is contained in the closure of its image.

3. Every Lipschitz function is uniformly continuous.

4. Every uniformly continuous function is continuous.

5. Every continuous function on a compact set is uniformly continuous and proper.

6. Let f be a *linear* mapping from a *normed* space E to a *normed* space F. The following conditions are equivalent:

    a. f is Lipschitz.

    b. $\|f\| = \sup\limits_{x \in E} \dfrac{\|f(x)\|_F}{\|x\|_E} < +\infty.$

    c. f is continuous at 0.

7. A *linear form* is *continuous* if and only if its *kernel* is *closed*.

8. Every *linear* mapping from a *finite* dimensional space to another is *continuous*.

9. Let f be a bilinear mapping from $E \times F$ to G. The following conditions are equivalent:

    a. f is continuous.

    b. f is continuous at $(0, 0)$.

    c. $\|f\| = \sup\limits_{x \in E, \, y \in F} \dfrac{\|f(x, y)\|_G}{\|x\|_E \|y\|_F} < +\infty.$

10. Let p be a *seminorm* defined on a normed space. The following conditions are equivalent:

    a. p is Lipschitz.

    b. $\sup\limits_{x \in E} \dfrac{|p(x)|}{\|x\|_E} < +\infty.$

    c. p is continuous at 0.

11. Every *convex* function defined on an *open* convex set of $\mathbb{R}^n$ is *continuous*.

12. Every *lower semicontinuous convex* function defined on an *open* subset of a *Banach* space is *continuous*.

13. The composition of two continuous functions is continuous and of two uniformly continuous functions is uniformly continuous.

14. Every linear combination of continuous mappings with values in a normed space is continuous. Similarly, the function $x \mapsto \|f(x)\|$ is continuous when f is continuous.

15. The product, the pointwise supremum, and the pointwise infimum of two continuous real-valued functions are continuous.

16. A real-valued function is continuous if and only if it is both upper and lower semicontinuous.

17. The *pointwise supremum* of an infinite family of real-valued continuous functions is lower semicontinuous.

18. Every function $f = \{f_1, \ldots, f_n\}$ from E to $\prod_{i=1}^{n} F_i$ is *continuous* if and only if the functions $f_i: E \mapsto F_i$ are *continuous*.

19. If E is a set and if the n mappings f_i from E to the metric spaces F_i are *collectively injective*, then f is *continuous* when E has the distance of the graph. If in addition, the family of the f_i's is *closed* and if the spaces F_i are *complete*, then E is *complete* for the distance of the graph.

20. Every *continuous* function from E to F remains continuous if the distance on E is replaced by a stronger distance and/or the distance on F is replaced by a weaker distance.

21. Let A be a *dense* subset of E and f a continuous mapping from A to F. There exists *at most* one continuous mapping from E to F, which is an extension of f.

22. Let A be a dense subset of E and f a *uniformly continuous* mapping from A to a *complete* space F. There exists at least one uniformly continuous mapping from E to F, which is an extension of f (extension by density).

23. Let E be a normed space, $E_0 \subset E$ a dense subspace and F a *Banach* space. Every continuous linear mapping from E_0 to F has a unique extension to a continuous linear mapping from E to $F: \mathscr{L}(E_0, F) = \mathscr{L}(E, F)$.

24. Let E and F be two normed spaces, $E_0 \subset E$ and $F_0 \subset F$ two dense subspaces and G a *Banach* space. Every continuous bilinear mapping from $E_0 \times F_0$ to G has a unique extension to a continuous bilinear mapping from $E \times F$ to $G: \mathscr{L}(E_0, F_0; G) = \mathscr{L}(E, F; G)$.

25. Let A and B be two nonempty closed disjoint subsets of E. There exists a continuous function f from E to $[0, 1]$ that takes the values 0 on A and 1 on B.

8. Homeomorphisms

1. Let f be a *continuous bijective* mapping from E onto F. If f is proper (in particular, if E is compact), then f is bicontinuous.

[If E and F are Banach spaces and if f is linear, continuous, and bijective, then f is bicontinuous.]

2. Every *contraction* from a complete metric space to itself has a *fixed point*. (Fixed point for contracting mappings.)

Every dissipative continuous map from a complete metric space to itself has a fixed point.

3. If E is a *Banach* space and if g is a *contraction* with proportionality factor λ from the ball $\mathring{B}(0, \beta)$ to E, then the mapping $f = 1 + g$ is a bijection from $B(0, \beta)$ onto its image C, which *contains the ball of radius* $\beta(1 - \lambda)$, and f^{-1} is Lipschitz with proportionality factor $1/(1 - \lambda)$.

4. If E and F are *Banach* spaces, if f is a *continuously differentiable* mapping from an open ball U of E to F such that $Df(0)$ is a *bicontinuous isomorphism* from E onto F, then f is a *homeomorphism* from a neighborhood V of 0 onto its image (inverse function theorem).

5. Every *continuous* mapping from a *compact convex* subset from $\mathbb{R}^n$ to itself has a *fixed point* (Brouwer).

6. If f is a *continuous* mapping from $\mathbb{R}^n$ to itself, which is *strictly monotone* (respectively, *coercive*), then f is *injective* (respectively, *surjective*).

7. Let f be a Lipschitz map from $\mathbb{R}^n$ to $\mathbb{R}^n$, with proportionality factor λ, satisfying $\langle f(x) - f(y), x - y \rangle \geq c \|x - y\|^2 \; \forall x, y \in \mathbb{R}^n$. For any positive definite $A \in \mathscr{L}(\mathbb{R}^n, \mathbb{R}^n)$, for any $\rho \in \;]0, 2c/\lambda \|A^{-1}\|[$, the sequence of solutions x_n of linear equations $Ax_n = Ax_{n-1} + \rho(y - f(x_{n-1}))$ converges to the unique solution x of equation $f(x) = y$.

8. Let E be a complete metric space and A be a mapping from E to $\mathscr{F}(E)$.

   a. If A is upper semicontinuous and weakly dissipative, it has a fixed point $x \in A(x)$.

   b. If A is lower semicontinuous and dissipative, it has a stationary point $x = A(x)$.

9. Convergence of Functions

1. If the functions f_n converge uniformly, they converge pointwise (but the converse is not true).

2. If continuous functions on $[-1, +1]$ converge uniformly, they converge in the mean of order $p(1 \leq p < +\infty)$.

3. If continuous functions on $[-1, +1]$ converge in the mean of order p, they converge in the mean of order q for $1 \leq q \leq p$.

4. The uniform limit of continuous functions is continuous.

5. If a sequence of continuously differentiable functions f_n on $[-1, +1]$ and the sequence of its derivatives f'_n converge uniformly to functions f and g, respectively, then $g = f'$.

6. Suppose that E is *compact* and that the sequence f_n of *continuous* functions is *increasing* and converges *pointwise* to a *continuous* function f. Then f_n converges *uniformly* to f (Dini).

7. Let $\mathcal{H}$ be an *equicontinuous* set of $\mathscr{C}(E)$, A a *dense* subset of E, and f_n a sequence of functions of $\mathcal{H}$.

  a. Then f_n converges *pointwise* to f *if and only if* $f_n(x)$ converges to $f(x)$ for every $x \in A$. In this case, f is *continuous*.
  b. If E is *compact*, every sequence of functions $f_n \in \mathcal{H}$, which converges pointwise to f, converges uniformly.

8. If a sequence of functions f_n of an equicontinuous set $\mathcal{H}$ converges uniformly to f and if the sequence of elements x_p converges to x, then $f_n(x_p)$ converges to $f(x)$.

10. Semicontinuity

1. Let f be a function from E to $]-\infty, +\infty[$, and let $x \in E$. The following conditions are equivalent:

  a. f is lower semicontinuous at x.
  b. for every sequence of elements x_n that converges to x, $f(x) \leq \lim\inf f(x_n)$
  c. $\forall \varepsilon > 0$, $\exists \eta > 0$ such that $f(x) \leq f(y) + \varepsilon$ when $d_E(x, y) \leq \eta$.

2. If f is a function from E to $]-\infty, +\infty[$, the following conditions are equivalent:

  a. f is lower semicontinuous.
  b. $\forall \lambda \in \mathbb{R}$, $F_\lambda = \{x \in E \text{ such that } f(x) \leq \lambda\}$ is closed.
  c. The epigraph of f is closed.

3. If f and g are lower semicontinuous functions, then $f + g$, $\inf(f, g)$, $\alpha f\,(\alpha > 0)$ are lower semicontinuous.

4. The pointwise supremum of an arbitrary family of lower semicontinuous functions is lower semicontinuous.

5. The *characteristic function* of a *closed* set is lower semicontinuous. The *gauge* of a *closed convex* set containing the origin is lower semicontinuous.

6. If E is a *complete* metric space and if f is a *lower semicontinuous* function defined on an *open* set U of E, there exists a *nonempty open* set $V \subset U$ on which f is *bounded above*.

7. If E is a *Banach* space and if f is a *convex lower semicontinuous* function on a nonempty open convex set U of E, then f is *continuous*.

8. Every lower *semicontinuous seminorm* on a *Banach* space is *continuous*.

11. Optimization Theory and Game Theory

1. The minimal set of a function f from U to $]-\infty, +\infty]$ is equal to
$$M^b(f) = \bigcap_{\lambda > \alpha} F_\lambda \qquad \text{where} \qquad F_\lambda = \{x \in U \text{ such that } f(x) \leq \lambda\}$$
where $\alpha = \inf_{x \in U} f(x)$.

2. The minimal set of f is *closed* if f is *lower semicontinuous*.

3. The minimal set of f is *compact* and *nonempty* if f is *lower semicontinuous* and *lower semicompact*.

4. The minimal set of f is *convex* if f is *convex*.

The minimal set of f contains *at most one point* if f is *strictly convex*.

5. Let E be complete and f a lower semicontinuous function from E to $\mathbb{R}_+$. $\forall \varepsilon > 0$, $\exists x_\varepsilon \in E$ such that $f(x_\varepsilon) = \min_{x \in E}[f(x) + \varepsilon d(x_\varepsilon, x)]$ (Ekeland's theorem).

6. Every real-valued *semicoercive* function on a subset of $\mathbb{R}^n$ is *lower semicompact*.

7. Let $E \subset U$, $f : E \mapsto \mathbb{R}$, and let f_E be defined by $f_E(x) = f(x)$ if $x \in E$ and $f_E(x) = +\infty$ if $x \notin E$. If E is *relatively compact*, then f_E is *lower semicompact*. If E is *closed* and if f is *lower semicontinuous*, then f_E is *lower semicontinuous*.

8. Consider a mapping $A : x \mapsto A(x)$ from a metric space E to the space $\mathscr{F}(F)$ of the *nonempty closed* subsets of a metric space F. Let f be a function from $E \times F$ to $\mathbb{R}$. Set $\alpha(x) = \inf_{y \in A(x)} f(x, y)$.

   a. If $x \mapsto A(x)$ is *upper semicontinuous* at x_0, if $A(x_0)$ is *compact* and if f is *lower semicontinuous* on $E \times F$, then α is *lower semicontinuous* at x_0.

   b. If $x \mapsto A(x)$ is a *lower semicontinuous* mapping at x_0 and if f is an *upper semicontinuous* function on $E \times F$, then α is *upper semicontinuous* at x_0.

9. If $f : E \times F \mapsto \mathbb{R}$, if E is *compact* and if the functions $x \mapsto f(x, y)$ are *lower semicontinuous* for all $y \in F$, then there exists $\bar{x} \in E$ such that

$$\sup_{y \in F} f(\bar{x}, y) = \inf_{x \in E} \sup_{y \in F} f(x, y).$$

Moreover, we obtain the equality

$$\inf_{x \in E} \sup_{y \in F} f(x, y) = \sup_{\{y_1, \ldots, y_n\}} \inf_{x \in E} \max_{i = 1, \ldots, n} f(x, y_i).$$

10. If we suppose *in addition* that F is *convex* and that the functions $y \mapsto f(x, y)$ are *concave*, then

$$\inf_{x \in E} \sup_{y \in F} f(x, y) = \sup_{D \in \mathscr{C}(E, F)} \inf_{x \in E} f(x, D(x))$$

$$= \inf_{C \in \mathscr{C}(F, E)} \sup_{y \in F} f(C(y), y).$$

11. If E is a *compact convex* set, if $f : E \times E \mapsto \mathbb{R}$ is *lower semicontinuous* with respect to the *first* variable and *concave* with respect to the *second* variable, then there exists $\bar{x} \in E$ such that

$$\sup_{y \in F} f(\bar{x}, y) \leq \sup_{y \in F} f(y, y) \text{(Ky-Fan)}.$$

12. Suppose that E and F are *compact* and *convex* and that $f: E \times F \mapsto \mathbb{R}$ is *convex* and *lower semicontinuous* with respect to the *first* variable and *concave and upper semicontinuous* with respect to the *second* variable. Then there exists a saddle point $\{\bar{x}, \bar{y}\} \in E \times F$:

$$\sup_{y \in F} f(\bar{x}, y) = f(\bar{x}, \bar{y}) = \inf_{x \in E} f(x, \bar{y}) \text{ (Von Neumann)}.$$

13. Suppose that the n sets E_i are convex and compact and that the n functions $f_i: \prod_{j=1}^{n} E_j \mapsto \mathbb{R}$ are *continuous* and satisfy:

$$y_i \mapsto f_i(x_1, \ldots, y_i, \ldots, x_n)$$

are *convex* for all $x_j \in E_j$ $(j \neq i)$. Then there exists a noncooperative equilibrium $x = \{x_1, \ldots, x_n\}$:

$$\forall i = 1, \ldots, n, \qquad \forall y_i \in E_i, \qquad f_i(x) \leq f_i(x_1, \ldots, y_i, \ldots, x_n).$$

14. Suppose that the I demand functions $\{p, r\} \mapsto f_i(p, r)$ and income function $p \mapsto r_i(p)$ are continuous and that the total income function $r = \sum_{i=1}^{I} r_i$ is convex. Then there exists a Walras equilibrium $p \in S^n$:

$$\forall q \in S^n, \qquad \langle q, f_i(p, r_i(p)) \rangle \leq r(q).$$

Exercises

Chapter 1

1. Show that in $]0, 1]$ the sequence $x_n = 1/n$ is a Cauchy sequence but does not converge.

2. **a.** Let $\{x_n\}_{n \in \mathbb{N}}$ be a sequence in a metric space E. Suppose that the subsequences

$$\{x_{2n}\}_{n \in \mathbb{N}}, \qquad \{x_{2n+1}\}_{n \in \mathbb{N}}, \qquad \text{and} \qquad \{x_{3n}\}_{n \in \mathbb{N}}$$

are convergent. Show that $\{x_n\}$ is convergent (but if we assume that only two of these three subsequences are convergent, the result is no longer true).

   b. Give an example of a sequence $\{x_n\}$ of real numbers that is not convergent but that has the property that for every integer $k \geq 2$ the subsequence $\{x_{kn}\}_{n \in \mathbb{N}}$ is convergent.

3. Let E be a metric space. Show that every Cauchy sequence in E is bounded.

4. Let E be a metric space and $\{a_{m,n}\}$ be a double sequence in E for which

$$\lim_{n \to \infty} a_{m,n} = a_m \qquad \text{and} \qquad \lim_{m \to \infty} a_m = a.$$

Show that there exists a subsequence $\{p_n\}$ such that $\lim_{n \to \infty} a_{n, p_n} = a$.

5. Let E be a set with a semidistance, that is, a mapping e from $E \times E$ to $\mathbb{R}_+$ possessing all the properties of a distance with the possible exception of

$$e(x, y) = 0 \Rightarrow x = y.$$

We define the relation $\mathscr{R}$ on E as follows:

$$x \mathscr{R} y \qquad \text{if and only if} \qquad e(x, y) = 0.$$

   a. Show that $\mathscr{R}$ is an equivalence relation.

b. Show that on $E/\mathscr{R}$ the semidistance e is compatible with the equivalence relation, that is, that

$$\left.\begin{array}{l} x_1 \mathscr{R} x_2 \\ y_1 \mathscr{R} y_2 \end{array}\right\} \Rightarrow e(x_1, y_1) = e(x_2, y_2).$$

c. Show that the mapping on $E/\mathscr{R}$ induced by e is a distance.

6. Given a set E and a mapping d from $E \times E$ to $\mathbb{R}_+$ such that

$$\begin{cases} d(x, y) = 0 \Leftrightarrow x = y, \\ d(x, z) \le d(x, y) + d(y, z) \qquad \forall x, y, z \in E, \end{cases}$$

show that δ defined by

$$\delta(x, y) = \sup\{d(x, y), d(y, x)\}$$

is a distance on E.

7. Let f be a continuous bounded function from $\mathbb{R}$ to $\mathbb{R}$. We define the mapping d from $\mathbb{R} \times \mathbb{R}$ to $\mathbb{R}_+$ by

$$d(a, b) = \sup_{x \in \mathbb{R}} | f(x - a) - f(x - b)|.$$

a. Show that d is a semidistance on $\mathbb{R}$ (see exercise 5). (One must first show that the range of d is in $\mathbb{R}_+$.)

b. Show that d is a distance if and only if f is not periodic.

8. We say that a metric space (E, d) is an ultrametric space if the distance d satisfies

$$\forall x, y, z \in E: d(x, z) \le \sup(d(x, y), d(y, z)).$$

a. Show that

$$[d(x, y) \ne d(y, z)] \Rightarrow d(x, z) = \sup\{d(x, y), d(y, z)\},$$

that is, that every triangle is isosceles, the two equal sides being at least equal to the third.

b. Show that every point of a ball is a center for this ball and that if two balls are not disjoint, one is contained in the other. (*Note*: This result is to be established for open balls *and* for closed balls.)

c. Show that in an ultrametric space $\{u_n\}_{n \in \mathbb{N}}$ is a Cauchy sequence if and only if $\lim_{n \to +\infty} d(u_n, u_{n+1}) = 0$.

d. Show that the elements of a Cauchy sequence that do not converge to a given point a remain at a constant distance from a after a certain index.

9. *The p-adic numbers.* Let p be a fixed prime number. We set $\forall r \in \mathbb{Q}$, $|r| = 1/p^\alpha$ where $r = p^\alpha(a/b)$ ($\alpha \in \mathbb{Z}$) for a and b relatively prime to p, and $|0| = 0$.

a. Show that $(\mathbb{Q}, |\cdot|)$ is a field with absolute value (see Section 1.4).

b. Show that the associated distance is ultrametric (see exercise 8).

10. Let E be a vector space and d a distance on E such that

(1) $\qquad\qquad d(x + t, y + t) = d(x, y) \qquad \forall x, y, t \in E$

(2) $\qquad\qquad d(\lambda x, \lambda y) = |\lambda| d(x, y) \qquad \forall x, y \in E \qquad \forall \lambda \in \mathbb{R}.$

Show that $\|x\| = d(x, 0)$ is a norm on E and that d is the associated distance (that is, verify that $d(x, y) = \|x - y\|$ for all x, y in E).

11. Let E be a vector space with the discrete distance defined by $d(x, y) = 1$ if $x \neq y$ and $d(x, y) = 0$ if $x = y$. Is it possible to define a norm on E for which d is the associated distance (in the sense that $d(x, y) = \|x - y\|$ for all $x, y \in E$)? (Use the preceding exercise.)

12. Let E be a vector space on $\mathbb{R}$ and C be a convex set that is symmetric about the origin (that is, $-x \in C$ if $x \in C$) and that contains no line but meets all the lines through the origin in at least one point other than the origin. Show that the gauge from E to $\mathbb{R}_+$ defined by

$$p(x) = \inf\{r \text{ such that } r \in \mathbb{R}_+ \text{ and } x/r \in C\}$$

is a norm on E (see Section 3.8).

13. Let $S = \{a_i\}_{i \in I}$ be a family of elements of a normed vector space E. We denote by $F(I)$ the set of finite subsets of I and set

$$S_J = \sum_{i \in J} a_i \qquad \text{where} \qquad J \in F(I).$$

We say that S is summable if there exists a in E such that

$$\forall \varepsilon > 0 \qquad \exists J_0 \in F(I) \qquad \forall J \in F(I), \qquad J \supset J_0 \Rightarrow \|a - S_J\| < \varepsilon.$$

Show the following:

a. If S is summable, it satisfies the following condition, which we call the Cauchy condition:

$$\forall \varepsilon > 0 \qquad \exists J_0 \in F(J) \qquad \forall J \in F(I), \qquad J \supset J_0 : \|S_J - S_{J_0}\| < \varepsilon.$$

b. If $E = \mathbb{R}$, S is summable if and only if the set of its finite sums is bounded.

c. If S is summable, it contains at most a countable number of non-zero elements. (First, prove this property in the case where $E = \mathbb{R}$ and where the a_i's are positive or zero, and then use the "Cauchy condition".)

d. If E is complete, S is summable if and only if it satisfies the "Cauchy condition."

e. If E is complete, every subset of a summable family is summable.

f. If E is complete and if the family $\{\|a_i\|\}_{i \in I}$ is summable in $\mathbb{R}_+$, then S is summable.

14. Let E be the set of continuous functions f from $[0, 1]$ to $\mathbb{R}$ such that $f(1) = 0$. Is the following function from E to $\mathbb{R}$ a norm:

$$q(f) = \left[\int_0^1 (|f(x)|)^{1/2} dx\right]^2 ?$$

(Show that q does not satisfy the triangle inequality.)

15. Let E_p be the vector space of real polynomials of degree less than or equal to p. Let $x_0, \ldots, x_p$ be $p + 1$ distinct real numbers.

   a. Show that the following function defines a norm on E_p:

$$P \to \sum_{i=0}^p |P(x_i)|.$$

   b. Let F be the following isomorphism from E_p to $\mathbb{R}^{p+1}$:

$$P(x) = a_0 + a_1 x + \cdots + a_p x^p \to (a_0, \ldots, a_n).$$

Calculate the norm induced by F from the preceding norm. We obtain many new examples of norms on $\mathbb{R}^n$ in this fashion.

16. Show that the set c_0 of sequences of real numbers $x = \{x_n\}_{n \geq 0}$ such that $\lim_{n \to \infty} x_n = 0$ is a Banach space for the norm $\|x\|_0 = \sup_{n \geq 0} |x_n|$.

17. Let X be a set and E a Banach space. Show that the space $\mathcal{U}(X, E)$ of bounded functions from X to E is a Banach space for the following norm:

$$\|f\|_\infty = \sup_{x \in X} \|f(x)\|_E.$$

18. Let X be a set and E a Banach space. Show that the space $\mathcal{C}_\infty(X, E)$ of continuous functions from X to E is a Banach space for the norm $\|\cdot\|_\infty$. (See exercise 17.)

19. Let $\mathcal{C}([a, b])$ be the space of continuous functions from the interval $[a, b]$ to $\mathbb{R}$.

   a. Show that the function $\int_a^b |f(x)| dx$ defines a norm on $\mathcal{C}([a, b])$.

   b. Show that $\mathcal{C}([a, b])$ is not a Banach space with this norm. To this end consider the sequence $\{f_n\}$ in $\mathcal{C}([a, b])$:

$$\begin{cases} f_n(x) = n^{1/2} & \text{if} \quad a \leq x \leq \left(a + \dfrac{1}{n}\right), \\[2ex] f_n(x) = \dfrac{1}{(x - a)^{1/2}} & \text{if} \quad \left(a + \dfrac{1}{n}\right) \leq x \leq b. \end{cases}$$

20. Let E and F be two normed vector spaces and $A \in \mathcal{L}(E, F)$ be a continuous linear operator. We set

$$a_1 = \sup_{x \neq 0} \frac{\|Ax\|}{\|x\|}, \qquad a_2 = \sup_{\|x\| = +1} \|Ax\|,$$

$$a_3 = \sup_{\substack{\|x\| < +1 \\ x \neq 0}} \|Ax\|, \qquad a_4 = \inf\{c > 0 \mid \|Ax\| < c\|x\|, x \in E\}.$$

Show that the four numbers a_1, a_2, a_3, and a_4 are equal.

21. Let E be a normed vector space. Show that every linear operator from $\mathbb{R}^n$ to E is continuous. (Compare this result with Proposition 1.10.3).

22. Let $a < c < b$ be three real numbers. Consider the function from $\mathscr{C}([a, b])$ to $\mathbb{R}$:

$$f \to f(c).$$

Show that this function is linear but not continuous when $\mathscr{C}([a, b])$ has the norm defined in exercise 19.

23. Let E be a normed vector space and X a vector subspace that is dense in E (that is, every point of E is the limit of a sequence of points of X). Let f be a linear mapping from X to $\mathbb{R}^n$ that is continuous. Show that there exists a *unique* continuous linear mapping g from E to $\mathbb{R}^n$ such that the restriction of g to X is equal to f and

$$\sup_{\|x\| \leq 1} \|g(x)\|_{\mathbb{R}^n} = \sup_{\{\|x\| \leq 1\} \cap X} \|f(x)\|_{\mathbb{R}^n}.$$

24. Consider the function from $\mathscr{C}_\infty([0, 1])$ to $\mathbb{R}$ defined by

$$f \to \int_0^1 f(x)dx.$$

Show that this function is linear and continuous, and calculate its norm.

25. Let E and F be two normed vector spaces, and let A be a linear mapping from E to F. We assume that if $\{x_n\}_{n \in \mathbb{N}}$ is a sequence in E converging to zero, then $\{A(x_n)\}_{n \in \mathbb{N}}$ is a bounded sequence in F. Show that under these conditions A is continuous.

26. Let E be a normed vector space.
  a. Show that

$$\forall A, B \in \mathcal{L}(E, E): \|A \circ B\|_{\mathcal{L}(E, E)} \leq \|A\|_{\mathcal{L}(E, E)} \cdot \|B\|_{\mathcal{L}(E, E)}.$$

b. Application: Let A and B be two elements of $\mathcal{L}(E, E)$ such that

$$(*) \qquad\qquad A \circ B - B \circ A = 1_E.$$

   i. Deduce from (*) that

$$\forall n \in \mathbb{N}, \ A \circ B^{n+1} - B^{n+1} \circ A = (n + 1)B^n.$$

   ii. Show that

$$\forall n \in \mathbb{N}, \ (n + 1) \cdot \|B^n\| \le 2\|A\| \cdot \|B\| \cdot \|B^n\|.$$

   iii. Deduce that B is null (using ii and then i).

Hence conclude that (*) was absurd.

27. Let E be the set of continuous functions from $[0, 1]$ to $\mathbb{R}$. Consider the following three norms on E:

$$\|f\|_\infty = \sup_{x \in [0, 1]} |f(x)|,$$

$$\|f\|_1 = \int_0^1 |f(t)| \, dt,$$

$$\|f\|_2 = \left[\int_0^1 |f(t)|^2 \, dt \right]^{1/2}.$$

If f and h are two elements of E, we define $h \cdot f$ by $(h \cdot f)(x) = h(x)f(x)$, and we consider the mapping φ from E to E:

$$f \to h \cdot f = \varphi(f)$$

for a fixed h in E.

a. Show that this mapping is linear and continuous for each of the three norms given above for E (the same norm "for domain and range").

b. Show that the three corresponding norms for φ in $\mathcal{L}(E, E)$ are less than or equal to $\|h\|_\infty$.

c. Show that these norms are equal to $\|h\|_\infty$.

28. Show that the bilinear form:

$$\langle f, g \rangle = \int_0^1 f(x)g(x) \, dx$$

is a scalar product on $\mathscr{C}_\infty([0, 1])$.

29. Show that the bilinear form:

$$((f, g))_1 = \int_0^1 f(x)g(x) \, dx + \int_0^1 \frac{d}{dx} f(x) \frac{d}{dx} g(x) \, dx$$

is a scalar product on $\mathscr{D}^{(1)}([0, 1])$.

30. Let $\mathscr{P}$ be the vector space of real polynomials. We set

$$\text{if} \qquad p \in \mathscr{P} : \|p\|_A = \sup_{x \in A} |p(x)|,$$

where A is a bounded subset of $\mathbb{R}$.

a. Show that $e(p, q) = \|p - q\|_A$ is a semidistance on $\mathscr{P}$. (See exercise 2.)

b. Show that this is a norm if and only if A is infinite.

c. If A is the interval $[0, 1]$, show that the linear mapping from $\mathscr{P}$ to $\mathbb{R}$:

$$p \to p(\alpha) \qquad \text{where} \qquad \alpha \in \mathbb{R}$$

is continuous if $\alpha \in [0, 1]$ and is not continuous if $\alpha \notin [0, 1]$. (Consider polynomials of the form $(kx)^n$ if $\alpha > 1$ and of the form $[k(1 - x)]^n$ if $\alpha < 0$.)

31. Let $\mathscr{F}$ be the set of functions from the interval $[a, b]$ to $\mathbb{R}$. For f in $\mathscr{F}$, we set

$$M_f = \sup_{\substack{s \neq t \\ s, t \in [a, b]}} \frac{|f(s) - f(t)|}{|s - t|^\alpha},$$

where α is a constant such that $0 < \alpha \leq 1$. We then define

$$\text{Lip } \alpha = \{f \in \mathscr{F} \mid M_f < +\infty\}.$$

a. Show that Lip α is a vector subspace of $\mathscr{F}$.

b. For f in Lip α, we set

$$\|f\|_1 = |f(a)| + M_f,$$
$$\|f\|_2 = \sup_{x \in [a, b]} |f(x)| + M_f.$$

Show that in this fashion we define two norms, $\|\cdot\|_1$ and $\|\cdot\|_2$, on Lip α.

c. Show that Lip α is a Banach space for these two norms.

d. Are the functions in Lip α continuous? differentiable? (Study this question according to the value of α).

32. We define the following function on $\mathscr{D}^{(1)}(]0, 1[)$ (cf. example 1.11.4):

$$\|f\| = \sup_{x \in]0, 1[} |f(x)| + \sup_{x \in]0, 1[} |f'(x)|.$$

a. Show that this is a norm. (First show that its range is in $\mathbb{R}_+$).

b. Is $\mathscr{D}^{(1)}(]0, 1[)$ a Banach space with this norm?

    c. Show that the linear form on $\mathscr{D}^{(1)}(]0, 1[)$

$$f \to f'(1/2)$$

is continuous for this norm. What is its norm?

    d. Show that this linear form is no longer continuous if $\mathscr{D}^{(1)}(]0, 1[)$ has the following norm:

$$\|f\|_\infty = \sup_{x \in]0, 1[} |f(x)|.$$

33. In $l^\infty(\mathbb{N})$ whose elements are denoted as $x = \{x_n\}_{n \in \mathbb{N}}$ we consider the operator τ_k from $l^\infty(\mathbb{N})$ to itself defined by

$$\tau_k(x) = (x_k, x_{k+1}, x_{k+2}, \dots, x_{k+n}, \dots)$$

or as

$$[\tau_k(x)]_n = x_{n+k}.$$

    a. Show that τ_k is a continuous linear operator with norm 1.

    b. If $(a_k)_{k \in \mathbb{N}}$ is an absolutely convergent series of real numbers, show that the operator A defined by

$$(Ax)_n = \sum_{k=0}^{\infty} a_k x_{n+k} \qquad \text{where} \qquad x = (x_n) \in l^\infty(\mathbb{N})$$

(with $(Ax)_n$ denoting the nth term of Ax) takes its values in $l^\infty(\mathbb{N})$. Show also that this is a continuous endomorphism of $l^\infty(\mathbb{N})$ and justify the notation $A = \sum_{k=0}^{\infty} a_k \tau_k$ (which is an equality in $\mathscr{L}(l^\infty(\mathbb{N}), l^\infty(\mathbb{N}))$.

    c. Show that

$$\|A\|_{\mathscr{L}(l^\infty, l^\infty)} = \sum_{k=0}^{\infty} |a_k|.$$

Chapter 2

1. Let $(E, \|\cdot\|)$ be a normed vector space. We denote by $\delta(B(a, r))$ the diameter of the ball $B(a, r) = \{x \in E \mid \|x - a\| \leq r\}$ and by $\delta(S(a, r))$ the diameter of the sphere $S(a, r) = \{x \in E \mid \|x - a\| = r\}$. Show that for all $a \in E$ and for all $r > 0$, $\delta(B(a, r)) = \delta(S(a, r))$. Give a counterexample to this proposition in the case where E is only a metric space.

2. Let (E, d) be a metric space. Show that a sequence $\{x_n\}$ is a Cauchy sequence if and only if $\lim_{n \to \infty} \delta(A_n) = 0$ where $\delta(A_n)$ denotes the diameter of the set $A_n = \{x_{n+p}\}_{p \geq 0}$.

3. Let E be a metric space. Show that a subset A of E is bounded if and only if there exist $a \in E$ and $r > 0$ such that A is included in the ball $B(a, r)$ with center a and radius r.

4. Let (E, d) be a metric space. Show that for all $a \in E$ and $r > 0$, the sphere $S(a, r) = \{x \in E \mid d(x, a) = r\}$ is closed in E.

5. Let (E, d) be a metric space and A an arbitrary subset of E. For all, $r > 0$ we set $\mathring{B}(A, r) = \{x \in E \mid d(x, A) < r\}$.
   a. Show that $\mathring{B}(A, r) = \bigcup_{x \in A} \mathring{B}(x, r)$.
   b. Show that $\overline{A} = \bigcap_{r > 0} \mathring{B}(A, r)$.

6. Show that in a real vector space two norms that define the same unit ball are identical.

7. Let E be a metric space. If A is a subset of E, we denote by A' the derived set of A where

$$A' = \{x \in E \mid x \in \overline{A \setminus \{x\}}\},$$

where $A \setminus B$ denotes $A \cap \complement[A \cap B]$. Show the following:
   a. A' is the set of those points x of E, which are the limit of a sequence of points of A, which *never assumes the value* x.
   b. For every subset A, A' is closed.
   c. The set A is closed if and only if it contains its derived set.

8. In a normed vector space, show that the closure of a vector subspace is a vector subspace

9. In $l^\infty(\mathbb{N})$, we denote by c the vector subspace of convergent sequences and by c_0 the vector subspace of sequences that converge to zero. Show that c and c_0 are closed in $l^\infty(\mathbb{N})$ (cf. exercise 16, chapter 1).

10. Show that the interior $\mathring{A}$ of a set A has the following three properties:

$$\mathring{A} \subset A; \qquad A \subset B \Rightarrow \mathring{A} \subset \mathring{B}; \qquad \widehat{A \cap B} = \mathring{A} \cap \mathring{B}.$$

11. Let $(E, \|\cdot\|)$ be a normed vector space.
   a. Show that the closed ball $B(a, r) = \{x \in E \mid \|a - x\| \leq r\}$ is the closure of the open ball $\mathring{B}(a, r) = \{x \in E \mid \|a - x\| < r\}$.
   b. Show that the open ball $\mathring{B}(a, r)$ is the interior of the closed ball $B(a, r)$.
   c. Show that the sphere $S(a, r) = \{x \in E \mid \|a - x\| = r\}$ is the boundary of the open ball $\mathring{B}(a, r)$ and of the closed ball $B(a, r)$. (The boundary of a subset $A \subset E$ is the set $\mathrm{Fr}(A) = \{x \in E \mid x \in \overline{A} \text{ and } x \notin \mathring{A}\}$).
   d. Give a counterexample to each of the three propositions a, b, and c in the case where E is only a metric space.

12. Let A be a subset of a metric space E. Show that the set $A \cup (\widehat{\complement A})$ is dense in E.

13. If A and B are two subsets of a metric space E, show that $\overset{\circ}{\widehat{A \cup B}} \supset \overset{\circ}{A} \cup \overset{\circ}{B}$. Give an example that shows that the opposite inclusion is false.

14. Let E be a metric space. Show that a subset A of E intersects every dense subset of E if and only if A has a nonempty interior.

15. Let A be a subset of a metric space E. Note that if $\bar{A} = \overset{\circ}{A}$, then A is open. Show that the contrary is false.

16. Let A be an arbitrary subset of $\mathbb{R}^n$. Setting

$$\mathbb{R}^n_+ = \{(x_1, \ldots, x_n) \in \mathbb{R}^n | x_i \geq 0, \qquad \forall i = 1, \ldots, n\},$$

show that $A + \overset{\circ}{\mathbb{R}}^n_+ = \overset{\circ}{\widehat{(A + \mathbb{R}^n_+)}}$.

17. Let E be a metric space.

    a. Show that for any subsets A and B of E:

$$\overline{A \cap B} \subset \bar{A} \cap \bar{B}.$$

    Show that the converse is false.

    b. Show that if A is open in E, then

$$\forall B \subset E : \bar{A} \cap \bar{B} \subset \overline{A \cap B}.$$

    c. Find an example in $\mathbb{R}$ of two intervals A and B such that $A \cap \bar{B}$ is not contained in $\overline{A \cap B}$.

    d. Find an example in $\mathbb{R}$ of two open sets A and B such that the following four sets are different:

$$A \cap \bar{B}, \qquad B \cap \bar{A}, \qquad \overline{A \cap B}, \qquad \bar{A} \cap \bar{B}.$$

18. Let U be an open set of $\mathbb{R}$ and x_0 a point of U.

    a. Show that the greatest lower bound b of the set $\complement U \cap [x_0, +\infty[$ is not in U and that $[x_0, b[\subset U$.

    b. By making the same reasoning for $]-\infty, x_0]$, show the existence of an interval $]b', b[$ such that

$$x_0 \in]b', b[; \qquad b \text{ and } b' \notin U; \qquad \text{and} \qquad]b', b[\subset U.$$

    c. *Application.* Show that the most general open set of the real line is the union of at most a countable number of open intervals that are mutually disjoint.

19. In a metric space E, if $x_1, \ldots, x_n$ are n distinct points, show that there exist n neighborhoods of these points $V_1, \ldots, V_n$ (V_i being the neighborhood of x_i), which are mutually disjoint.

20. In a metric space show that the intersection of the neighborhoods of an arbitrary point reduces to this point.

21. We call an isolated point of a subset B of E a point b of B for which there exists a neighborhood V of b such that $V \cap B = \{b\}$. If a subset A of a metric space E has no isolated point, show that the same is true for $\bar{A}$.

22. In a metric space E let $\{x_n\}_{n \in \mathbb{N}}$ be a sequence that converges to an element x. Show that the set

$$\left[\bigcup_{n \in \mathbb{N}} \{x_n\} \right] \cup \{x\}$$

is compact in E.

23. **a.** Consider

$$A = \left\{ n + \frac{1}{n} \,\middle|\, n \in \mathbb{N}, \qquad n \geq 2 \right\} \subset \mathbb{R}$$

$$B = \{-n \,|\, n \in \mathbb{N}, \qquad n \geq 2\} \subset \mathbb{R}$$

Show that A and B are closed in $\mathbb{R}$ but that $A + B$ is not closed.

b. Let A and B be the following two subsets of $\mathbb{R}^2$:

$$A = \{(x, y) \,|\, x \geq 0, \qquad x \cdot y \geq 1\},$$

$$B = \{(x, y) \,|\, x \geq 0, \qquad x \cdot y \leq -1\}.$$

Show that A and B are closed and that $A + B$ is open.

c. Show that if A is compact and B is closed, then $A + B$ is closed. Give an example to show that $A + B$ is not in general compact under these conditions.

24. We recall that the compact subsets of $\mathbb{R}^n$ are those subsets that are closed and bounded.

a. Show that this property does not hold in an infinite dimensional normed vector space. (For example, look for a sequence that does not have a cluster point in the unit ball of $l^\infty(\mathbb{N})$.)

b. In the Hilbert space $l^2(\mathbb{N})$ whose elements are denoted $x = \{x_n\}_{n \in \mathbb{N}}$ we consider the subset

$$Q = \left\{ x \in l^2(\mathbb{N}) \,|\, \forall n \in \mathbb{N} \,|\, x_n| \leq \frac{1}{n} \right\}.$$

Show that Q is not contained in any finite dimensional vector subspace of $l^2(\mathbb{N})$. Show that Q is compact.

c. Let $\{x_n\}_{n \in \mathbb{N}}$ be a sequence in a normed vector space E. Suppose that the following condition, denoted (H), is satisfied:

$$(H): \quad \begin{cases} \forall x \in E \text{ the sequence of positive numbers} \\ \|x_n - x\| \text{ converges.} \end{cases}$$

Show that if E is of finite dimension, we deduce from (H) that $\{x_n\}_{n \in \mathbb{N}}$ is a convergent sequence in E. Give an example of a sequence that satisfies (H) in $E = l^1(\mathbb{N})$ but that, nevertheless, does not converge in $l^1(\mathbb{N})$.

25. Let E be a real vector space.
    a. Show that the intersection and the sum of a finite family of convex sets is convex.
    b. Show that every vector subspace of E is a convex set.

26. Let E be a normed vector space. Show that the closed balls and the open balls of E are convex sets.

27. Give an example of a closed set A in a normed space E whose convex hull is not closed. (For example, $A = F \cup G$ where

$$F = \{(x, y) \in \mathbb{R}^2 \,|\, x \geq 0, \quad \text{and} \quad xy \geq 1\},$$
$$G = \{(x, y) \in \mathbb{R}^2 \,|\, x \geq 0, \quad \text{and} \quad xy \leq -1\}).$$

28. **a.** Let K be a compact convex subset of $\mathbb{R}^n$ that does not contain the origin, and let $\tilde{K} = \bigcup_{\lambda \geq 0} \lambda K$ be the convex cone generated by K. Show that $\tilde{K}$ is closed.
    b. Show that the preceding proposition is false without the hypothesis that $0 \notin K$. (Consider, for example, $K = \{(x, y) \in \mathbb{R}^2 \,|\, (x - 1)^2 + y^2 \leq 1\}$.)

29. Let E be an ultrametric space. (See exercise 8, Chapter 1.) Show that every ball, open or closed, is both open and closed in E.

30. In a metric space E a subset A is called topologically free if $\forall x \in A : x \notin \overline{A - \{x\}}$. In $l^\infty(\mathbb{N})$, we define the sequence $(x_n)_{n \in \mathbb{N}}$ by

$$\text{for } n = 0: \quad (x_0)_i = 1 \qquad i = 1, 2, \dots.$$

$$\text{for } n \geq 1: \quad (x_n)_i = 1 \qquad \text{if} \qquad i \neq n.$$

$$(x_n)_n = 1 + \frac{1}{n}.$$

Show that for every p the set

$$\{x_n \,|\, 0 \leq n \leq p\}$$

is topologically free, but that the union of these sets

$$\{x_n \,|\, n \in \mathbb{N}\}$$

is not topologically free.

31. Let E be a metric space and A a subset of E. We recall that the boundary of A is defined by

$$\text{Fr}(A) = \overline{A} \backslash \mathring{A}$$

    a. Show that the following formula holds:

$$\text{Fr}(A) = [A \cap \complement \overline{A}] \cup [\overline{A} \backslash A].$$

b. Show that

$$\mathrm{Fr}(\bar{A}) \subset \mathrm{Fr}(A).$$

Give a counterexample showing that, in general, the equality $\mathrm{Fr}(\bar{A}) = \mathrm{Fr}(A)$ does not hold.

Chapter 3

1. Show that the image of a Cauchy sequence under any uniformly continuous function is a Cauchy sequence.
2. Let E and F be two metric spaces. Show that the following four properties for a function f from E to F are equivalent:

   i. f is continuous.

   ii. $\forall A \subset E : f^{-1}(\mathring{A}) \subset \widehat{f^{-1}(A)}$.

   iii. $\forall A \subset E : \overline{f^{-1}(A)} \subset f^{-1}(\bar{A})$.

   iv. $\forall A \subset E : f(\bar{A}) \subset \overline{f(A)}$.

3. Let (E, d) be a metric space. Show that the mapping $d : (x, y) \to d(x, y)$ from $E \times E$ to $\mathbb{R}$ is uniformly continuous.
4. Show that the extended reals $\bar{\mathbb{R}} = [-\infty, +\infty]$ (furnished with the distance $\bar{d}$ defined in Section 1.3) is a compact metric space.
5. Let (E, d) be a metric space in which the closed balls are compact. Show that for every nonempty closed subset A of E and for every $x \in E$, there exists a in A such that $d(x, A) = d(x, a)$.
6. Let (E, d) be a metric space, A a nonempty closed subset of E, and B a nonempty compact subset of E such that $A \cap B = \varnothing$. We define $d(A, B) = \inf_{\substack{a \in A \\ b \in B}} d(a, b)$. Show that $d(A, B) > 0$.
7. Let (E, d) be a metric space, A a compact subspace of E, and U an open subspace of E containing A. For $\varepsilon > 0$ we set

$$\mathring{B}(A, \varepsilon) = \{x \in E \mid d(x, A) < \varepsilon\}.$$

a. Show that there exists $\varepsilon > 0$ such that

$$\mathring{B}(A, \varepsilon) \subset U.$$

b. Show that this result is not true in general if we suppose that A is closed rather than compact. (Consider, for example,

$$E = \mathbb{R}^2, \qquad A = \{(0, y) \mid y \in \mathbb{R}\}) \qquad \text{and} \qquad U = \{(x, y) \mid xy < 1\}.$$

8. Show that the space $\mathbb{R}$ of real numbers is homeomorphic to the following sets:
    a. The interval $\mathbb{R}_+ =]0, +\infty[$.
    b. An open interval $]a, b[$ $(a < b)$.
    c. The subset $\{(x, 1/x) | x > 0\}$ of $\mathbb{R}^2$.
    d. The set $C \backslash \{A\}$ where C is the subset $\{(x, y) | x^2 + y^2 = 1\}$ of $\mathbb{R}^2$ and A is the point $(0, 1)$.

9. Show that the following function from $\mathbb{R}^2$ to $\mathbb{R}$ is continuous:

$$f(x, y) = e^{-1/|x \cdot y|} \qquad \text{if} \qquad x \cdot y \neq 0.$$

$$f(x, y) = 0 \qquad \text{if} \qquad x \cdot y = 0.$$

10. The vector space of quadratic forms on $\mathbb{R}^n$ is isomorphic to $\mathbb{R}^N$ where $N = n(n - 1)/2$. Show that the set of positive definite quadratic forms is open in $\mathbb{R}^N$.

11. In the set of M of $n \times n$ matrices with real coefficients show that the set U of those matrices that have an inverse is open. (We may give M any norm for a finite dimensional vector space.) Show as well that the mapping $A \rightarrow A^{-1}$ is a homeomorphism from U to itself.

12. Let E, F, G be three metric spaces and f and g continuous mappings from E to F and from F to G, respectively. We suppose that f is surjective and that $g \circ f$ is a homeomorphism from E to G. Show that f is a homeomorphism from E to F and that g is a homeomorphism from F to G.

13. Let E be a metric space and A and B be two nonempty subsets of E such that

$$A \cap \bar{B} = \bar{A} \cap B = \varnothing.$$

Show that there exist two open sets U and V in E such that

$$U \supset A, \qquad V \supset B, \qquad \text{and} \qquad U \cap V = \varnothing.$$

(We may consider the function: $x \rightarrow d(x, A) - d(x, B)$.)

14. Let f be a continuous function from $[0, 1]^2$ to $\mathbb{R}$. For x in $[0, 1]$, we set

$$\varphi(x) = \sup_{0 \leq y \leq x} f(x, y).$$

Show that φ is a continuous function on $[0, 1]$.

15. We consider the following function from l^∞ to $\mathbb{R}$:

$$f(\{x_n\}) = \sup_{n \in \mathbb{N}} (x_n + x_{n+1}).$$

If l^∞ has the usual norm $\| \cdot \|_\infty$, show that f is continuous.

16. Let E and F be two metric spaces. We say that a mapping f from E to F is open if the image under f of every open set in E is open in F, and

similarly that a mapping is closed if the image of every closed set is closed.

a. Show that if the mapping f is bijective then f is open if and only if f is closed.

b. Let A be a subset of E and i the canonical injection from A to E. Show that i is open if and only if A is open in E and that i is closed if and only if A is closed in E.

17. Let E and F be two metric spaces and f a mapping from E to F.

a. Show that f is an open mapping if and only if for every subset A of X:

$$f(\mathring{A}) \subset \mathring{\widehat{f(A)}}.$$

b. Show that f is a closed mapping if and only if for every subset A of X:

$$\overline{f(A)} \subset f(\overline{A}).$$

18. Let f be the projection of $\mathbb{R}^2$ onto $\mathbb{R}$ defined by $f(x, y) = x$ for all $x, y \in \mathbb{R}$.

a. Show that f is an open mapping.

b. Show that f is not a closed mapping.

19. Let E be a real normed vector space. Show that the vector subspaces of E of finite dimension are closed.

20. Let E and F be two normed spaces and f a mapping from E to F such that

    i. $f(x + y) = f(x) + f(y)$ for all x, y in E.

    ii. $\exists K > 0$ $\forall x \in E,$ $\|x\| \leq +1 \Rightarrow \|f(x)\| \leq K.$

a. Show that $f(\lambda x) = \lambda f(x)$ for all x in E and for every rational number λ.

b. Show that for every sequence $\{x_n\}$ of E such that $\lim_{n \to +\infty} x_n = 0$, we have $\lim_{n \to +\infty} f(x_n) = 0$.

c. Deduce from the above that f is linear and continuous.

21. Let f be a continuous mapping from E to F and g a continuous mapping from F to G. Show that

a. If f and g are proper, then $g \circ f$ is proper.

b. If $g \circ f$ is proper, then f is proper.

c. If $g \circ f$ is proper and if f is surjective, then g is proper.

22. Show that $\mathbb{R}$ and the following subset of $\mathbb{R}^2$:

$$\{(x, y) | x^2 + y^2 = 1\}$$

are not homeomorphic.

23. Let L and X be two compact metric spaces and F a continuous mapping from $L \times X$ to Y. Let y_0 be a fixed element of Y.

    a. Show that

$$L_0 = \{\lambda \in L \mid \exists x \mid y_0 = F(\lambda, x)\}$$

    is a closed set of L.

    b. Suppose that for every λ in L_0 the mapping F_λ from X to Y:

$$x \to F_\lambda(x) = F(\lambda, x)$$

    is bijective. Show that the solutions $x = \varphi(\lambda)$ of the equation $y_0 = F(\lambda, x)$ is continuous in L_0.

24. Let f and g be two lower semicontinuous mappings from a metric space E to $\mathbb{R}$ such that $f(x) > 0$ and $g(x) > 0$ for all x in E. Show that $f \cdot g$ is lower semicontinuous.

25. Let f and g be two functions that are lower semicontinuous at a point x. Show that $f + g$ and $\inf(f, g)$ are also lower semicontinuous at x.

26. Let $\{f_n\}_{n \in \mathbb{N}}$ be a sequence of positive real-valued functions defined in the same interval I of $\mathbb{R}$. We suppose that the series $g(x) = \sum_{n=1}^{\infty} f_n(x)$ converges. One of the following propositions is true, the other is false:

    i. If each f_n is upper semicontinuous, then the function $g(x) = \sum_{n=1}^{+\infty} f_n(x)$ is upper semicontinuous.

    ii. If each f_n is lower semicontinuous, then g is lower semicontinuous.

27. Let E be a normed space.

    a. Show that the mapping $x \to \|x\|$ from E to $\mathbb{R}$ is convex.

    b. Show that the mapping $(x, y) \to \|x - y\|$ from the product space $E \times E$ (defined in Chapter 4) to $\mathbb{R}$ is convex.

    c. Let A be a nonempty convex subset of E. Show that the mapping $x \to d(x, A) = \inf_{y \in A} \|x - y\|$ from E to $\mathbb{R}$ is convex.

28. Let E be a real vector space, X a convex subset of E, and f a mapping from X to $\mathbb{R}$. We set

$$\mathcal{E}\!p(f) = \{(x, y) \in X \times \mathbb{R} \mid y \geq f(x)\}$$

$$\mathcal{E}\!p'(f) = \{(x, y) \in X \times \mathbb{R} \mid y > f(x)\}.$$

    a. Show that the following three propositions are equivalent:

        i. f is a convex function.

        ii. $\mathcal{E}\!p(f)$ is a convex subset of $X \times \mathbb{R}$.

        iii. $\mathcal{E}\!p'(f)$ is a convex subset of $X \times \mathbb{R}$.

    b. We say that a function f from X to $\mathbb{R}$ is "quasiconvex" if for every real λ the sets $s(\lambda) = \{x \in X \mid f(x) \leq \lambda\}$ are convex. Show that a convex function is quasiconvex.

    c. Give an example of a quasiconvex function that is not convex.

29. Let E and F be two real vector spaces, A a convex subset of $E \times F$ and f a convex function that is bounded below on A. Let X be the image of A under the projection of $E \times F$ onto E. For x in X, we set

$$g(x) = \inf_{\substack{y \in F \\ (x,\, y) \in A}} f(x, y)$$

Show that g is a convex function on X.

30. Let $\{(E_n, d_n)\}_{n \in \mathbb{N}}$ be a sequence of complete and bounded metric spaces. We suppose that there exists $\delta > 0$ such that

$$\forall n \in \mathbb{N} \qquad \forall x, y \in E_n, \qquad d_n(x, y) \leq \delta.$$

For each n, let h_n be a mapping from E_{n+1} to E_n. We suppose that h_n is an α_n-contraction (that is, $d_n(h_n(x), h_n(y)) \leq \alpha_n \cdot d_{n+1}(x, y)$). Finally, we suppose that

$$\forall p \in \mathbb{N}, \ \lim_{p \to +\infty} \left(\prod_{n \geq p} \alpha_n \right) = 0.$$

Show that there exists a unique sequence $\{x_n\}_{n \in \mathbb{N}}$ such that $x_n \in E_n$ $\forall n \in \mathbb{N}$, $x_n = h_n(x_{n+1})$.

31. Let X be a set and $\mathscr{U}(X)$ the Banach space of bounded functions from X to $\mathbb{R}$. Given a bounded function g from $X \times X$ to $\mathbb{R}$ and a real number $\alpha(0 \leq \alpha < 1)$, we consider the mapping φ from $\mathscr{U}(X)$ to itself:

$$(\varphi u)(x) = \sup_{y \in X} \{(1 - \alpha) \cdot g(x, y) + \alpha \cdot u(y)\}.$$

Show that φ is a contraction with proportionality factor α and that the ball with center 0 and radius $\|g\|_\infty$ is invariant under φ. Hence deduce that φ has a fixed point in this ball.

32. Let A be a compact subset of $\mathbb{R}^n$. Show that the function $\psi : x \to \max_{y \in A} \sum_{i=1}^{n} y_i \cdot x_i$ from $\mathbb{R}^n$ to $\mathbb{R}$ is continuous.

33. Let E be a normed vector space and $\mathring{B}(a, r)$ the open ball with center a and radius $r(r > 0)$.

    a. Show that $\mathring{B}(a, r)$ is homeomorphic to $\mathring{B}(0, r)$.

    b. Show that the mapping $x \to (r/(1 + \|x\|)) \cdot x$ from E to itself is a homeomorphism from E onto $\mathring{B}(0, r)$. Calculate the inverse homeomorphism.

    c. Hence conclude that $\mathring{B}(a, r)$ is homeomorphic to E.

34. Show that a mapping φ is a homeomorphism from $[0, 1]$ to itself if and only if φ satisfies **i** or **ii**:

   i. φ is continuous, strictly increasing, and

$$\varphi(0) = 0; \qquad \varphi(1) = 1.$$

   ii. φ is continuous, strictly decreasing, and

$$\varphi(0) = 1; \qquad \varphi(1) = 0.$$

35. Let E be a compact metric space and f a continuous mapping from E to itself. Let $\{K_n\}_{n \in \mathbb{N}}$ be a decreasing sequence of nonempty closed subsets of E. Show that

$$f\left(\bigcap_{n=1}^{+\infty} K_n\right) = \bigcap_{n=1}^{+\infty} f(K_n).$$

36. Show that if f and g are two real-valued lower semicontinuous functions such that $f + g$ is continuous, then f and g are continuous.

37. Let E be a metric space and f a continuous mapping from E to itself. Show that the set I of fixed points of f ($I = \{x \in E \mid f(x) = x\}$) is a closed (possibly empty) subset of E. Show that if E is compact and if I is empty then there exists $k > 0$ such that

$$\forall x \in E \qquad d(x, f(x)) \geq k.$$

Chapter 4

1. Let E be the set of continuous functions from $[0, 1]$ to $\mathbb{R}$. We recall the following definitions for f in E:

$$\|f\|_\infty = \sup_{x \in [0, 1]} |f(x)|$$

$$\|f\|_1 = \int_0^1 |f(x)| \cdot dx.$$

a. Show the following inequality for all f in E:

$$\|f\|_1 \leq \|f\|_\infty.$$

b. Show that there exists no $k > 0$ such that for all f in E:

$$\|f\|_\infty \leq k \cdot \|f\|_1.$$

(Consider the sequence of functions $f_n(t) = t^n$.) Hence conclude that the norms $\|\cdot\|_1$ and $\|\cdot\|_\infty$ are not equivalent.

c. Give another proof of the result of **b** using exercise 19 of Chapter 1.

2. Under the hypotheses and notation of exercise 7 of Chapter 1, show that if f is Lipschitz (Section 3.1), the corresponding distance is weaker than the usual distance.

3. We define the following function d on $\mathbb{R} \times \mathbb{R}$:

$$d(x, y) = |x^3 - y^3|.$$

   a. Show that d is a distance on $\mathbb{R}$.
   b. Show that the Cauchy sequences are not the same for this distance and the usual distance.
   c. Show that d defines the same open sets as does the usual distance.

4. Let E be a metric space and X and Y two subsets of E such that $E = X \cup Y$. Show that if $M \subset X \cap Y$ is an open (respectively, closed) subset of the subspace X of E *and* of the subspace Y of E, then M is open (respectively, closed) with respect to E.

5. Let E be a metric space and X and Y two subsets of E such that $E = X \cup Y$. Show that if X and Y are open, then the following proposition is true: A real function defined on E whose restrictions to X and to Y are continuous for the induced distance is continuous on E. Show by means of a counterexample that this property does not hold for arbitrary X and Y.

6. Let E be a metric space, F a subspace of E, and A a subspace of F.
   a. Show that

$$\text{Int}_E(A) \subset \text{Int}_F(A)$$

$$\text{Fr}_F(A) \subset \text{Fr}_E(A)$$

   where $\text{Int}_E(A)$ and $\text{Fr}_E(A)$ (respectively, $\text{Int}_F(A)$ and $\text{Fr}_F(A)$) denote the interior and the boundary of A in the space E (respectively, in the space F).
   b. Show that the inverse inclusions are in general not true. Consider, for example, $E = \mathbb{R}$; $F = [-1, +1]$, and $A = [0, 1]$.

7. Let E be a metric space and A and B two subsets of E such that $B \subset A \subset E$. Show that $\mathring{B}_1 \subset \mathring{B}_2$ where the subscript 1 corresponds to the metric space E and the subscript 2 to its subspace A. What can one say about $\bar{B}_1$ and $\bar{B}_2$? Show that $\text{Fr}_2(B) \subset A \cap \text{Fr}_1(B)$ but that the equality is not in general true. Consider $E = \mathbb{R}$, $A = B = {}^{`}\mathbb{Q}$.

8. Show that the subspaces of a complete metric space are not necessarily complete.

9. Consider the following distance for $\mathbb{Z}$:

$$d(x, y) = |x - y| \qquad \text{for} \qquad x, y \quad \text{in} \quad \mathbb{Z},$$

   where $|\cdot|$ denotes the absolute value in $\mathbb{R}$. Show that $(\mathbb{Z}, d)$ is a complete metric space.

10. Let E and F be two metric spaces and A and B subsets of E and F, respectively. Show that

$$\mathring{A} \times \mathring{B} = \widehat{A \times B}^{\circ}.$$

11. Let E be a metric space. Show that the diagonal

$$D = \{(x, x) \in E \times E \mid x \in E\}$$

is closed in the product space $E \times E$.

12. Let E and F be two metric spaces and e and f two arbitrary points of E and F, respectively. We define the following subspaces of the product space $E \times F$:

$$\mathscr{E} = \{(x, f) \mid x \in E\}$$

$$\mathscr{F} = \{(e, y) \mid y \in F\}$$

and the mappings $u : x \to (x, f)$ from E to $E \times F$ and $v : y \to (e, y)$ from F to $E \times F$. Show that u is an isometry from E onto $\mathscr{E}$ and that v is an isometry from F onto $\mathscr{F}$.

13. Show that the metric space E is isometric to the diagonal

$$D = \{(x, x) \mid x \in E\}$$

of the product space $E \times E$.

14. Let E and F be two metric spaces and A and B two subsets such that $A \subset E$ and $B \subset F$ Show that

$$\mathrm{Fr}(A \times B) = (\mathrm{Fr}(A) \times \bar{B}) \cup (\bar{A} \times \mathrm{Fr}(B)).$$

15. Let X, Y, and Z be three metric spaces. Let A be an open subset of $X \times Y$ and B an open subset of $Y \times Z$. Show that the subset C of $X \times Z$ defined by

$$C = \{(x, z) \in X \times Z \mid \exists y \in Y \mid (x, y) \in A \quad \text{and} \quad (y, z) \in B\}$$

is open.

16. Let E be a normed vector space and $K_1, \ldots, K_n$ nonempty compact subsets of E.

a. Show that the mapping $\varphi : (x_1, \ldots, x_n) \to \sum_{i=1}^{n} x_i$ from $K_1 \times \cdots \times K_n$ to E is continuous and conclude from this that the set $\sum_{n=1}^{n} K_i$ is compact.

b. We suppose that the sets K_i are convex and denote by $S^n = \{(\lambda_1 \cdots \lambda_n) \mid \lambda_i \geq 0, \sum_{i=1}^{n} \lambda_i = +1\}$, the simplex of $\mathbb{R}^n$. Show that the mapping $\psi : (\lambda_1, \ldots, \lambda_n, x_1, \ldots, x_n) \to \sum_{i=1}^{n} \lambda_i x_i$ from $S \times K_1 \times \cdots \times K_n$ to E is continuous. Conclude from this that the convex hull of the union of the sets K_i is compact.

17. Let $\{(E_n, d_n)\}_{n \in \mathbb{N}}$ be a sequence of metric spaces. We suppose that

$$\forall n \in \mathbb{N}, \forall x, y \in E_n \qquad d_n(x, y) \leq 1.$$

Show that the following function is a distance on $E = \prod_{n=1}^{+\infty} E_n$:

$$d(\{x_n\}, \{y_n\}) = \sum_{n=1}^{+\infty} \frac{1}{2^n} d_n(x_n, y_n).$$

18. Let M be the normed vector space with real coefficients. Show that the map $(A, B) \mapsto A \cdot B$ is continuous from $M \times M$ to M.

19. Let f be a uniformly continuous mapping from $\mathbb{R}$ to $\mathbb{R}$. We define for all real a the function $f_a : x \to f(x - a)$ from $\mathbb{R}$ to $\mathbb{R}$. Show that the family of functions $\{f_a | a \in \mathbb{R}\}$ is uniformly equicontinuous.

20. Let f_n be a sequence of increasing functions from an interval $[a, b]$ to $\mathbb{R}$. Suppose that the sequence f_n converges pointwise on $[a, b]$ to a *continuous* function f. Show that the convergence is uniform.

21. Let E be a compact metric space and k a strictly positive real number. We denote by I_k the set of functions from E to a fixed interval $[a, b]$ of $\mathbb{R}$ that are Lipschitz with proportionality factor k. Show that I_k is a compact convex subset of $\mathscr{C}_\infty(E)$.

22. Let $\mathscr{D}$ be the space of functions on $[0, 1]$ that possess derivatives of all orders. We denote by $f^{(p)}$ the pth derivative of $f \in \mathscr{D}$, and we set

$$\|f\|_p = \sup_{x \in [0, 1]} |f^{(p)}(x)|.$$

a. Show that $e(f, g) = \|f - g\|_p$ defines a semidistance on $\mathscr{D}$ (see exercise 5 of Chapter 1).

b. Let a_n $(n = 0, 1, 2, \ldots)$ be a sequence of strictly positive real numbers whose series converges. For f in $\mathscr{D}$ we define

$$\delta(f) = \sum_{n=0}^{+\infty} a_n \frac{\|f\|_n}{1 + \|f\|_n}.$$

Show that $d(f, g) = \delta(f - g)$ is a distance on $\mathscr{D}$.

c. Let n be an integer and ε a positive number and set

$$V(n, \varepsilon) = \left\{ f \in \mathscr{D} \,|\, \sup_{j \leq n} \|f\|_j < \varepsilon \right\}.$$

Show that every neighborhood of 0 in $\mathscr{D}$ (for the distance d) contains a neighborhood $V(n, \varepsilon)$. Deduce from this that the particular choice of the sequence $\{a_n\}$ has no influence on the family of open sets defined by d.

d. We shall say that a subset H of $\mathscr{D}$ is pseudobounded if it satisfies the following property:

$$\forall n \in \mathbb{N} \qquad \exists M \in \mathbb{R} \qquad \forall f \in H : \|f\|_n \le M.$$

Show, using Ascoli's theorem, that every pseudobounded sequence in $\mathscr{D}$ has a convergent subsequence.

e. Hence conclude that the compact subsets of $\mathscr{D}$ are those subsets that are closed and pseudobounded.

23. We denote by $\mathscr{C}^1$ the space of continuously differentiable functions from $[0, 1]$ to $\mathbb{R}$ with the norm:

$$\|f\| = \sup_{x \in [0, 1]} \max(|f(x)|, |f'(x)|)$$

(where f' denotes the derivative of f). We consider the set A:

$$A = \{f \in \mathscr{C}^1 \,|\, f \text{ has a first and second derivative on } [0, 1] \text{ and}$$

$$|f(0)| \le 1; \qquad |f'(0)| \le 1; \qquad \sup_{x \in [0, 1]} |f''(x)| \le 1\}.$$

Show that A is relatively compact in $\mathscr{C}^1$.

24. Consider the vector space $\mathscr{P}$ of real polynomials with the norm

$$\|P\| = \sup_{x \in [0, 1]} \max(|P(x)|, |P'(x)|).$$

Show that the completion of $\mathscr{P}$ is the space $\mathscr{C}^1$ defined in exercise 23.

25. Show by means of examples that the Hausdorff distance between two discs in the plane can be less than, equal to, or greater than the distance between their centers.

26. Let E be a bounded metric space and $A, B, C,$ and D four nonempty closed subsets of E. Show that

$$h(A \cup B, C \cup D) \le \max(h(A, C), h(B, D))$$

where h is the Hausdorff distance.

27. Let (E, d) be a bounded metric space and $\mathscr{F}(E)$ the set of nonempty closed subsets of E with the Hausdorff distance. Show that the mapping $(x, A) \to d(x, A)$ from $E \times \mathscr{F}(E)$ to $\mathbb{R}$ is uniformly continuous.

28. Let X and E be two metric spaces and A a mapping from X to the set $\mathscr{F}(E)$ of the nonempty closed subsets of E. Show that if $A(x)$ reduces to a single point for all x in X, then the following statements are equivalent:

a. A is continuous when considered as a mapping from X to E.

b. A is continuous from X to $\mathscr{F}(E)$.

   c. A is upper semicontinuous from X to $\mathscr{F}(E)$.

   d. A is lower semicontinuous from X to $\mathscr{F}(E)$.

29. Let X and E be two metric spaces and A a mapping from X to the set $\mathscr{F}(E)$ of the nonempty closed subsets of E. Let V be a subset of E, and set:

$$A^+(V) = \{x \in X \mid A(x) \subset V\}$$
$$A^-(V) = \{x \in X \mid A(x) \cap V \neq \varnothing\}.$$

   a. Denoting by $\complement V$ the complement of the subset V of X or of E, show that

$$A^+(\complement V) = \complement[A^-(V)] \qquad \text{for all} \qquad V \subset E.$$

   b. Suppose that for every open set V of E the set $A^+(V)$ is open in X. Show that A is upper semicontinuous.

   c. Suppose that A is upper semicontinuous and that $A(x)$ is compact for all x in X. Show that for every open set V in E the set $A^+(V)$ is open in X. Also show that for every closed set F in E the set $A^-(F)$ is closed in X.

30. Let X and E be two metric spaces and A a mapping from the set X to the set $\mathscr{F}(E)$ of the nonempty closed subsets of E. Let $A^+(V)$ and $A^-(V)$ be the sets defined in the preceding exercise.

   a. Suppose that A is lower semicontinuous. Show that for every open set V in E the set $A^-(V)$ is open in X, and for every closed set F in E the set $A^+(F)$ is closed in X.

   b. Suppose that $A(x)$ is compact for every x in X and that for every open set V in E the set $A^-(V)$ is open in X. Show that A is lower semicontinuous.

31. Let X and E be two metric spaces and A an upper semicontinuous mapping from X to the set $K(E)$ of nonempty *compact* subsets of E. Suppose that X is compact. Show that the set $A(X) = \bigcup_{x \in X} A(x)$ is compact. (Consider an open covering $\{V_i\}_{i \in I}$ of $A(X)$ and show that for every x in X there exists an open set W_x that is a finite union of open sets V_i and such that $x \in A^+(W_x)$ (see exercise 30).)

32. Let X and E be two metric spaces and A a mapping from X to the set $\mathscr{F}(E)$ of the nonempty closed subsets of E. We call the graph of A the subset $G(A) = \{(x, y) \in X \times E \mid y \in A(x)\}$.

   a. Suppose that A is upper semicontinuous and show that $G(A)$ is closed in $X \times E$. Show that the converse is not in general true.

   b. Suppose in addition that E is compact, and show that if the set $G(A)$ is closed in $X \times E$ then A is upper semicontinuous.

33. Let (E, d) be a metric space, $\mathscr{F}(E)$ the set of the nonempty closed subsets of E, $\overline{\mathscr{F}}(E) = \mathscr{F}(E) \cup \{\varnothing\}$, and $\mathscr{C}(E)$ the set of continuous functions from E to $\mathbb{R}$.

a. Let i be the mapping from $\overline{\mathscr{F}}(E)$ to $\mathscr{C}(E)$ that associates to a closed set A of E the mapping

$$i(A):x \to \frac{d(x, A)}{1 + d(x, A)}$$

(with the convention that $d(x, \varnothing) = +\infty$ for all x in E and $\infty/1 + \infty = +1$). Show that i is injective.

b. We say that a sequence A_n of $\overline{\mathscr{F}}(E)$ converges pointwise to a closed set A of E if the functions $i(A_n)$ converge pointwise to $i(A)$ in $\mathscr{C}(E)$. Show that A_n converges pointwise to A in $\overline{\mathscr{F}}(E)$ if and only if $\lim_{n \to +\infty} d(x, A_n) = d(x, A)(\leq +\infty)$ for all x in E.

c. Show that the subset $i(\overline{\mathscr{F}}(E))$ of $\mathscr{C}(E)$ is uniformly equicontinuous. Conclude from this that a sequence A_n converges pointwise to A if and only if $\lim_{n \to +\infty} d(x, A_n) = d(x, A)(\leq +\infty)$ for all x in a dense subset D of E.

34. Let (E, d) be a metric space and A_n a sequence of closed (possibly empty) subsets of E. We call the upper limit of the sequence A_n the set, denoted $L_s A_n$, of points x of E for which every neighborhood of x has a nonempty intersection with an infinite number of the sets A_n.

a. Show that x belongs to $L_s A_n$ if and only if $\liminf_{n \to +\infty} d(x_1 A_n) = 0$.

b. Show that

$$L_s A_n = \bigcap_{n=1}^{+\infty} \overline{\bigcup_{k \geq 0} A_{n+k}}.$$

35. Let (E, d) be a metric space and A_n a sequence of closed (possibly empty) subsets of E. We call the lower limit of the sequence A_n the set, denoted $L_i A_n$, of points x of E for which for every neighborhood V of x there exists an integer n_0 such that $V \cap A_n$ is nonempty for every $n \geq n_0$.

a. Show that x belongs to $L_i A_n$ if and only if $\lim_{n \to +\infty} d(x, A_n) = 0$.

b. Using the definitions of the preceding exercise, show that $L_i A_n \subset L_s A_n$.

36. Let A_n be a sequence of closed subsets of a metric space E that converges "pointwise" (see exercise 33) to a closed subset A of E. Show that $A = L_s A_n = L_i A_n$ (see exercises 34 and 35).

37. Let (E, d) be a metric space in which the closed balls are compact. Show that a sequence A_n of closed sets of E converges pointwise (see exercise 33) to a closed subset A of E if and only if $A = L_s A_n = L_i A_n$ (use the preceding exercise).

38. Let A be a countable subset of $[0, 1]$ and α a mapping from A to
$]0, +\infty[$ such that

$$\sum_{t \in A} \alpha(t) < +\infty.$$

For f in $\mathscr{C}_\infty([0, 1])$ we set

$$\|f\| = \sum_{t \in A} \alpha(t) f(t).$$

   a. Show that $e(f, g) = \|f - g\|$ defines a semidistance on $\mathscr{C}_\infty([0, 1])$
       (see exercise 5, Chapter 1).
   b. Show that this is a norm if and only if A is dense in $[0, 1]$.
   c. If A is dense in $[0, 1]$, is the norm $\|\cdot\|$ equivalent to the uniform
       convergence norm?
   d. If A is dense in $[0, 1]$, show that the norms associated with (A, α)
       and with (A, α') are equivalent if and only if the functions α/α'
       and α'/α are bounded on A.

39. Let (E, d) be a compact metric space and $\mathscr{C}$ the set of continuous
mappings from E to itself. Consider the distance δ on $\mathscr{C}$:

$$\delta(f, g) = \sup_{x \in E} d(f(x), g(x)).$$

Let G be the set of isometries from E to E:

$$G = \{f \in \mathscr{C} \mid \forall x, y \in E \qquad d(x, y) = d(f(x), f(y))\}.$$

   a. Let a be an element of E and f an isometry. Show that a is a cluster
       point for the sequence $\{f^n(a)\}_{n \in \mathbb{N}}$.
   b. Conclude from this that $f(E) = E$.
   c. Show that G is closed in $\mathscr{C}$.
   d. Show that G is compact in $\mathscr{C}$. To this end we need to use the follow-
       ing generalization of the Ascoli theorem (which may also be proved
       as an additional exercise): Every uniformly equicontinuous subset
       of $\mathscr{C}$ is relatively compact.

Chapter 5

  1. Let E be a metric space.
    a. Let A and B be two locally compact subspaces of E. Show that
       $A \cap B$ is locally compact.
    b. Give an example of two locally compact subspaces of a metric
       space whose union is not locally compact.

2. Let E_1 and E_2 be two locally compact metric spaces. Show that $E_1 \times E_2$ is locally compact.

3. Show that the subspace $\mathbb{Q}$ of the rational numbers in the space $\mathbb{R}$ is not locally compact.

4. Let Ω be an open subset of $\mathbb{R}^p$. We denote by d the Euclidean distance in $\mathbb{R}^p$, by $\complement\Omega$ the complement of Ω in $\mathbb{R}^p$, and by $0 = (0, \ldots, 0)$ the origin in $\mathbb{R}^p$. For every integer n, we define the set:

$$U_n = \left\{ x \in \mathbb{R}^p \mid d(x, 0) < n \quad \text{and} \quad d(x, \complement\Omega) > \frac{1}{n} \right\}.$$

Show that the sets U_n are open and relatively compact in the space Ω and verify:

   i. $\bar{U}_n \subset U_{n+1}$.
   ii. $\Omega = \bigcup_n U_n$.

5. Show that in an infinite dimensional normed space every compact subset has an empty interior.

6. Let (E, d) be a separable locally compact metric space and $\mathscr{C}(E)$ the set of continuous functions from E to $\mathbb{R}$. Show that the set $\mathscr{C}(E)$ with the topology of uniform convergence on every compact subset of E is a metrizable space (that is, show there exists a distance δ on $\mathscr{C}(E)$ such that every sequence f_n of $\mathscr{C}(E)$ converges uniformly on every compact subset of E to a function f of $\mathscr{C}(E)$ if and only if $\lim_{n \to +\infty} \delta(f_n, f) = 0$).

7. Let (E, d) be a separable locally compact metric space.
   a. Show that the set $\bar{\mathscr{F}}(E)$ with "pointwise convergence" (see exercise 33, Chapter 4) is a metrizable space (see exercise 6).
   b. Show that the mapping $d: (x, A) \to d(x, A)$ from $E \times \mathscr{F}(E)$ to $\mathbb{R}$ is continuous. (Use the equicontinuity of $i(\bar{\mathscr{F}}(E))$. See exercise 33, Chapter 4.)

8. Show that the space $\mathbb{R}$ with the distance $d(x, y) = 0$ if $x = y$ and $d(x, y) = 1$ if $x \neq y$ is not separable.

9. Show that every subspace of a separable metric space is separable.

10. a. Show that the product of two separable metric spaces is separable.
    b. Show that the continuous image of a separable metric space is separable.

11. Show that the Banach spaces l^1 and c_0 (see section 1.7 and exercise 9, Chapter 2) are separable.

12. Let E be a real vector space and p and q two norms for E. Suppose that there exists $k > 0$ such that $p(x) \leq k \cdot q(x)$ for all x in E. Show that if the normed space (E, q) is separable, then the space (E, p) is also separable.

13. Let E be a metric space.
    a. Suppose that for all $\varepsilon > 0$ there exists a finite number of points $x_1, \ldots, x_n$ in E such that $E \subset \bigcup_i B(x_i, \varepsilon)$. Show that E is separable.
    b. Deduce from this that a compact metric space is separable.
14. Let E be a separable metric space. Show that every open covering of E has a countable subcovering.
15. Let E be a compact metric space and (F, d) a separable metric space. We denote by $\mathscr{C}(E, F)$ the set of continuous functions from E to F, and we set $\delta(f, g) = \sup_x d(f(x), g(x))$ for f, g in $\mathscr{C}(E, F)$. Show that $(\mathscr{C}(E, F), \delta)$ is a separable metric space.
16. Let E be a separable locally compact metric space and $\mathscr{C}(E)$ the metrizable space (see exercise 6) of continuous functions from E to $\mathbb{R}$ with uniform convergence on the compact sets of E. Show that the space $\mathscr{C}(E)$ is separable.
17. Let E and F be two compact metric spaces and f a continuous mapping from $E \times F$ to $\mathbb{R}$. Show that for every $\varepsilon > 0$ there exist an integer n, n continuous functions u_i from E to $\mathbb{R}$, and n continuous functions v_i from F to $\mathbb{R}$ such that

$$\max_{(x, y) \in E \times F} \left| f(x, y) - \sum_{i=1}^{n} u_i(x) v_i(y) \right| \leq \varepsilon.$$

(Apply the Stone–Weierstrass theorem to the algebra generated by $\mathscr{C}(E) \cup \mathscr{C}(F)$).
18. If a complete metric space is the union of a countable sequence of closed sets, show that at least one of these closed sets has a nonempty interior.
19. Show that every open subspace of a Baire space is a Baire space.
20. Let E be a complete metric space. Show that the intersection $U \cap F$ of an open set U of E and a closed set F of E is a Baire space.
21. Let E be a complete metric space and f a lower semicontinuous function from E to $\mathbb{R}$. Show that f is continuous on a G_δ dense set of E where a G_δ set is a set that is a countable intersection of open sets of E.
22. Let E be a complete metric space. We shall call any subset of E that is contained in a countable union of closed sets of E with empty interior a "meager set." Let f_n be a sequence of continuous functions from E to $\mathbb{R}$. Suppose that the sequence f_n converges to a function f from E to $\mathbb{R}$ for every point of E. Show that f is continuous on the complement of a meager set.
23. Show that the product of a finite number of connected spaces is a connected space.
24. Let E be a metric space, A an arbitrary subspace of E, and B a connected subspace of E. Show that if B has nonempty intersections with the

interior of A and with the exterior of A (the complement in E of the closure of A), then B has a nonempty intersection with the boundary of A.

25. **a.** Let m be an arbitrary point of $\mathbb{R}^2$. Show that $\mathbb{R}^2 \setminus \{m\}$ is connected.

   b. Prove from this that the spaces $\mathbb{R}$ and $\mathbb{R}^2$ are not homeomorphic.

26. Let E be a metric space. The "connected component" of a point x of E, denoted $C(x)$, is the union of all the connected subsets of E containing x.

   a. Show that $C(x)$ is connected and closed.

   b. Show that the set of distinct connected components in E forms a partition of E.

   c. Show that E is connected if and only if $C(x) = E$ for every x of E.

   d. Let $\mathbb{Q}$ be the subspace of $\mathbb{R}$ consisting of the rational numbers. Show that the connected component $C(x)$ of any point x of $\mathbb{Q}$ is the set $\{x\}$.

27. We say that a metric space E is locally connected if for every open set G of E and for every x of G there exists a connected open set U contained in G and containing x. Show that a space E is locally connected if and only if the connected components of every point of E (see definition in exercise 26) are open.

28. We say that a metric space E is arcwise connected if for every x, y of E there exists a continuous mapping f from the interval $[0, 1]$ to E such that $f(0) = x$ and $f(1) = y$.

   a. Show that an arcwise connected space is connected.

   b. Let $A = \{(x, y) \in \mathbb{R}^2 \mid y = \sin 1/x, x \in {]}0, 1]\}$. Show that the closure of the subset A of $\mathbb{R}^2$ is connected but not arcwise connected.

   c. Show that every convex subset of a normed vector space is arcwise connected.

29. Let (E, d) be a connected metric space. Show that for every pair (x, y) of points of E and for every $\varepsilon > 0$ there exists a finite sequence $x_1, \ldots, x_n$ of points of E such that $x_1 = x$ and $x_n = y$ and $d(x_i, x_{i+1}) \le \varepsilon$ for all $1 \le i \le n - 1$. (Consider the subset $E(x, \varepsilon)$ of those points y of E for which there exists a sequence $x_1, \ldots, x_n$ satisfying the given condition, and show that it is open and closed.)

30. **a.** Show that the converse of the preceding exercise is false. (Take $E = \mathbb{Q}$.)

   b. Suppose that E is a compact metric space, and show that the converse of the preceding exercise is now true.

31. Show that the intersection of a decreasing sequence of connected compact sets of a metric space is connected and compact.

32. **a.** Let (E, d) be a metric space, A a closed subset of E, and f a continuous bounded mapping from A to $\mathbb{R}$. Show that there exists a continuous bounded mapping g from E to $\mathbb{R}$ such that

  i. $g(x) = f(x)$ for all x in A.
  ii. $\sup_{x \in E} g(x) = \sup_{x \in A} f(x)$.
  iii. $\inf_{x \in E} g(x) = \inf_{x \in A} g(x)$.

(Suppose without loss of generality that $\sup_{x \in A} f(x) = 2$ and $\inf_{x \in A} f(x) = 1$ and define g by $g(x) = f(x)$ if $x \in A$ and $g(x) = 1/d(x, A) \inf_{y \in A} f(y) \cdot d(x, y))$ if $x \notin A$.)

b. Deduce from the preceding result the theorem on the separation of two disjoint closed sets (Proposition 5.5.1).

33. Let E be a normed vector space, K a nonempty compact convex set of E, and F a mapping from K to the set of nonempty subsets of K. We suppose that

  i. $F(x)$ is convex for all x of K.
  ii. $F^-(y) = \{x \in K \mid y \in F(x)\}$ is open in K for all $y \in K$.

a. Show that there exists a finite subset $y_1, \ldots, y_n$ of K such that $K \subset \bigcup_{i=1}^{n} F^-(y_i)$.

b. Consider a continuous partition of unity β_i, $i = 1, \ldots, n$ subordinate to the open covering $F^-(y_i)$, $i = 1, \ldots, n$. Show that the function $f : x \to \sum \beta_i(x) y_i$ is continuous from K to K, and verify that $f(x) \in F(x)$ for all x in K.

34. Let f and g be two mappings from a compact metric space E to $\mathbb{R}$. Suppose that f is lower semicontinuous, that g is upper semicontinuous, and that they satisfy $g(x) < f(x)$ for all x in E. Show that there exists a continuous function φ from E to $\mathbb{R}$ such that $g(x) < \varphi(x) < f(x)$ for all x in E. (Consider the open covering of E by subsets

$$\{x \in E \mid f(x) > \alpha \quad \text{and} \quad g(x) < \alpha\} \quad \text{for} \quad \alpha \quad \text{in} \quad \mathbb{Q}).$$

35. Let E be a compact *convex* subset of a Banach space V and F an arbitrary subset of the Banach space V. We denote by $\mathscr{C}(E, F)$ (respectively, $\mathscr{C}(F, E)$) the set of continuous functions from E to F (respectively from F to E). Let φ be an arbitrary mapping from $E \times F$ to $\mathbb{R}$.

a. Show that for all $D \in \mathscr{C}(E, F)$ and for all $C \in \mathscr{C}(F, E)$

$$\varphi^b(D) = \inf_{x \in E} \varphi(x, D(x)) \leq \sup_{y \in F} \varphi(C(y), y) = \varphi^{\#}(C).$$

(Apply the Brouwer theorem to the function $C \circ D$).

b. Deduce from this that

$$\sup_{y \in F} \inf_{x \in E} \varphi(x, y) \leq \sup_{D \in \mathscr{C}(E, F)} \varphi^b(D) \leq \inf_{C \in \mathscr{C}(F, E)} \varphi^\#(C) \leq \inf_{x \in E} \sup_{y \in F} \varphi(x, y).$$

c. Suppose in addition that F is convex, that the mappings $y \to \varphi(x, y)$ are concave for all x of E, and that the mappings $x \to \varphi(x, y)$ are lower semicontinuous for all y of F. Deduce Theorem 5.6.2, from Theorem 5.6.1 and the preceding results.

36. Let E a compact space, F a convex subset of a Banach space V, and f a mapping from $E \times F$ to $\mathbb{R}$. We denote by $K(E, F)$ the set of correspondences D from E to F whose values are nonempty closed convex sets and whose graph $G(D) = \{(x, y) \in E \times F \,|\, y \in D(x)\}$ is closed in $E \times F$. We denote by $\mathscr{C}(E, F)$ the set of continuous functions from E to F. For $D \in K(E, F)$, we set

$$\varphi^b(D) = \inf_{x \in E} \inf_{y \in D(x)} \varphi(x, y) = \inf_{(x, y) \in G(D)} \varphi(x, y).$$

a. Show that for all $D \in K(E, F)$:

$$\varphi^b(D) \leq \inf_{x \in E} \sup_{y \in F} \varphi(x, y).$$

b. Suppose that the hypotheses of Theorem 5.6.1, are satisfied. Deduce from the inclusion $\mathscr{C}(E, F) \subset K(E, F)$ that

$$\sup_{D \in K(E, F)} \varphi^b(D) = \inf_{x \in E} \sup_{y \in F} \varphi(x, y).$$

c. Let $\bar{D}$ be the correspondence from E to F defined by

$$\bar{D}(x) = \left\{ y \in F \,|\, \varphi(x, y) = \max_{u \in F} \varphi(x, u) \right\} \qquad \text{for all} \qquad x \in E.$$

Show that if in addition we suppose that F is compact and that φ is continuous from $E \times F$ to $\mathbb{R}$ then $\bar{D} \in K(E, F)$.

d. Show that under the hypotheses assumed in b and c we have

$$\varphi^b(\bar{D}) = \max_{D \in K(E, F)} \varphi^b(D) = \inf_{x \in E} \sup_{y \in F} \varphi(x, y).$$

37. Let E and F be two nonempty compact convex subsets of a Banach space V and f a mapping from $E \times F$ to $\mathbb{R}$. We denote by $\mathscr{C}(E, F)$ the set of continuous functions from E to F.

a. Suppose that the mappings $x \to f(x, y)$ are convex for all y in F and that the mappings $y \to f(x, y)$ are upper semicontinuous for all x in E. Show that

$$\sup_{y \in F} \inf_{x \in E} f(x, y) = \sup_{D \in \mathscr{C}(E, F)} \inf_{x \in E} f(x, D(x)).$$

(Apply Theorem 5.6.2 to $-f$).

b. Suppose in addition that the mappings $x \to f(x, y)$ are lower semicontinuous for all y in F and that the mappings $y \to f(x, y)$ are concave for all x in E. Using Theorem 5.6.1, show that

$$\sup_{y \in F} \inf_{x \in E} f(x, y) = \inf_{x \in E} \sup_{y \in F} f(x, y)$$

and deduce from this the Von Neumann theorem (Theorem 5.7.2).

38. Let $X = \{x_1, \ldots, x_n\}$ and $Y = \{y_1, \ldots, y_p\}$ be two finite subsets of strategies of $\mathbb{R}^k$, and let f be a mapping from $X \times Y$ to $\mathbb{R}$. We define the sets of "mixed" strategies E and F by

$$E = \left\{ \sum_{i=1}^{n} \lambda_i x_i \in \mathbb{R}^k \,|\, \lambda_i \geq 0 \quad \text{and} \quad \sum_{i=1}^{n} \lambda_i = +1 \right\},$$

$$F = \left\{ \sum_{j=1}^{p} \mu_j y_j \in \mathbb{R}^k \,|\, \mu_j \geq 0 \quad \text{and} \quad \sum_{j=1}^{p} \mu_j = +1 \right\}.$$

Thus we have $X \subset E$ and $Y \subset F$, and we define the extension $\hat{f}$ of f to $E \times F$ by

$$\hat{f}(u, v) = \sum_{j=1}^{p} \sum_{i=1}^{n} \lambda_i \mu_j f(x_i, y_j)$$

for $u = \sum_{i=1}^{n} \lambda_i x_i \in E$ and $v = \sum_{j=1}^{p} \mu_j y_j \in F$. Show that the mapping f has a saddle point $(\bar{u}, \bar{v}) \in E \times F$ and that we have the following result as well:

$$\sup_{j} \inf_{i} f(x_i, y_j) \leq \hat{f}(\bar{u}, \bar{v}) \leq \inf_{i} \sup_{j} f(x_i, y_j).$$

Problems

1. In the Banach space l^∞ (see Section 1.7), we denote by e_x the constant sequences:
$$\forall x \in \mathbb{R}, \qquad e_x = (x, x, \ldots, x, \ldots).$$

   Let H be a continuous linear form on l^∞ such that $H(e_x) = x$ for all $x \in \mathbb{R}$.
   a. Show that the norm of H is greater than or equal to 1.
   b. We say that H is positive if
$$\forall n \in \mathbb{N}, \qquad x_n \geq 0 \Rightarrow H((x_n)) \geq 0.$$

   Show that the norm of H is equal to 1 if and only if H is positive.
   c. We denote by a^i the following element of l^∞:
$$\begin{cases} a_n^i = 0 & \text{if} \quad n \neq i. \\ a_i^i = 1. \end{cases}$$

   Suppose that H is positive and that $H(a^i) = 0$ for every integer i. Show that for all $\{x_n\}_{n \in \mathbb{N}} \in l^\infty$ we have
$$\liminf_{n \to +\infty} x_n \leq H(\{x_n\}) \leq \limsup_{n \to +\infty} x_n.$$

2. We denote by $\mathscr{C}$ the space of continuous functions from $[0, 1]$ to $\mathbb{R}$ and by φ a fixed element of $\mathscr{C}$. Consider the linear form:
$$F : \mathscr{C} \to \mathbb{R}$$
$$f \to \int_0^1 \varphi(t) \cdot f(t)dt.$$

   a. If we consider $\mathscr{C}$ with the norm
$$\|f\|_1 = \int_0^1 |f(t)|dt,$$

   show that the norm of F is $\|\varphi\|_\infty = \sup_{x \in [0, 1]} |\varphi(x)|$.

b. If we consider $\mathscr{C}$ with the norm $\| \cdot \|_\infty$, show that the norm of F is $\|\varphi\|_1$.

3. We denote by c the Banach space of real convergent sequences with the norm

$$\forall x \in c, \qquad \|x\|_\infty = \sup_{n \in \mathbb{N}} |x_n|.$$

(See exercise 9, Chapter 2.) If x is an element of l^1 (Section 1.7), we set

$$\forall t \in c \qquad g(x)(t) = x_1 \cdot \left[\lim_{n \to +\infty} t_n \right] + \sum_{n=1}^{+\infty} x_{n+1} \cdot t_n.$$

a. Show that in this fashion we define a mapping g from l^1 to $c^* = \mathscr{L}(c, \mathbb{R})$.

b. Show that g is linear and an isometry (where c^* has the norm for $\mathscr{L}(c, \mathbb{R})$ defined in Section 1.10).

c. Show that g is surjective and, therefore, is an isomorphism of the Banach space. Use the fact that a sequence $x = \{x_n\}$, which *tends to zero*, can be expressed in c by

$$x = \sum_{n=1}^{+\infty} x_n e_n$$

where the e_n's are defined by

$$(e_n)_i = 0 \quad \text{if} \quad i \neq n \quad \text{and} \quad (e_n)_n = 1.$$

4. *Criterium for normally convergent series.* Let E be a Banach space and $\{u_n\}_{n \in \mathbb{N}}$ a sequence of elements of E.

a. Show that if the series $\sum_{n=1}^{+\infty} \|u_n\|$ converges in $\mathbb{R}$, then the series $\sum u_n$ converges in E, and, moreover,

$$\left\| \sum_{n=1}^{+\infty} u_n \right\| \leq \sum_{n=1}^{+\infty} \|u_n\|.$$

b. Let E be a normed vector space in which $\sum_{n=1}^{+\infty} \|u_n\| < +\infty$ implies that $\sum_{n=1}^{+\infty} u_n$ converges in E. Show that E is a Banach space.

5. *Summation techniques.* Given a double sequence $(a_{ij})_{i,\,j \in \mathbb{N}}$, let us consider the transformation:

$$x = \{x_n\} \in l^\infty \to Ax = \{y_n\} : y_n = \sum_{j=1}^{+\infty} a_{nj} x_j.$$

We denote by c the Banach space of convergent sequences (see problem 3). The transformation A, mentioned previously is called a summation technique if A is an element of $\mathscr{L}(c, c)$ and if, in addition,

$$\forall x \in c \qquad \lim_{n \to +\infty} (Ax)_n = \lim_{n \to +\infty} x_n.$$

Show that A is a summation technique if and only if

$$\forall j \in \mathbb{N} \qquad \lim_{i \to +\infty} a_{ij} = 0$$

$$\sup_{i \in \mathbb{N}} \sum_{j=1}^{+\infty} |a_{ij}| < +\infty.$$

$$\lim_{i \to +\infty} \sum_{j=1}^{+\infty} a_{ij} = 1.$$

If these conditions are satisfied, show that A belongs to $\mathcal{L}(l^\infty, l^\infty)$ and calculate its norm.

EXAMPLE.

1. $a_{ij} = \begin{cases} \dfrac{1}{i} & 1 < j \le i \\ 0 & i < j \end{cases}$ (Cesaro mean)

2. $a_{ij} = (1 - r_i)r_i^j$ with $0 < r_i < 1$ and $r_i \to 1$.

In each of these two examples, find a nonconvergent sequence whose image under A is convergent.

6. *Invertible mappings in Banach spaces.* Let E and F be two Banach spaces.
    a. Given $u \in \mathcal{L}(E, F)$ whose inverse u^{-1} exists and belongs to $\mathcal{L}(F, E)$, show that

$$\|u^{-1}\|_{\mathcal{L}(F, E)}^{-1} \le \|u\|_{\mathcal{L}(E, F)}.$$

b. Let v be an element of $\mathcal{L}(E, F)$ such that

$$\|v\|_{\mathcal{L}(E, F)} < \|u^{-1}\|_{\mathcal{L}(F, E)}^{-1}.$$

Show that $(u + v)$ is invertible and that

$$\|(u + v)^{-1}\| \le \frac{1}{\|u^{-1}\|^{-1} - \|v\|}.$$

c. Deduce from this that the set $\mathcal{U}_{E, F}$ of elements of $\mathcal{L}(E, F)$ that are invertible in $\mathcal{L}(E, F)$ is an open set in $\mathcal{L}(E, F)$ and that the mapping $u \to u^{-1}$ is a homeomorphism $\mathcal{U}_{E, F} \to \mathcal{U}_{F, E}$. (Compare with exercise 7, Chapter 3.)

7. Let A be a subset of a complete metric space E. Let $\{u_n\}_{n \in \mathbb{N}}$ be a sequence of mappings from A to E. Suppose that $u_n(A)$ is relatively compact in E for every n and that $\{u_n\}$ converges uniformly on A to a function u. Show that $u(A)$ is relatively compact.

8. Let φ be a continuous function from $\mathbb{R}$ to $\mathbb{R}$ such that the set $\{x \in \mathbb{R} \mid \varphi(x) \neq 0\}$ is bounded. Let f be a bounded function from $\mathbb{R}$ to $\mathbb{R}$, which is continuously differentiable and whose derivative is bounded on $\mathbb{R}$. For $h > 0$, we denote

$$F_h(f) = \int_{\mathbb{R}} \frac{1}{h} (f(x + h) - f(x)) \cdot \varphi(x)dx,$$

$$F_0(f) = \int_{\mathbb{R}} f'(x) \cdot \varphi(x)dx.$$

a. Let E be the following Banach space (see exercise 32, Chapter 1):

$$E = \{f \in \mathscr{C}_\infty(\mathbb{R}) \mid f \quad \text{is continuously differentiable}$$
$$\text{and} \quad \|f\| = \max(\|f\|_\infty, \|f'\|_\infty) < +\infty\}.$$

Show that for all $h \geq 0$, F_h is a continuous linear form on E.

b. Show that if f is twice continuously differentiable, then

$$\lim_{h \to 0} F_h(f) = F_0(f).$$

c. Show that this property still holds if f belongs to E, but that F_h does not converge to F_0 in $\mathscr{L}(E, \mathbb{R})$.

9. Let S be a compact metric space and φ a continuous mapping from S to itself. We say that a subset A of S is stable under φ if $\varphi(A) = A$.

a. Show that every union of subsets that are stable under φ is stable under φ. Deduce from this that there exists a maximal subset in S that is stable under φ and that we shall denote by $\tilde{S}$.

b. Show that $\tilde{S}$ is compact.

c. If s belongs to S, we set

$$0^i(x) = \{\varphi^j(x)\}_{j \geq i}$$

and

$$\omega(x) = \bigcap_{i=0}^{+\infty} \overline{0^i(x)}.$$

We call $\omega(x)$ the limit set of x. Show that it is a nonempty compact subset of S.

d. Show that the limit set of x is stable under φ.

e. EXAMPLE.   Let S be the following set:

$$S = \{(x, y) \in \mathbb{R}^2 \mid x^2 + y^2 = 1\}.$$

We define the mappings φ_α by

$$\forall \alpha \in \mathbb{R}, \qquad \varphi_\alpha(e^{i\pi t}) = e^{i\pi(t+\alpha)}.$$

Show that $\omega(x)$ is finite when $\alpha \in \mathbb{Q}$ and that $\omega(x) = S$ (for all x) when α is irrational.

10. If f is a continuous function from $[0, 1]$ to $\mathbb{R}$, we shall say that the real number $\varepsilon \geq 0$ is associated with f when the set $\{x \in [0, 1] \mid |f(x)| \geq \varepsilon\}$ can be covered by a finite family of intervals the sum of whose lengths is bounded above by ε. We set

$$p(f) = \inf\{\varepsilon \geq 0 \mid \varepsilon \text{ is associated with } f\}.$$

   a. Show that $p(-f) = p(f)$ and $p(f + g) \leq p(f) + p(g)$ for all f and g in $\mathscr{C}_\infty([0, 1])$. Show, nevertheless, that p is not a norm on $\mathscr{C}_\infty([0, 1])$.
   b. Show that the relation

$$d(f, g) = p(f - g)$$

defines a distance on $\mathscr{C}_\infty([0, 1])$.
   c. If B is a ball with center 0 for the distance d, show that the convex hull of B is all of $\mathscr{C}_\infty([0, 1])$.
   d. Show that the only continuous linear form for this distance is the linear form that is identically zero.

11. Let E be a compact metric space and $\mathscr{C} = \mathscr{C}_\infty(E)$ the Banach space of continuous functions on E. For every subset M of $\mathscr{C}$ and for every subset A of E we set

$$V(M) = \{x \in E \mid \forall u \in M, \qquad u(x) = 0\},$$

$$J(A) = \{u \in \mathscr{C} \mid \forall x \in A, \qquad u(x) = 0\}.$$

   a. Show that $V(M)$ is closed in E and that $J(A)$ is closed in $\mathscr{C}$.
   b. A function e of $\mathscr{C}$ is said to be idempotent if $e^2(x) = e(x)$ for all x in E. Show that the mapping $e \to V(\{e\})$ is a bijection from the set of idempotent functions onto the set of open and closed subsets of E. Deduce from this that if E is connected, the only idempotent elements of $\mathscr{C}$ are the two constant functions $e = 0$ and $e' = 1$.
   c. Show that if A is closed in E, then for every element x in the complement of A, there exists a function u of $\mathscr{C}$, which is null on A, non-negative on $\complement A$ and for which $u(x) \neq 0$. Hence deduce the formula

$$\forall A \subset E : V(J(A)) = \bar{A}.$$

12. *Metric spaces with compact closed balls.* We denote by (compact closed balls) CCB those metric spaces (E, d) in which the closed balls of E are compact.

I.a. Show that a metric space E is CCB if and only if the closed and bounded subsets of E are compact.

I.b. Show that the space $\mathbb{R}^n$ is a CCB space for the Euclidean distance d (or any distance derived from a norm), but that it is not a CCB space for the equivalent distance $d/(1 + d)$.

I.c. Show that a normed space is a CCB space if and only if it is finite dimensional.

I.d. Show that every closed subspace of a CCB space is a CCB space. Show that an open subspace of a CCB space is not in general a CCB space.

I.e. Show that a CCB space is complete, locally compact, and separable.

II. Let (E, d) be a separable locally compact metric space.

a. Let U_n be a sequence of relatively compact open sets in E such that $\overline{U}_n \subset U_{n+1}$ and $E = \bigcup_{n=1}^{\infty} U_n$. Show that there exists a continuous mapping f from E to $\mathbb{R}$ such that

$$f(x) \leq n \quad \text{if} \quad x \in \overline{U}_n \quad \text{and} \quad f(x) \geq n \quad \text{if} \quad x \notin \overline{U}_n.$$

b. Show that the distance $d'(x, y) = d(x, y) + |f(x) - f(y)|$ is topologically equivalent to the distance d (that is, that the identity mapping from (E, d) to (E, d') is a homeomorphism).

c. Show that the closed balls of the metric space (E, d') are compact.

III. Let (E, d) be a CCB metric space.

a. Show that for every closed set A of E and for every $x \in E$ there exists $a \in A$ such that $d(x, A) = d(x, a)$.

b. Let A be a compact subset of E and B a closed subset of E. We define

$$d(A, B) = \inf_{\substack{a \in A \\ b \in B}} d(a, b).$$

Show that there exist $a \in A$ and $b \in B$ such that

$$d(A, B) = d(a, b).$$

IV.a. Let (E, d) be a CCB metric space and $\{x_n\}$ a sequence in E. Show that the following three statements are equivalent:

i. For every compact set K of E there exists n_0 such that $x_n \notin K$ for all $n \geq n_0$.

ii. $\forall a \in E \qquad \lim_{n \to +\infty} d(a, x_n) = +\infty.$

iii. $\exists a \in E \qquad \lim_{n \to +\infty} d(a, x_n) = +\infty.$

If one of these conditions is satisfied, we say that the sequence x_n "approaches infinity."

IV.b.   Let E and F be two CCB metric spaces and f a continuous mapping from E to F. Show that the following two statements are equivalent:

   i.   For every compact set K of F, $f^{-1}(K)$ is compact in E.

   ii.   For every sequence x_n of E that "approaches infinity" the sequence $f(x_n)$ of F "approaches infinity."

INDEX

261